ADVANCES IN MOLECULAR AND CELL BIOLOGY

Volume 27 • 1998

INTRACELLULAR PROTEIN DEGRADATION

ADVANCES IN MOLECULAR AND CELL BIOLOGY

INTRACELLULAR PROTEIN DEGRADATION

Series Editor: E. EDWARD BITTAR
Department of Physiology
University of Wisconsin
Madison, Wisconsin

Guest Editor: A.J. RIVETT
Department of Biochemistry
University of Bristol
Bristol, England

VOLUME 27 • 1998

JAI PRESS INC.

Stamford, Connecticut *London, England*

100 Prospect Street
Stamford, Connecticut 06902-1640

JAI PRESS LTD.
38 Tavistock Street
Covent Garden
London WC2E 7PB
England

ISBN: 0-7623-0387-5

Transferred to digital printing 2006

CONTENTS

LIST OF CONTRIBUTORS

Didier Attaix — INRA de Theix
Unité de'Etude du Métabolisme Azoté
Theix, France

Enrique Jose Andreu — Fundación Valenciana de Investigaciones Biomédicas
Instituto de Investigaciones Citológicas
Valencia, Spain

Jill Beyette — Department of Molecular Genetics
University of Cincinnati
Cincinnati, Ohio

Jeffrey L. Brodsky — Department of Biological Sciences
University of Pittsburgh
Pittsburgh, Pennsylvania

Wolfgang Dubiel — Institute of Biochemistry
Humboldt University
Berlin, Germany

Erwin Knecht — Fundación Valenciana de Investigaciones Biomédicas
Instituto de Investigaciones Citológicas
Valencia, Spain

Jose Javier Martin de Llano — Fundación Valenciana de Investigaciones Biomédicas
Instituto de Investigaciones Citológicas
Valencia, Spain

Grant G.F. Mason — Department of Biochemistry
University of Bristol
Bristol, England

Michael R. Maurizi — Laboratory of Cell Biology
National Cancer Institute
Bethesda, Maryland

Ardythe A. McCracken — Biology Department
University of Nevada
Reno, Nevada

Isabel Moreno Miralles — Fundación Valenciana de Investigaciones Biomédicas
Instituto de Investigaciones Citológicas
Valencia, Spain

Martin Rechsteiner — Department of Biochemistry
University of Utah School of Medicine
Salt Lake City, Utah

Daniel Taillandier — INRA de Theix
Unité d'Etude du Métabolisme Azoté
Theix, France

Nobuyuki Tanahashi — Institute of Enzyme Research
University of Tokushima
Tokushima, Japan

Keiji Tanaka — The Tokyo Metropolitan Institute of Medical Science
Tokyo, Japan

Michael Thumm — Institut für Biochemie
Universität Stuttgart
Stuttgart, Germany

Eric D. Werner — Biology Department
University of Nevada
Reno, Nevada

Keith D. Wilkinson — Biochemistry Department
Emory University School of Medicine
Atlanta, Georgia

Dieter H. Wolf — Institut für Biochemie
Universität Stuttgart
Stuttgart, Germany

PREFACE

This volume brings together a set of reviews that provide a summary of our current knowledge of the proteolytic machinery and of the pathways of protein breakdown of prokaryotic and eukaryotic cells. Intracellular protein degradation is much more than just a mechanism for the removal of incorrectly folded or damaged proteins. Since many short-lived proteins have important regulatory functions, proteolysis makes a significant contribution to many cellular processes including cell cycle regulation and transciptional control. In addition, limited proteolytic cleavage can provide a rapid and efficient mechanism of enzyme activation or inactivation in eukaryotic cells.

In the first chapter, Maurizi provides an introduction to intracellular protein degradation, describes the structure and functions of bacterial ATP-dependent proteases, and explores the relationship between chaperone functions and protein degradation. Many of the principles also apply to eukaryotic cells, although the proteases involved are often not the same. Interestingly, homologues of one of the bacterial proteases, lon protease, have been found in mitochondria in yeast and mammals, and homologues of proteasomes, which are found in all eukaryotic cells (see below), have been discovered in some eubacteria.

Studies of proteolysis in yeast have contributed greatly to the elucidation of both lysosomal (vacuolar) and nonlysosomal proteolytic pathways in eukaryotic cells. Thumm and Wolf (chapter 2) describe studies that have elucidated the functions of proteasomes in nonlysosomal proteolysis and the contributions of lysosomal proteases to intracellular protein breakdown. Proteins can be selected for degradation by a variety of different mechanisms. The ubiquitin system is one complex and highly regulated mechanism by which eukaryotic proteins are targetted for degradation by proteasomes. In chapter 3, Wilkinson reviews the components and functions of the ubiquitin system and considers some of the known substrates for this pathway which include cell cycle and transcriptional regulators.

The structure and functions of proteasomes and their regulatory components are described in the two subsequent chapters by Tanaka and Tanahashi and by Dubiel and Rechsteiner. Proteasomes were the first known example of threonine proteases. They are multisubunit complexes that, in addition to being responsible for the turnover of most short-lived nuclear and cytoplasmic protein, are also involved in antigen processing for presentation by the MHC class I pathway. Recent studies reviewed by McCracken and colleagues (chapter 6) lead to the exciting conclusion that some ER-associated proteins are degraded by cytosolic proteasomes.

Lysosomes are responsible for the degradation of long-lived proteins and for the enhanced protein degradation observed under starvation conditions. In chapter 7, Knecht and colleagues review the lysosomal proteases and describe studies of the roles of lysosomes and the mechanisms for protein uptake into lysosomes. Methods of measuring the relative contribution of different proteolytic systems (e.g., ubiquitin–proteasome pathway, calcium-dependent proteases, lysosomes) to muscle protein degradation, and the conclusions from such studies, are reviewed by Attaix and Taillinder in the following chapter.

Finally, proteases play an important role in signaling apoptosis by catalyzing the limited cleavage of enzymes. Mason and Beyette review the role of the major players, caspases, which are both activated by and catalyze limited proteolysis, and also consider the involvement of other proteolytic enzymes in this pathway leading to cell death.

A. Jennifer Rivett
Guest Editor

BIOCHEMICAL PROPERTIES AND BIOLOGICAL FUNCTIONS OF ATP-DEPENDENT PROTEASES IN BACTERIAL CELLS

Michael R. Maurizi

Advances in Molecular and Cell Biology
Volume 27, pages 1-41.

ISBN: 0-7623-0387-5

I. INTRODUCTION

Intracellular protein degradation, the cleavage of many peptide bonds in a protein leading to complete loss of function, has come to be recognized as an important posttranslational regulatory mechanism on a par with reversible covalent modification. Limited proteolysis, the cleavage of one or a few peptide bonds to produce an altered but functional form of the protein, is also an important metabolic process but will not be discussed in this review. While the majority of cellular proteins are stable for several generations, some proteins are degraded nearly as fast as they are synthesized and have half-lives of less than one minute. For such proteins, changes in the rate of synthesis will lead to expansion or depletion of the intracellular pool of the protein within a fraction of the cellular generation time, even in rapidly dividing bacterial cells. As pointed out by Schimke more than 25 years ago (Schimke and Doyle, 1970), the rate at which a new steady-state is achieved is dependent only on the degradation rate under the new conditions—the higher the degradation rate, the faster the new steady-state is reached. Changes in degradation rates in response to regulatory signals will also lead to rapid increases or decreases in protein pools.

The regulatory functions of protein degradation involve essentially every vital process in the cell, because many of the rapidly degraded proteins are themselves important regulatory proteins, including transcription factors, replication proteins, cell division inhibitors, and other proteins involved in multimeric, multifunctional enzyme complexes. Thus, in different organisms, protein degradation is involved in the timing of cell division, heat shock and other stress responses, DNA repair and mutagenesis, developmental changes, and response to nutrient depletion. This review will describe some of the biological functions of proteases in bacterial cells (see Table 1). Examples of regulatory degradation in *Escherichia coli* have been well-known for some years, but recently a number of exciting examples have come from work in other bacteria, such as *Caulobacter* and *Myxococcus*.

Protein degradation also serves in protein quality control, an important role often classified as a housekeeping function. Structurally abnormal and otherwise nonfunctional proteins are wasteful and potentially harmful to the cell and must be either repaired by chaperone-assisted refolding or destroyed by the proteolytic machinery. The amount of rapid protein degradation that goes on in cells under normal growing conditions is still not known with certainty, and there have been no recent efforts to confirm or refine earlier estimates that, of newly synthesized polypeptides, between 5–15% in *E. coli* and ~20% in animal cells are degraded with half-lives of <20 minutes. Because the number of intrinsically unstable native proteins appears to be small, it is likely that much of the rapidly degraded protein under normal growth conditions represents incomplete proteins arising from errors in transcription or translation. Protein degradation is increased by conditions, such as starvation or treatment with certain heavy metal ions, that produce increased errors in biosynthesis, and also by conditions, such as heat or other stress, that lead to unfolding or misfolding of proteins. This review will address recent data obtained with *E. coli* ATP-dependent proteases that suggest that chaperones and proteases are partners in a system of protein quality control, and that the ability of a targeted protein to either adopt a stable conformation or assemble into a macromolecular complex is probably a major determinant of whether it is refolded or degraded.

Because this review will focus on intracellular protein degradation, only cytoplasmic proteases or membrane-bound proteases with active sites within the cytoplasm will be discussed. Also, only the eubacterial proteolytic systems, particularly those of *E. coli*, will be discussed in detail. The Archaea proteasome is discussed elsewhere in this volume and will be mentioned only for purposes of comparison. Finally, because most if not all cytoplasmic protein degradation is energy-dependent, this review will address ATP-dependent proteases exclusively (see Table 1). These once esoteric and poorly understood enzymes have in recent years been shown to be elegant and intriguing multifunctional complexes, designed to carry out the seemingly mundane process of peptide bond cleavage with fidelity, discernment, and efficiency. *E. coli* proteases are found in the cytoplasm, in the inner membrane, and within the periplasmic space. Although the latter have not been shown to be dependent on ATP or other energy sources, protein quality control and proteolysis in the periplasm are important processes and influence expression or activity of cytoplasmic proteases and chaperones by as yet unknown mechanisms. Readers are referred to recent general reviews by others on the subject of *E. coli* proteases (Miller, 1996).

The ATP-dependent proteases of *E. coli* (see Table 2) can be divided into two categories: those, including Lon and FtsH, composed of a single polypeptide chain carrying both proteolytic activity and ATP-dependent protein remodeling activity, and those, including ClpAP, ClpXP, and ClpYQ, in which the proteolytic and protein remodeling activities are present on separate polypeptide chains. In both types of proteases the active form of the protease is a complex with multiple active sites of both kinds. The basic structure of ATP-dependent proteases, based on analysis of ClpAP, has a modular design in which sites with similar function are grouped together into structural domains, which then interact with and communicate with other functional domains. Thus, the proteolytic subunits associate to form a degradative center which is shielded from the surrounding medium. The protein-binding and remodeling domains form a separate structural domain that interacts with the proteolytic domain, further blocking indiscriminate access to the degradative center while simultaneously facilitating the transfer of select proteins to the proteolytic site. In more complex proteases, such as the 26 S proteasome, additional functional domains are present, although the number of such domains and their functions are still incompletely understood.

One of the intriguing aspects of protein degradation is the mechanism of selectivity—how are appropriate substrates recognized by proteases and how do the proteases avoid degrading normal proteins? Why are specific intrinsically unstable proteins targeted primarily by a single protease, at least to the extent that the functional half-life of the protein is dependent on one protease? Eukaryotic and prokaryotic cells appear to diverge significantly in their mechanisms of substrate selection. In bacteria, most protein substrates interact directly with the chaperone component of the protease which must recognize a feature of the protein that identifies it as an appropriate substrate for degradation. At least five different ATP-dependent proteases exist in *E. coli*, and each apparently has specific targets that it degrades from three to 15 times as fast as do other proteases. Most highly unstable proteins are degraded at significant rates in the absence of the protease that carries out the rate-limiting steps in its degradation, suggesting that there is some overlap in specificity between the proteases. Also, the proteases are at least partially redundant for degradation of the broad classes of abnormal or unfolded proteins. One explanation for these findings is that unstable proteins have, in addition to a specific motif recognized by a specific protease, general motifs that are recognized by any number of proteases. Such an explanation demands that the proteases themselves have multiple sites for binding protein substrates, some specific for a unique motif and some specific for more general degradation motifs.

In contrast, a majority of, if not all, proteins degraded in the cytosol and nucleus of eukaryotic cells are first modified by ubiquitination, and it is the ubiquitinated proteins that are recognized by the major ATP-dependent protease, the 26 S proteasome. The decision to ubiquitinate different proteins is made by a complex set of ubiquitin conjugating enzymes (E2s) and ubiquitin protein ligases (E3s) of which there are many with varying specificities in the cell. Thus, signal recognition and degradation are functionally and spatially separated in most eukaryotic cell compartments. Interestingly, a number of the eubacterial proteases are conserved in eukaryotic cells, where they occur in the mitochondria of animal cells and in the chloroplasts of plants and other photosynthetic organisms. Thus, the fidelity and efficiency of degradation by the eubacterial systems appears to suffice in highly evolved prokaryotes and eukaryotes, and the selective advantage of the ubiquitin-proteasome pathway for protein degradation remains to be elucidated.

Recent data suggests that bacterial cells also have a system for tagging proteins for degradation. Unlike the ubiquitin system, the tagging activity acts on incompletely synthesized proteins. Because the tag is added cotranslationally, this system cannot be used to mark completed proteins for degradation. Nevertheless, the tagged proteins are degraded by ATP-dependent proteases and thus provide the first direct clue regarding a universal recognition motif in substrates for one or more *E. coli* proteases. This review will describe this system in more detail and discuss what is known about the role of subunit dissociation and protein unfolding in exposing additional degradation motifs that reside in inaccessible regions of most stable cellular proteins.

II. LON PROTEASES

Lon protease from *E. coli* (also called protease La) was the first enzyme shown to have proteolytic activity that depended on ATP. Several reviews of Lon structure and activity have appeared (Goldberg, 1992; Gottesman and Maurizi, 1992; Maurizi, 1992; Goldberg et al., 1994). *E. coli* Lon is the product of a single gene encoding 784 amino acids (Chin et al., 1988). Estimates of the size of the purified protein from gel filtration and from sucrose density gradients suggest that Lon is a tetramer in solution and sometimes forms octamers. *In vitro*, Lon has protease and peptidase activity and has an active protein-stimulated ATPase activity (Waxman and Goldberg, 1986; Menon and Goldberg, 1987). Lon also binds to DNA, both nonspecifically and to a TG-rich sequence found in human immunodeficiency virus (called the peri-ets site) (Zehnbauer et al., 1981; Fu et al., 1997). The affects of this interaction on Lon activities *in vivo* are not clear, although there is at least one report of activation of Lon activity by DNA *in vitro* (Chung and Goldberg, 1982), and protein substrates weaken interaction between Lon and nonspecific DNA (Sonezaki et al., 1995). Lon binding to DNA may have additional significance in light of the reported tendency of yeast *lon* mutants to accumulate mitochondrial DNA deletions (Suzuki et al., 1994; Van Dyck et al., 1994).

Lon genes have been identified in a number of bacteria, in Archaea, in yeast, and in mammalian cells (Gottesman et al., 1995). Sequence analysis identified a Walker-type ATP binding consensus sequence in the center of the coding region and a serine protease active site located in the carboxy-terminal 25% of Lon (Chin et al., 1988). Ser679 is believed to be the active site residue, and is conserved in all Lon proteins. The protein is very well conserved, with 36% identity between the human and *E. coli* amino acid sequences (Wang et al., 1993). In eukaryotic cells, Lon is encoded in the nucleus but the protein is found apparently exclusively in mitochondria (Wang et al., 1993). A number of *in vivo* activities have been found for Lon proteases from different organisms, but very little is known about the *in vitro* properties of the proteases other than *E. coli* Lon.

A. Enzymatic Properties of Lon

ATP-Dependence of Lon Activity

In vitro, Lon will hydrolyze a variety of peptides and proteins, with unfolded or "denatured" proteins being preferred substrates (Waxman and Goldberg, 1986). Lon shows a preference for cleaving in regions of proteins with aliphatic side chains in the P1 position and hydrophilic residues in P2 or P3, but factors other than the sequence surrounding the cut site are equally or more important (Waxman and Goldberg, 1985; Maurizi, 1987). Turnover of most proteins is rather slow, about two peptide bonds per minute per active site of Lon (assuming one active site on each subunit). With a purified physiological substrate, λ N protein, Lon cleaved much faster, 60 peptide bonds per minute, but the K_m for N protein (~13 μM) was about four times that of α-casein (Maurizi, 1987). That Lon did not have a higher affinity for N protein was surprising in light of the very rapid turnover rate of N protein *in vivo* ($t_{1/2}$ ~ 1–2 minutes) (Gottesman et al., 1981). Another physiological Lon substrate, CcdA, had an even weaker binding affinity for Lon (K_m >200 μM) (Van Melderen et al., 1996). Interestingly, CcdA is degraded much more slowly *in vivo* ($t_{1/2}$ ~ 30 minutes) (Van Melderen et al., 1994), so that the K_ms for these two substrates are proportional to their degradation rates *in vivo*. The data suggest that purified N protein and CcdA are not in optimum conformations for recognition by Lon or that a site for specific substrates is not accessible on Lon under *in vitro* conditions.

Proteolytic activity of Lon on many high molecular weight substrates requires ATP hydrolysis. A point mutation in invariant Lys at the ATP-binding site inactivates the ATPase and proteolytic activity of Lon (Fischer and Glockshuber, 1994). Lon is also allosterically activated by nucleotide binding, and peptidase activity of Lon is highly stimulated by non-hydrolyzable analogues of ATP (Waxman and Goldberg, 1982). For protein degradation, ATP hydrolysis is needed continuously, and with saturating protein substrate and nucleotide, an average of 2 ATP molecules are hydrolyzed per peptide bond cleaved (Menon et al., 1987). Degradation by Lon is processive or recursive, meaning that the bound protein is cleaved at many sites in continuous order without release of high molecular weight intermediates. The products of degradation are peptides of 5–15 amino acids (Maurizi, 1987). Studies showing that ATPase activity of Lon was stimulated by protein substrates but not by non-substrate proteins led to a proposal that ATP hydrolysis is coupled to peptide bond cleavage during protein degradation by Lon (Waxman and Goldberg, 1982; Menon et al., 1987). However, several lines of evidence indicate that this model is not explicitly correct. First, a number of proteins, including N protein and α-casein, can be degraded without ATP hydrolysis (although a non-hydrolyzable analogue of ATP is needed as an activator) (Goldberg and Waxman, 1985; Maurizi, 1987). Second, a mutant of Lon in which Ser679 has been changed to Ala lacks proteolytic activity, but the ATPase activities of the mutant and wild-type Lon are stimulated equally by protein substrates (Fischer and Glockshuber, 1993). These results suggest that the ATPase and peptide bond cleaving activities are linked but mechanistically separated.

Protein Remodeling by Lon

Recent results suggest that ATP hydrolysis by Lon and other ATP-dependent proteases may be required for unfolding or remodeling protein substrates. Studies with a yeast homologue of Lon protease suggest that Lon can display chaperone-like activities *in vivo*. Multi-copy *Saccharmyces cerevisiae* Lon corrected a defect in assembly of ATP synthase in a mutant lacking the wild-type chaperones that normally promote this process (Rep et al., 1996b). The activity was ATP-dependent but did not require a functional proteolytic active site. Since other studies have shown that Lon can also degrade ATP synthase *in vivo* (Suzuki et al.,

Table 1. Activities of ATP-Dependent Proteases *In Vivo*

Biologically Active Species	*Substrate/Type of Activity*	*Phenotype in Absence of Activity*
ClpP alone	Degradation of CRAG	Artificial substrate, no known phenotypes associated with ClpP alone
ClpA alone	No substrates identified	
ClpAP	Degradation of MazE	Cell death due to action of MazF
	Degradation of β-Gal fusion proteins	Increased β-galactosidase
ClpX alone	Disassembly of MuA/DNA strand transfer complex	No propagation of phage Mu
ClpXP	Degradation of λ O protein	Slight defect of lambda growth
	Degradation of phage P1 Phd	Cell death after loss of phage P1 caused by action of Doc
	Degradation of RpoS	Inappropriate expression of stationary phase genes
ClpYQ	Sigma E	Not known
Lon	Degradation of SulA	Sensitivity to DNA damage due to inhibition of septation
	Degradation of RcsA	Overproduction of capsular polysaccharide
	Degradation of CcdA	Cell killing by CcdB after loss of F factor
C. crecentus Lon	Degradation of CcrM	Improper timing of cell division
B. subtilis Lon	Degradation of sigma G	Inappropriate expression of sporulation genes
S. cerevisiae Lon	Degradation of ATP synthase subunits	Respiratory defects and no growth on nonfermentable carbon sources, accumulation of mitochondrial DNA deletions
FtsH	Degradation of sigma 32	Temperature sensitive growth, constitutive expression of heat shock genes
	Degradation of λ CII	Reduced lytic growth of phage λ
S. cerevisiae YTA-10 (FtsH homolgue)	Degradation and assembly of membrane protein complexes	Respiratory defects, no growth on nonfermentable carbon sources

1994), these results imply that Lon can alter the conformation of proteins with which it interacts. A completely different type of experiment also points to Lon's ability to alter protein conformation. *In vitro* studies of CcdA degradation by *E. coli* Lon indicate that the presence of secondary structure in a substrate determines the need for ATP hydrolysis (Van Melderen et al., 1996). CcdA which has stable α-helical and β-sheet structures requires ATP hydrolysis to be degraded by Lon, whereas a truncated mutant protein, CcdA41, which lacks stable secondary structure is degraded in the presence of non-hydrolyzable analogues of ATP. Moreover, incubating CcdA with Lon at 50 °C which is the melting temperature for the α-helices, allows degradation in the presence of non-hydrolyzable ATP analogues (Van Melderen et al., 1996).

The correlation between substrate secondary structure and a requirement for ATP hydrolysis and the demonstration that proteins can be released from Lon with altered conformations suggest that Lon carries out two consecutive steps in degrading proteins—a protein remodeling reaction that prepares the protein for degradation and a peptide bond cleavage. Presumably, the first step is rate-limiting for most proteins, and thus most protein degradation is dependent on ATP hydrolysis. A relatively constant ratio of ATP hydrolyzed to peptide bonds cleaved is expected if the energy cost for unfolding and translocating a segment of protein is proportional to the size of the segment and if, on the average, there are a constant number of susceptible bonds per length of polypeptide.

General Chaperone Contribution to Lon Activity

In vivo there is an overlap between the substrates degraded by Lon and some of the proteins whose degradation is affected by molecular chaperones (Gottesman et al., 1997). The largest groups of such proteins studied are canavanyl proteins and nascent polypeptides truncated with puromycin. Rates of degradation of these abnormal proteins are reduced about 50% in *lon* mutants and about 50–80% in mutants lacking DnaJ, a component of a major chaperone system (Straus et al., 1988). Also, at least one specific Lon substrate, RcsA, is not degraded in the absence of DnaJ (Jubete et al., 1996). Since many Lon substrates are degraded equally in wild-type and *dnaJ* mutant cells, chaperones are obviously not required for all Lon degradation (Jubete et al., 1996). *In vitro*, Lon activity, although much slower than activity *in vivo*, is not affected by purified DnaJ and DnaK (Maurizi, unpublished observations), indicating that Lon can act directly on protein substrates without help from chaperones. When chaperones appear to be needed for degradation of a protein, the effect may be indirect. For example, canavanyl proteins have a tendency to aggregate and in the absence of DnaJ are recovered in the insoluble fraction of cell extracts (Jubete et al., 1996). Similar results were obtained in studies with yeast mitochondria; chaperone mutations interfered with Lon-dependent degradation of abnormal proteins and the substrates were found to have aggregated (Wagner et al., 1994). Thus, chaperone effects on degradation can be explained by their maintaining proteins in a soluble form that is accessible to Lon rather than to any direct interaction between the protease and the chaperone..

B. *In vivo* Activities of Lon Proteases

E. coli *Lon*

Functions of Lon have been identified on the basis of the effects seen in knockout mutants, which presumably lose both putative chaperone activity as well as proteolytic activity. Lon has activity *in vivo* directed at specific proteins and at a variety of misfolded proteins. Mutants lacking Lon have 50% reduced rates of degradation of abnormal proteins, such as canavanine-containing proteins and puromycyl fragments (Maurizi et al., 1985). Lon also degrades a number of foreign proteins, unassembled components of complexes, and temperature sensitive mutant proteins (Gottesman, 1996). Lon is expressed constitutively, but synthesis increases under heat shock and other stress, probably in response to the presence of misfolded proteins (Baker et al., 1984; Goff et al., 1984).

Lon also targets several specific proteins and thereby exerts a number of regulatory effects in *E. coli*. Lon substrates are among the most rapidly degraded proteins, with half-lives from less than one minute for RcsA (Torres-Cabassa and Gottesman, 1987) to about 30 minutes for CcdA (Van Melderen et al., 1994). The specificity of degradation is indicated by the change in degradation rate of these proteins when *lon* is mutated. In general, the most rapidly degraded proteins are the most specific substrates; thus, the half-lives of RcsA and SulA increase 10–20 fold in *lon* mutants (Jubete et al., 1996), whereas the half-life of CcdA increases only threefold (Van Melderen et al., 1994).

Lon Activity in UV Damaged Cells. *lon* mutants of *E. coli* are sensitive to UV irradiation and other DNA damaging agents (Gottesman, 1987). Induction of LexA-controlled genes by DNA damage leads to production of SulA (also called SfiA), a protein that inhibits septation by binding to FtsZ (Jones and Holland, 1985). Accumulation of SulA is a lethal event, because the cells continue to grow without putting down septa and dividing. Eventually the long filaments of undivided cells lyse and die. In wild-type cells, irreversible filamentation is avoided by rapid degradation of SulA by Lon (Mizusawa and Gottesman, 1983). When synthesis of SulA is shut off following DNA repair, the SulA is depleted and septation proceeds.

Recently, another role for Lon during cellular response to DNA damage was reported. Two proteins, UmuC and UmuD, components of an error-prone DNA repair system, are targets for Lon protease (Nohmi et al., 1988; Frank et al., 1996). UmuD is a precursor of UmuD' which is formed by cleavage of UmuD by activated RecA following DNA damage. UmuD' forms a homodimeric complex and binds to UmuC making the active mutagenic DNA repair enzyme. Lon degradation appears to be aimed at keeping the concentrations of the precursors of the mutagenic com-

Table 2. ATP-Dependent Proteases in *E. coli*

Component	*Subunit Molecular Weight*	*Oligomeric State*	*Activities*	*Protease*	*Selected Homologues*
ClpP	21,600	Heptamer	Peptidase		Chloroplast ClpP
ClpA	84,600	Hexamer	ATPase, chaperone	$ClpA_6ClpP_{14}ClpA_6$	Chloroplast ClpC
ClpX	46,400	Oligomeric ring	ATPase, chaperone	Complex of $ClpX_6ClpP_1$ $ClpX_6$	
ClpQ	19,100	Hexamer	Peptidase		Archaea proteasome β subunit
ClpY	49,600	Hexamer and heptamer	ATPase	Complex of $ClpY_6$ or Y_7 and $ClpQ_{12}$	Pasteurella LapA
ClpB	95,500	Hexamer	ATPase (protein activated)	None known	HSP104
Lon	87,400	Tetramer and octamer	Peptidase, ATPase, ATP-dependent protease	Both activities in same subunit	Matrix of mitochondria
FtsH/HflB	70,700	Oligomeric	ATPase, ATP-dependent protease	Both activities believed to be in same subunit	Membrane of mitochondria, plant thylakoid membrane

plex as low as possible. UmuD' is less susceptible to degradation by Lon, but whether the difference is due to the removal of the N-terminal segment from UmuD or to consequent conformational changes has not been reported.

Degradation of a Transcriptional Regulator, RcsA. *E. coli* strains carrying the genes for the capsular polysaccharide (cps), colanic acid, become mucoid when *lon* is mutated (Gottesman, 1987). *cps* expression is under control of a positive transcriptional regulator, RcsA, which is also a substrate for Lon protease. RcsA is apparently synthesized under most growth conditions, but degradation by Lon ($t_{1/2}$ 1–5 minutes) keeps the amount of RcsA too low to activate *cps* transcription (Torres-Cabassa & Gottesman, 1987; Stout et al., 1991). RcsA is protected from degradation in the presence of the activated form of another positive regulator, RcsB, with which it probably forms a complex (Stout et al., 1991), suggesting that sites recognized by Lon may be in the region of RcsA that interacts with RcsB. RcsA has a tendency to aggregate into a form insoluble in cell extracts, and accumulation of an active soluble form of RcsA requires the DnaK/DnaJ chaperone system (Jubete et al., 1996). In DnaJ mutants, RcsA is somewhat more stable, but the protein that survives is inactive and insoluble. Excess RcsB will also help keep RcsA soluble even in the absence of DnaJ (Jubette, Y., Maurizi, M.R., and Gottesman, S., unpublished observations). These results are consistent with a model in which Lon, chaperones, and activated RcsB all interact with the same form of RcsA, which otherwise has a tendency to rapidly aggregate. The rates at which RcsA interacts with the others or itself are in the order: RcsB ≥ chaperones > aggregation > Lon. As proposed for degradation of abnormal proteins, chaperones would facilitate RcsA degradation by maintaining the protein in a soluble form that is more accessible to Lon.

Lambda: Benefiting From a Degrading Experience

Nowhere have host cell's proteases been exploited more impressively than with phage lambda. The list of unstable lambda protein includes those involved in most of the critical functions of the lambda life cycle, starting with the repressor, CI (Roberts and Roberts, 1975), and continuing with the replication proteins, O and P (Gottesman et al., 1981), the lytic/lysogenic switch protein, CII (Gottesman et al., 1981; Kornitzer et al., 1991), a coregulator of CII, CIII, the antitermination protein, N (Gottesman et al., 1981), and the excision protein, Xis (Weisberg and Gottesman, 1971). A number of these substrates are degraded by the ATP-dependent proteases of *E. coli* and will be discussed later. The regulatory consequences of degradation of these proteins have not been worked out in most cases.

Lon degrades the λ N protein and recent data suggests that it may also degrade Xis. The half-life of N protein increases from 1–2 minutes when Lon is present to 20–30 minutes in a *lon* mutant (Gottesman et al., 1981). The consequences of N protein degradation for lambda have not been defined. Lambda does have a phenotype on *lon* mutant strains, having lower frequencies of lysogenization; however, this phenotype is unrelated to N protein degradation. Degradation of λ CII protein (see below) is actually faster in *lon* mutants, which could account for the inhibition of lysogenization. The effect of *lon* mutations on CII degradation underscore the interrelationships of the ATP-dependent proteolytic systems. Different proteases have both unique substrates and common substrates, proteins that act within a common pathway may be targeted by different proteases, and expression and activities of different proteases may be regulated by common signals. Thus, metabolic signals aimed at or generated by a specific protease will have pleiotropic effects on other proteases.

Post-Segregational Killing. Lon activity has also been exploited by low copy number plasmids, including the lysogenic form of phage P1, to help maintain plasmid stability in a population. Cells that have harbored certain low copy number plasmids have a tendency to die if the plasmid is lost or segregated, a phenomenon known as postsegregational killing (Jensen et al., 1995). Because the cell appears to become dependent on the plasmid, the process has also been called plasmid addiction (Yarmolinsky, 1989). In each case the plasmid-encoded components of the system are two proteins, one of which is a toxin and the other of which is an antidote. In the systems relevant to this discussion, the toxin is stable and the antidote protein is unstable but produced in sufficient excess to neutralize the toxin as long as the antidote can be continuously synthesized. If a plasmid is lost, synthesis of the antidote is no longer possible in that cell, and degradation of the antidote leaves the toxin to kill the cell. A phage P1 postsegregational killing system involves the toxin, CcdB, a DNA gyrase inhibitor, and CcdA, the antidote (de Feyter et al., 1989). CcdA is a substrate for Lon and is degraded with a half-life of about 30 minutes (Van Melderen et al., 1994). Loss of a miniP1 carrying CcdA and CcdB kills about 90% of wild-type cells, but *lon* mutants, in which CcdA is stable, are not killed.

The IncFII plasmids R1 and R100 also have a postsegregational killing system made up of a toxin (PemK) and an unstable antidote (PemI) (Tsuchimoto et al., 1992). PemK is a substrate for Lon protease. There is little similarity in sequence between the Lon substrates CcdA and PemI, although the proteins may fold in a similar manner.

Developmental Roles for Bacterial Lon Proteases

Myxococcus xanthus is a gram-negative bacterium that grows vegetatively but can be induced by nutrient deprivation to undergo a developmental change to form a mass of round myxospores called a fruiting body. *M. xanthus* has two *lon* genes that encode proteins with more than 50% homology to *E. coli* Lon (Gill et al., 1993; Tojo et al., 1993a,b). The two proteins are differentially expressed and are apparently required at different times of growth and development. LonV is essential for vegetative growth, whereas LonD is required for myxospore formation. The targets for these proteases have not been identified, but mutants in LonD are arrested at an early stage of development.

In *Caulobacter crescentus* Lon is responsible for degradation of an adenine DNA methyltransferase, CcrM (Wright et al., 1996). CcrM is essential for viability but is present only during part of the cell division cycle. CcrM is synthesized in predivisional cells and must be degraded before the cells divide. In *lon* mutants, CcrM is stable and therefore present throughout the cell cycle, leading to a number of defects, including incorrect timing of DNA replication.

Lon proteases may also target transcriptional factors in bacteria and affect developmental pathways. *Vibrio parahaemolyticus* undergoes developmental changes from a cell type called a swimmer to one called a swarmer, and this differentiation involves changes in expression of many genes. Lon mutations in *V. parahaemolyticus* are constitutive for swarmer cell gene expression but showed no defect in swimming motility (Stewart et al., 1997). The target for Lon has not been identified but may be a protein that positively regulates swarmer cell gene expression. A cloned *V. parahaemolyticus* Lon complemented an *E. coli lon* mutant for both UV resistance and capsular polysaccharide gene expression. A mutation in the *lon* gene of *Bacillus subtilis* leads to inappropriate expression of genes under control of a specialized sporulation sigma factor, sigma G (Schmidt et al., 1994). The properties of the mutant suggest that sigma G itself may be a target for degradation but direct evidence for such degradation has not been reported. Sigma G positively regulates its own expression, and in wild-type cells Lon may help modulate its activity.

Lon Activity in Yeast Mitochondria

The eukaryotic Lon protease homologue is synthesized in the cytosol with a mitochondrial matrix targeting sequence and taken up into mitochondria (Wang et al., 1993; Suzuki et al., 1994; Van Dyck et al., 1994). Recombinant human and yeast Lon proteases have ATP-dependent protein degrading activity. Yeast mutants lacking Lon protease (also known as PIM1) have respiration defects and do not grow on non-fermentable carbon sources (Suzuki et al., 1994; Van Dyck et al., 1994). The mitochondrial genome in a yeast *lon* mutant is nonfunctional and accumulates DNA deletions. Yeast *lon* mutants degrade several mitochondrial matrix proteins, including the b subunit of F1 ATPase, at a reduced rate and accumulate dense inclusion bodies within the matrix.

Yeast Lon protease also displays an apparent chaperone activity by promoting assembly of ATP synthase in yeast cells lacking either of two membrane-associated protease/chaperones, YTA-10 and YTA-12 (Rep et al., 1996b). Although this chaperone activity is more readily seen with a proteolytically inactive mutant of Lon, it is also seen with active Lon that is capable of degrading the ATP synthase subunits. Lon thus appears to function at a branchpoint in assembly of ATP synthase and possibly other multimeric complexes: it either promotes interaction between the components and helps in assembly or it retains unassembled subunits and degrades them. Analysis of the mechanism of discrimination between the pathways and the kinetics of the decision process should provide a fascinating insight into protein quality control in cells. A cloned *E. coli* Lon protease targeted to mitochondria complemented the yeast *lon* mutation for degradation of misfolded proteins and helped to maintain mitochondrial genome integrity and respiratory sufficiency (Teichmann et al., 1996). *E. coli* Lon lacking proteolytic activity was unable to restore mitochondrial function to the *lon* mutant, confirming that proteolytic activity is at least one essential aspect of Lon's contribution to mitochondrial function.

III. FtsH: MEMBRANE-BOUND ATP-DEPENDENT PROTEASES

One of the most exciting recent developments in the protease field has been the identification of a membrane-bound ATP-dependent proteolytic system in *E. coli* and other organisms. FtsH (also known as HflB) is an integral membrane protein of 644 amino acids (Tomoyasu et al., 1993). It has two membrane-spanning domains, and the carboxy-terminal 530 amino acids are in the cytoplasm. FtsH appears to form an oligomeric complex in the membrane but its structure is not known (Akiyama et al., 1995). Sequence analysis reveals a consensus ATP binding sequence starting at Gly192 and a Zn-metalloprotease motif about 220 amino acids from the carboxy-terminus. *ftsH* transcription is partially under heat shock control. Purified FtsH displayed a significant ATP-dependent proteolytic activity against the purified *E. coli* protein, σ^{32}, purified λ CII protein (Shotland et al., 1997), and several membrane proteins (Tomoyasu et al., 1995).

FtsH homologues are widespread in bacteria and eukaryotic cells. As with Lon, the eukaryotic homologues are synthesized in the cytosol and imported into mitochondria where they are located in the inner membrane (Langer and Neupert, 1996). An FtsH homologue has also been found in the thylakoid membrane of plant cells (Lindahl et al., 1996). FtsH belongs to the AAA family of ATPases and has an extended domain surrounding the ATP consensus sequence that resembles comparable regions of a number of ATPases with diverse functions. Notable among these are ATPase subunits of the 26 S proteasome which share >40% identical amino acid residues over a region of >220 amino acids. This degree of homology suggests that there is likely to be structural or mechanistic similarities for the ATPase domains of these protease components.

A. Enzymatic Properties of FtsH

Purified *E. coli* FtsH degrades the *E. coli* sigma factor, σ^{32} *in vitro* (Tomoyasu et al., 1995). Degradation requires ATP hydrolysis. The rate of degradation increased about two-fold between 37 °C and 42 °C and there was about 1 molecule of σ^{32} degraded per minute per subunit of FtsH at the higher temperature. Similar to Lon, FtsH proteolytic activity is also activated by CTP (and possibly by UTP with some substrates), but other nucleotide triphosphates are ineffective, as are non-hydrolyzable analogues of ATP. Proteolytic activity is activated by Zn, and high concentrations of EDTA or *o*-phenanthroline are inhibitory. Lack of complete inhibition by chelators suggests that either the Zn is not essential for proteolytic activity or the metal ion is very tightly bound and not completely removed by the chelating agents. Basal ATPase activity is quite low (<2 μmol/min/mg FtsH) and is dependent on Mg. Purified FtsH is also able to degrade two membrane proteins, SecY and subunit a of H^+-ATPase, both of which are substrates for the protease *in vivo* (Akiyama et al., 1996a,b). FtsH with a mutation in the ATP consensus sequence (K198N) was proteolytically inactive, while a mutant in which His414 of the Zn binding motif was changed to Lys had reduced activity (Akiyama et al., 1996b).

B. *In vivo* Activities of *E. coli* FtsH

FtsH is essential in *E. coli*, but whether specific activities of FtsH are required for cell growth or loss of those activities causes pleiotropic effects that interfere with growth is not known. Several different FtsH functions have been identified on the basis of the phenotypes of conditional mutants or partially defective mutants. The *ftsH* gene was originally identified as *hflB*, a gene that affected the frequency of lambda lysogenization (Banuett et al., 1986). The decision to grow lytically or to form lysogens is determined for lambda by the amounts of a protein called CII, which is rapidly degraded in wild-type cells. Mutations in *hflB* led to higher frequencies of lambda lysogenization because of stabilization of λCII . The original *hflB* mutation was only partially defective in FtsH activity and did not cause defects in any other FtsH functions. λCII degradation is inhibited by another lambda protein, CIII (Kornitzer et al., 1991). Overproduction of CIII also leads to stabilization of the heat shock sigma factor, σ^{32} (Bahl et al., 1987), providing the first hint that one protease might degrade both CII and σ^{32}. Temperature-sensitive *ftsH* mutants accumulate σ^{32} at the nonpermissive temperature; these mutants also cannot grow at the higher temperature. The half-life of σ^{32} is extended by a factor of >10 under conditions that destroy functional FtsH, indicating that FtsH is responsible for the rate-limiting degradation of σ^{32} (Herman et al., 1993, 1995b, 1997). High concentrations of Fe (II), Ni (II), Mn (II), and Co (II) suppress the temperature-sensitivity of growth of *ftsH*(ts) mutants and restore degradation of σ^{32} (Herman et al., 1995a). Thus, divalent metal ions bind to FtsH *in vivo*, stabilize its conformation, and may activate its functions. σCIII is also degraded by FtsH; CIII competes for FtsH with the other substrates, thereby slowing their degradation (Kornitzer et al., 1991; Herman et al., 1997).

A number of phenotypes of *ftsH* mutants reflect problems with assembly of membrane protein complexes and other activities dependent on proper structure and organization of the cytoplasmic membrane. Two potential targets for degradation have been identified in *E. coli* membranes. SecY, an integral membrane protein that normally associates with SecE, is degraded *in vivo* when it is not complexed with SecE (Kihara et al., 1995). Degradation is blocked in cells lacking functional FtsH (Akiyama et al., 1996b). FtsH with a mutation in the ATPase site can be chemically cross-linked to SecY *in vivo* demonstrating that these proteins can form a complex in the membrane (Akiyama et al., 1996b). Another integral membrane component, subunit a of the H^+-ATPase, was degraded in an FtsH-dependent reaction *in vivo* when it was produced in excess (Akiyama et al., 1996a). Although SecY and subunit a were overproduced to generate sufficient amounts of uncomplexed protein for the degradation assays, these data suggest that FtsH may be involved in regulating the proper stoichiometry of the components of membrane-protein complexes.

The targets of FtsH, and even the nature of the activity, responsible for producing certain phenotypes at the nonpermissive temperatures are not known. For example, a mutation in His418 in the putative Zn binding site is responsible for the tolerance to colicins and the inability to grow on non-fermentable carbon sources in *tolZ* (an allele of FtsH) mutants (Qu et al., 1996). *ftsH*(ts) mutants have effects on protein assembly and transport through the membrane (Akiyama et al., 1994a).

In other *ftsH* mutants, so-called stop transfer-defective (*std*) strains, PhoA fusion proteins that normally remain in the cytoplasm are allowed to be exported to the periplasm (Akiyama et al., 1994b). These latter phenotypes can be suppressed by overproduction of chaperones (Shirai et al., 1996), suggesting that some functions of *ftsH* may require the protein interaction/remodeling activity and not the proteolytic activity.

C. Yeast Mitochondrial FtsH

The yeast mitochondrial inner membrane contains two FtsH homologues, YTA-10 (also called AFG3) and YTA-12 (also called Rca1) (Guelin et al., 1994; Tauer et al., 1994). Both proteins are synthesized in the cytosol and imported into mitochondria. The proteins associate to form a complex in the membrane with their active sites facing opposite sides of the membrane (Arlt et al., 1996; Langer and Neupert, 1996). YTA-10 and YTA-12 express ATP-dependent proteolytic activity in yeast cell extracts.

Mutants lacking both proteins have defects in assembly of membrane complexes, degradation of endogenous mitochondrial proteins, and respiration-dependent growth. *In vivo* degradation of unassembled ATPase synthase subunits was partially blocked in YTA-10–12 mutants and could be restored by bringing in either protein on a multi-copy plasmid (Rep et al., 1996a). When YTA-10–12 carrying a mutation in the Zn-metalloprotease site was reintroduced, no degradation was observed. Instead, assembly of ATP synthase complexes was accelerated. These data indicate that, as seen for Lon, the protein interaction/chaperone domain of these FtsH homologues can function independently of the proteolytic active site in promoting assembly of membrane complexes. With wild-type FtsH, subunits that cannot assemble rapidly or with the correct structure will be degraded.

IV. HflA: A G-PROTEIN COMPLEX OF UNKNOWN ACTIVITY

The *hflA* locus was also identified because it affected lambda lysogenization (Hoyt et al., 1982). There are three components encoded within *hflA*: HflX, HflK, and HflC (Banuett and Herskowitz, 1987). Unlike *hflB*, *hflA* is not essential in *E. coli*; mutations in *hflA* lead to stabilization of the CII protein of lambda (Hoyt et al., 1982). A number of other *E. coli* proteins may also be targeted by HflA. HflK and HflC may be integral membrane proteins (Cheng and Echols, 1987). HflC has a sequence that is reminiscent of the ClpP proteolytic site motif, but otherwise the protein sequence is not similar to that of ClpP or any known proteases. HflX has sequence similarity to GTP-binding proteins (Noble et al., 1993).

HflK and HflC were purified and shown to have a proteolytic activity against purified λCII (Cheng et al., 1988). The activity was not dependent on ATP or other nucleotides. No reports of the effects of HflX on this or other activities have appeared. Whether these proteins constitute a GTP-dependent protease or any protease at all is still not clear. Recently, mutational studies of the effect of HflK-HflC on SecY degradation have suggested that HflC/HflK, rather than contributing to the degradation of proteins targeted by FtsH, actually antagonizes the activity of FtsH (Kihara et al., 1996). Two lines of experiments provided evidence for interaction between HflC-HflK and FtsH: cross-linking and immunoprecipitation showed an association between the proteins, and addition of HflC-HflK to reaction mixtures inhibited FtsH degradation of SecY. Thus, at this point, whether HflC-HflK is a negative modular of FtsH, a protease, or both is not known.

V. THE CLP FAMILY

Members of the Clp family were first identified and isolated in *E. coli* (Katayama-Fujimura et al., 1987). The Clp name was given in reference to the proteolytic activity of ClpAP, which "clips" proteins into seemingly random peptides 5–15 amino acids in length (Katayama et al., 1988). The name now applies to two separate groups of proteins, the ATPases ClpA, ClpB, ClpX, and ClpY (HslU), and the proteases ClpP and ClpQ (HslV). The Clp ATPases are related by sequence similarity (Gottesman et al., 1995), whereas the proteases, ClpP (Maurizi et al., 1990c) and ClpQ (Chuang et al., 1993), have no sequence similarity to each other or to the ATPases but are functionally related by interacting with different members of the Clp ATPase family. Biochemical and genetic data have shown that ClpA and ClpX interact with and activate ClpP (Maurizi, 1992), while ClpY interacts with and activates ClpQ (Kessel et al., 1996; Missiakas et al., 1996; Rohrwild et al., 1996; Yoo et al., 1996). ClpB stands alone in having no known activity as a component of a protease.

The amino acid sequences of ClpA and ClpB are most closely related, with 42% identity and 63% similarity (Gottesman et al., 1990b, 1995). The homology between ClpA and ClpB is interrupted by an insertion in ClpB of 119 amino acids between the two regions containing the ATPase sites. ClpX and ClpY are slightly more closely related to each other (32% identity, 55% similarity) than they are to ClpA or ClpB (21% identity, $\leq$ 50% similarity to both) (Gottesman et al., 1995). ClpA (758 amino acids) and ClpB (857 amino acids) are also much larger than ClpX (423 amino acids) and ClpY (442 amino acids). ClpA and ClpB have two ATP consensus sequences located in otherwise nonhomologous regions of the protein, suggesting that each protein is a fusion of two separate ATPases. ClpX and ClpY each have a single consensus ATP binding sequence.

The Clp ATPases are highly conserved in other bacteria and in eukaryotes, although it is not clear whether the four classes retain their identities in all organisms. The widespread occurrence of all the Clp ATPases indicates that the separate *E. coli* proteins evolved in a common ancestral organism, which also accounts for the considerable divergence between them. A striking difference between ClpB and the other Clp ATPases is that ClpB apparently does not function as a component of a protease. ClpB has a number of functions that can be explained by its acting as a chaperone, possibly by promoting protein disaggregation (for a review see Schirmer et al., 1996). Since other Clp ATPases also have chaperonelike activities, this property of ClpB does not preclude its acting as part of a proteolytic system also, but resolution of this issue awaits identification of a cognate protease.

ClpP and ClpQ are also found in many organisms. In *E. coli* ClpP the proteolytic active site is composed of a catalytic triad of Ser111, His136, and Asp185 and these residues are universally conserved (Maurizi et al., 1990a,b; Wang et al., 1997). ClpP has been found in most prokaryotes and eukaryotes but not yet in Archaea. In plants, ClpP is encoded in the chloroplast genome and the protein has been shown to be expressed in chloroplasts (Ostersetzer et al., 1996). To date, plants remain the only eukaryotic organisms in which the proteolytic activity of Clp has been demonstrated. Plants express ClpA homologues (sometimes referred to as ClpC) with chloroplast transit sequences attached, and an active ATP-dependent protease is formed between ClpA and ClpP within the chloroplast (Shanklin et al., 1995). ClpP has also been found in humans, where it is encoded in the chromosome but imported into mitochondria (Bross et al., 1995; Maurizi et al., 1998). No Clp ATPase has yet been found in human mitochondria, although the gene for a yeast mitochondrial ClpX has been identified (Ni, B. and Mauriji, M.P., unpublished observations). A recombinant form of the human ClpP has been purified after expression in *E. coli* and was shown to be similar in structure to the *E. coli* protein (Maurizi et al., 1998). The activity and function of human ClpP have not been characterized.

ClpQ, also known as HslV, belongs to the proteasome family (Chuang et al., 1993). ClpQ has 27% identity and 50% similarity in amino acid sequence to the β subunit of the Archaea proteasome, including the catalytic, amino terminal threonine residue.

A. The ClpAP and ClpXP Modular Proteases

The first Clp protease was isolated from cell extracts of an *E. coli lon* mutant as an ATP-dependent protease that could degrade casein. That enzyme, ClpAP (also referred to as protease Ti) is a complex of two proteins, ClpA and ClpP, which are separated during purification and must be reconstituted to make the active protease (Katayama-Fujimura et al., 1987; Hwang et al., 1988; Katayama et al., 1988). Subsequently, it was found that another protein, ClpX, could also combine with ClpP to form an ATP-dependent protease with a different substrate specificity (Gottesman et al., 1993; Wojtkowiak et al., 1993). ClpA and ClpX are both ATPases and enable ClpP to degrade higher molecular weight protein substrates. In addition to their activity in combination with ClpP, ClpA and ClpX on their own display an ATP-dependent protein structure-remodeling activity similar to that of molecular chaperones (Wickner et al., 1994; Levchenko et al., 1995; Wawrzynow et al., 1995). In at least one instance, this chaperone activity is sufficient for a biologically important function (Levchenko et al., 1995; Kruklitis et al., 1996).

Proteolytic Activity and Specificity of ClpAP and ClpXP

In the absence of other proteins, ClpP has peptidase activity against short (<10 amino acids) peptides (Woo et al., 1989; Thompson & Maurizi, 1994). Degradation of longer polypeptides and proteins requires the complex of ClpP and ClpA (Thompson et al., 1994). As with Lon, ClpP shows no strict requirement for residues near the peptide bond cleaved but cuts preferentially after non-polar residues (Thompson and Maurizi, 1994; Thompson et al., 1994). Degradation of longer polypeptides and proteins requires the ClpAP complex, and degradation is dependent on ATP binding and, in most cases, hydrolysis. The current model is that substrates with significant secondary and tertiary structure have limited access to the active sites in ClpP alone. Formation of the ClpAP complex produces an allosteric change in ClpP which allows polypeptides or short proteins with little structure greater access to the active sites (Thompson and Maurizi, 1994). Cleavage of such substrates occurs very rapidly without ATP hydrolysis. Non-hydrolyzable analogues of ATP (ATPγS and AMPPNP) are as effective as ATP in activating degradation of polypeptides, but degradation of large proteins by ClpAP requires continuous ATP hydrolysis. Turnover rates for cleavage of peptide bonds in proteins (10–20 min^{-1}) are considerably lower than those seen with polypeptide substrates (≤ 800 min^{-1}) (Maurizi et al., 1994), suggesting that the rate-determining step is structural disruption of the substrate and its translocation to the active sites. The products of degradation by ClpP or ClpAP are peptides of five to 15 amino acids and no exopeptidolytic activity is seen.

ClpA has an active ATPase activity which is stimulated 20–40% in the presence of various protein and peptide substrates. With wild-type ClpA, neither binding of ClpP nor substrate degradation increases ATPase activity significantly. Site-directed mutations in either of the ATPase sites of ClpA affect both ATPase and protein degradation activities (Singh and Maurizi, 1994; Seol et al., 1995), but mutations in the carboxy terminal domain (called domain II) have the largest effects (> 90%), suggesting that domain has the major role in activating degradation by ClpP. Some mutations in the ATPase site of the amino terminal region (called domain I) interfere with self-association of ClpA subunits indicating that domain I is important in assembly of oligomeric ClpA and of the ClpAP complex (Singh and Maurizi, 1994).

ClpAP activity is usually measured *in vitro* using casein as a substrate. ClpAP has activity with a variety of other unstable proteins *in vitro*; most Lon substrates (e.g., N protein and CcdA) are also degraded by ClpAP (Maurizi, unpublished observations). ClpAP also degrades the RepA protein of P1 phage in a purified system (Wickner et al., 1994), although RepA does not appear to be a substrates *in vivo*.

Conversely, several β-Gal fusion proteins degraded by ClpAP *in vivo* have been tested as substrates *in vitro*, but no degradation has been observed (Maurizi, unpublished observations). ClpAP does show specificity for a peptide that corresponds to the amino terminal region of ClpP itself (Thompson and Maurizi, 1994), which undergoes autocatalyzed processing *in vivo*.

ClpX was identified as a protein that activated ClpP when purified λ O protein was used as a substrate (Wojtkowiak et al., 1993). ClpXP degrades λ O protein into small peptides, and degradation is dependent on ATP hydrolysis. *In vivo*, ClpXP appears to be the major protease responsible for λ O protein degradation (Gottesman et al., 1993), and *in vitro* ClpX largely retains that fidelity. ClpXP has little or no activity against casein and activity with peptide substrates has not been reported. ClpAP, on the other hand, does have activity with λO protein (Wojtkowiak et al., 1993, Maurizi, unpublished observations), although degradation occurs at < 25% of the rate with ClpXP. ClpX has a much higher K_m for ATP (> 1 mM) than does ClpA (< 0.2 mM) in activating degradation by ClpP, but whether this difference reflects differences in regulation *in vivo* remains to be determined. ClpX also activated ATP-dependent degradation of purified MuA protein in the presence of ClpP, but no details of the kinetics of this degradation are available (Levchenko et al., 1995).

Chaperone Activities of ClpA and ClpX

ClpA interacts strongly with the phage P1 replication protein, RepA, and converts the inactive dimer of this protein into an active monomer (Wickner et al., 1994). This remodeling of RepA structure was previously demonstrated *in vitro* for the classical DnaK/DnaJ chaperone system (Wickner et al., 1991), which also appears to be the enzyme that activates RepA *in vivo*. The ability of ClpA to mimic DnaK/DnaJ indicates that it has an intrinsic protein remodeling activity. When ClpP is present with ClpA, the RepA protein is degraded, suggesting that the primary function of the remodeling activity of ClpA is to partially unfold proteins so that they can enter the ClpP active sites. ClpA also binds unfolded luciferase and prevents its irreversible denaturation *in vitro* (Wickner et al., 1994). Thermally inactivated luciferase can be reactivated by the molecular chaperones, DnaK/DnaJ, but the efficiency of reactivation is dependent on binding to chaperones during the heating step. The presence of ClpA during heating allows luciferase to be reactivated when the other chaperones are added, although ClpA itself cannot catalyze the refolding.

The chaperone activity of ClpX has been shown in two systems. The λ O protein has a tendency to aggregate under various conditions *in vitro*, and ClpX can prevent this aggregation (Wawrzynow et al., 1995). Prevention of aggregation does not require ATP hydrolysis, suggesting that binding of ClpX to λ O protein is sufficient for this activity. Previously aggregated λ O protein can be disaggregated by ClpX, but this reaction requires ATP-hydrolysis. ClpX has been shown to have a very specific chaperone activity with MuA (Levchenko et al., 1995; Kruklitis et al., 1996). During replication and transposition of Mu, a tetramer of the replication protein, MuA, forms a very tight strand transfer complex on the DNA. This complex is extremely stable and resists heating to 65 °C and exposure to urea. Addition of ClpX and ATP to the MuA/DNA complex released MuA in less than five minutes. The release required ATP hydrolysis, and the released form of MuA was completely active in binding DNA and catalyzing strand transfer.

Structure of ClpA, ClpP, and ClpAP

ClpP is processed after synthesis by cleavage after Met14 producing the mature 193 amino acid form of the protease (Maurizi et al., 1990a). Processing of ClpP is dependent on an active ClpP but does not require ClpA or ClpX. Electron microscopy, and recently X-ray crystallography (Wang et al. 1997), has shown that ClpP is arranged in rings of seven subunits each, and the native protein is composed of two such rings axially aligned (Kessel et al., 1995). The interface of the two rings produces an interior aqueous cavity to which there is limited access. As seen for the Archaea proteasome, which has a similar arrangement for the β subunits (Lowe et al., 1995), the proteolytic active sites of ClpP line the surface of the aqueous cavity forming a degradative chamber (Wang et al., 1997). Cleavage of proteins into multiple peptides of a minimum length in a processive or recursive manner should be the consequence of the close packing of 14 active sites within the cavity. The range of product lengths, 5–15 amino acids, suggests that there is not a discrete molecular ruler employed to select sites for cleavage. The occurrence of hydrophobic amino acids periodically in protein sequences, the density of proteolytic sites in ClpP, and the rate-limiting character of the substrate unfolding and translocation steps are sufficient to explain the frequency and efficiency of cleavage.

In the absence of nucleotide, purified ClpA exists as a mixture of monomers and dimers (Maurizi et al., 1997a). Upon binding ATP or a non-hydrolyzable analogue of ATP, ClpA assembles into stable oligomers, which exchange very slowly with free subunits (Singh, S.K. and Maurizi, M.R., unpublished). Top views of ClpA in the electron microscope have sixfold symmetry indicating that ClpA subunits form rings of six subunits (Kessel et al., 1996). Side views of ClpA shows the ring to be equatorially bifurcated, indicating that ClpA subunits are bilobed. The two lobes, which are of approximately equal size, appear to correspond to the two ATPase domains deduced from sequence analysis. ClpA thus is composed of two covalently joined, but structurally and possibly functionally distinct, ATPases.

The ClpA hexamer has high affinity for ClpP, whereas there is no interaction between unassembled ClpA and ClpP. The ClpAP complexes are similar in appearance whether formed in the presence of ATP or ATPγS and show a barrellike molecule with six bands (Kessel et al., 1996). The two shorter central bands correspond to side views of ClpP, and the two longer bands on either side represent side views of ClpA bound to the two exposed faces ClpP (Figure 1). The ClpAP complex

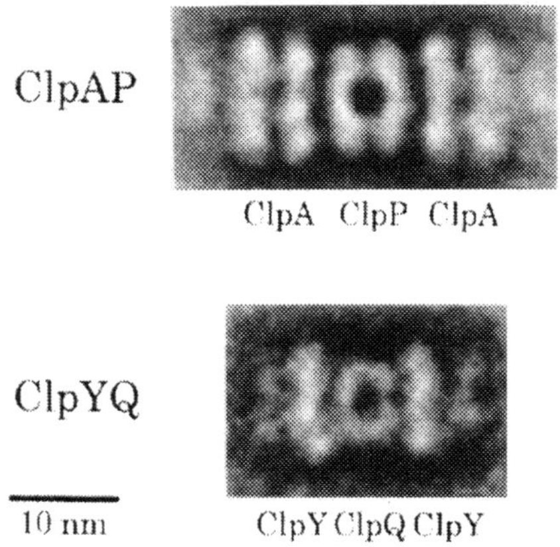

Figure 1. Electron micrographs of negatively stained ClpAP and ClpYQ. The ClpAP image first appeared in Kessel et al., 1995. The ClpYQ image was obtained by M. Kessel, NCI and NIAMS, NIH.

is about 13 nm in diameter (slightly less in the center where ClpP is) and 45 nm long. Solution studies indicate that, in the presence of nucleotide, ClpA and ClpP have a high affinity for each other ($K_d < 10$ nM) and that dissociation of ClpP from the complex is very slow (Singh, S.K. and Maurizi, M.R., unpublished).

In the presence of ATP or an analog of ATP, ClpX subunits also associate to form six-membered rings (Grimaud et al., 1998). ClpX and ClpA have similar affinities for ClpP, and ClpX hexamers bind on each face of the double-ringed structure of ClpP producing a stacked-ring complex similar to that formed between ClpA and ClpP. ClpX also allosterically activates oligopeptidase activity of ClpP to the same extent as does ClpA, suggesting that the two ATPases have similar modes of interaction with ClpP.

In vivo *Activities of ClpAP, ClpXP, and the ATPases Alone*

Abnormal Protein Degradation. ClpA, ClpX, and ClpP alone or together are not essential for growth of *E. coli* under any conditions tested. The *clpP* and *clpA* genes map in separate locations on the *E. coli* chromosome (Katayama et al., 1988). *clpP* and *clpX* are found in a bicistronic operon with *clpP* upstream, and the two genes are cotranscribed (Maurizi et al., 1990a). The 5'-upstream region of *clpP/clpX* has an σ^{32}-dependent promoter and transcription is enhanced five- to ten-fold upon heat shock (Kroh and Simon, 1990). ClpA transcription also increases at elevated temperatures, but ClpA is not under σ^{32} control and does not appear to be coordinately regulated with ClpP (Katayama et al., 1988).

ClpAP has a lower activity than Lon in degrading abnormal proteins in *E. coli*. There appears to be overlap in the abnormal proteins targeted by Lon and either ClpAP or ClpXP, because the effects of *clpP* mutations are more obvious in a *lon* mutant background than in *lon*$^+$ cells (Katayama et al., 1990).

ClpAP Substrates. Several β-galactosidase fusion proteins are degraded by ClpAP *in vivo*. The first one described was found in a random screen and has the amino terminal portion of ClpA fused to LacZ (Gottesman et al., 1990a). Another set of fusions were generated from variants of ubiquitin-LacZ fusions by enzymatic removal of the ubiquitin *in vivo* with ubiquitin carboxy terminal hydrolase, leaving different amino acids at the N-terminus of products (Tobias et al., 1991). Surprisingly, stability of the LacZ fusions followed an N-end rule similar to that seen in eukaryotic cells. Fusions with Leu, Phe, and Try at the amino terminus were very unstable ($t_{1/2} < 5$ minutes), and their degradation was dependent on active ClpA and ClpP in the cell. While the specificity of degradation of these proteins is quite striking, other substrates for ClpAP *in vitro* and *in vivo* do not show the same restrictive specificity (Thompson and Maurizi, 1994).

The only known naturally occurring substrate for ClpAP is an 81-amino acid protein, MazE (Aizenman et al., 1996). The MazE protein is the product of an endogenous *E. coli* gene but shows homology to the antidote protein of plasmid R100, PemI (Masuda et al., 1993). MazE neutralizes the cell-killing activity of another *E. coli* protein called MazF. Expression of MazE and MazF is under stringent control, and the proteins may have some role in adaptation to starvation conditions, but nothing is know about their specific function or how ClpAP-dependent degradation of MazE affects that function.

Proteolytic Activities of ClpXP In Vivo. Several proteolytic activities of ClpXP acting *in vivo* on physiological targets have been identified. During exponential growth of *E. coli* ClpXP degrades the starvation sigma factor, σ^S, which is needed for expression of specific genes during stationary phase (Schweder et al., 1996). The half-life of σ^S is ~ 2 minutes, and the protein is thus maintained at low levels in exponentially growing cells. As the cells enter stationary phase cells, the σ^S becomes resistant to degradation and its intracellular concentration increases. σ^S degradation dependent on another protein, RssB (Muffler et al., 1996; Pratt and Silhavy, 1996), which has sequence similarity to response regulators of *E. coli* . RssB is apparently required to modify σ^S in some manner to make it a substrate for ClpXP. Whether ClpXP recognizes the modified form of σ^S or the modification alters other properties of σ^S (e.g., binding to core RNA polymerase) making it more accessible to ClpXP remains to be determined.

ClpXP also degrades λ O protein (Gottesman et al., 1993; Wojtkowiak et al., 1993), the major replication protein of phage λ, *in vivo*. Although the half-life of l O protein is ~ 1 minute in wild-type cells, λ O protein is not limiting for lambda replication and the consequence of its degradation is not easy to see. It has been proposed that degradation serves only to remove λ O protein synthesized in excess (Szalewska et al., 1994). When λ O protein expression was attenuated by synthesis of a poorly suppressed amber mutant, λ O protein was limiting for replication and lambda burst size was increased in *clpP* and *clpX* mutants (Gottesman et al., 1993).

ClpXP is a regulator of a postsegregational killing system of phage P1 (Lehnherr et al., 1993). Phd is an antidote to Doc, which kills *E. coli* by an unknown mechanism. Wild-type cells that lose P1 are killed because Phd is degraded, but both *clpP* and *clpX* mutants survive loss of P1, because Phd is stable (Lehnherr and Yarmolinsky, 1995). Certain mutant forms of the phage Mu repressor protein are also degraded by ClpXP *in vivo* (Laachouch et al., 1996). The mutant proteins are altered in the carboxy terminus containing a new sequence that appears to be recognized by ClpXP (see below).

Chaperone Activity of ClpX In Vivo. ClpX is required for lytic growth of phage Mu, but ClpP is not, suggesting that ClpX can function independently of ClpP *in vivo* (Mhammedi-Alaoui et al., 1994). In ClpX mutants, Mu replication is blocked after formation of the strand transfer complex and before assembly of the replication forks. The explanation for this defect was obtained from *in vitro* work showing that ClpX is required for disassembly of MuA/DNA strand transfer complex (Levchenko et al., 1995; Kruklitis et al., 1996). Under normal circumstances *in vivo*, ClpP is present and MuA would be expected to be degraded. Mutational changes in MuA that affect the ability of the ClpX chaperone to act on it also affect its degradation by ClpXP in vivo, suggesting that the chaperone activity is a partial reaction in the degration pathway.

In vivo *Activities of ClpP.* An abnormal fusion protein, CRAG, is degraded in wild-type cells but is stable in *clpP* mutant cells (Kandror et al., 1994). Curiously, CRAG is degraded in *clpA, clpX,* and *clpB* mutants, suggesting that ClpP activity on this protein is independent of Clp ATPases. CRAG degradation is also blocked in *groEL* mutants and in cells with reduced activity of trigger factor, a peptidylproyl isomerase (Kandror et al., 1995). How these protein remodeling enzymes can render a protein susceptible or accessible to ClpP is not yet known. Another group of proteins degraded by ClpP independently of ClpA and ClpX are certain membrane proteins synthesized during carbon starvation (Damerau and St. John, 1993). These findings suggest that there may be alternative pathways for presentation of substrates to ClpP.

B. ClpYQ: A Union of the Clp and the Proteasome Families

ClpYQ is the latest Clp protease to be identified. The *clpQ* (*hslV*) and *clpY* (*hslU*) genes are in an operon originally identified because it is under σ^{32}-dependent heat shock control (Chuang et al., 1993). The amino acid sequence of ClpY proved to be similar to that of ClpX and highly homologous to a previously noted protein from Pasteurella (Gottesman et al., 1993). The ClpQ sequence showed similarity to that of the proteasome β subunits, and the presence of a catalytically active amino terminal threonine residue, confirmed its identity as a proteasome. A clone of *clpYQ* was independently isolated from an *E. coli* library as a multi-copy suppressor of mutations in *htrC*, which is involved in protein quality control in the periplasm (Missiakas et al., 1996).

Enzymatic Properties of ClpYQ

ClpY and ClpQ were purified in several laboratories either from clones overexpressing both proteins together (Rohrwild et al., 1996; Yoo et al., 1996) or from clones of the individual proteins (Kessel et al., 1996; Missiakas et al., 1996). In the former case, ClpY and ClpQ separated from each other during purification. ClpQ catalyzes cleavage of fluorogenic peptides, and this activity is stimulated 10-fold by ClpY in the presence of ATP (Rohrwild et al., 1996; Yoo et al., 1996). Peptide cleavage was also activated by non-hydrolyzable analogues of ATP. HslU (ClpY) has ATPase activity which is activated fivefold by HslV (ClpQ). Treatment of ClpY with sulfhydryl reagents inhibited both ATPase activity the ability to activate ClpQ peptidase activity (Yoo et al., 1996). ClpY and ClpQ together displayed an ATP-dependent proteolytic activity against casein (Kessel et al., 1996). Degradation of [^{3}H]casein occurred at ~20% of the rate seen with ClpAP.

Mutations in Thr2 resulted in ClpQ that no longer had suppressing activity *in vivo* and had lost peptidase activity *in vitro* (Missiakas et al., 1996). Mutations in the ATP binding consensus sequence blocked ATPase activity and eliminated ATP-dependent peptidase activity (Shin et al., 1996). The ATPase site mutant also did not form oligomers.

Structure of ClpYQ

Electron micrographs of ClpQ showed *en face* views of sixfold symmetric rings (Kessel et al., 1996; Rohrwild et al., 1997). ClpQ thus differs from the homologous proteasome and from ClpP in having only six subunits per ring. The molecular weight of ClpQ (220,000) indicates that it is composed of two rings. ClpY is in a dissociated state in the absence of nucleotide and forms an oligomer in the presence of ATPγS or ATP. Micrographs of the oligomeric ClpY indicate that it is also a ring and, in most preparations, is predominantly hexameric (Kessel et al., 1996) ; some preparations did give predominantly sevenfold symmetric particles. Rohrwild et al. (1997) used scanning transmission electron microscopy (STEM) analysis of individual particle masses to show nearly equal amounts of $ClpY_6$ and $ClpY_7$. Thus, ClpY can apparently associate in two oligomeric forms. Whether this is an artifact of reconstitution *in vitro* or an important property that expands the versatility of the enzyme is not yet known.

ClpY and ClpQ form a complex that resembles ClpAP and the proteasome in its overall arrangement of components (Rohrwild et al., 1997; Kessel, M. and Maurizi, M.R., unpublished). Side views of the complex show a double ring of ClpQ with a single ring of ClpY axially aligned on each side (Figure 1). In the images of ClpY in the complex, an additional but lighter density is seen on the outside surface of the ClpY rings. This mass might represent a domain of ClpY that undergoes a large conformation change or may simply be loosely associated with the remainder of the protein.

Activity of ClpYQ In Vivo

ClpY and ClpQ are not essential for *E. coli*. A knockout mutation in the *clpY* gene made the cells slightly thermosensitive and blocked about 50% of the degradation of puromycyl fragments (Missiakas et al., 1996). ClpY thus appears to be quantitatively as important as Lon in degrading these abnormal proteins. Transcription from σ^E promoters, which are expressed during acute heat shock, was also enhanced in ClpY mutants (Missiakas et al., 1996).

Multi-copy plasmids carrying *clpQ* and *clpY* under their own promoter could complement an *E. coli lon* mutation both for sensitivity to DNA damage and for capsular polysaccharide production (Wu, W.F., Maurizi, M.R., and Gottesman, S., unpublished). *lon* mutants become relatively UV resistant at 42 °C, in part because of increased degradation of SulA by a previously unknown protease. Cells remain UV sensitive in *lon clpY* or *lon clpQ* double mutants, suggesting that ClpYQ might degrade SulA at 42 °C. Direct measurements of SulA half-life in *clpY* and *clpQ* mutants confirmed that ClpYQ is responsible for SulA degradation at 42 °C.

VI. DEGRADATION SIGNALS AND PROTEASE RECOGNITION MOTIFS

It is useful to distinguish degradation signals and recognition motifs, although in some cases they may be biochemically identical. A degradation signal should be a recognizable feature of a protein or other cellular component associated with the protein that signals that the protein is an appropriate target for degradation. Examples of degradation signals from eukaryotic cells are "destruction boxes" (Glotzer et al., 1991), N-end degrons (Varshavsky, 1992), and possibly PEST sequences (Rogers et al., 1986), although the latter are more likely to reflect underlying chemical properties of proteins that result in exposure of degradation signals. These degradation signals or equivalent ones might also exist in *E. coli*. In eukaryotic cells, degradation signals give rise to ubiquitination, and the polyubiquitinated protein is the target for the protease. In *E. coli*, the degradation signal in some cases will serve as a direct recognition motif for a protease or will result in exposure of a recognition motif.

As defined here, recognition motifs are simply what proteases, chaperones, and their auxiliary proteins bind to. Recognition motifs are commonly encountered in protein-DNA, receptor-agonist, and antibody-antigen interactions. For simple proteases, such as trypsin, the recognition motif is a basic residue. The recognition motif for ATP-dependent proteases of *E. coli* have not been well defined. Moreover, the proteases may not have specificity for unique motifs. If, as was suggested earlier, *E. coli* proteases must directly recognize appropriate substrates, then they should have the ability to interact with weak and diffuse motifs as well as with highly specific ones.

It is also important to consider that selectivity on the part of proteases may be influenced by extrinsic factors such as localization or the participation of auxiliary proteins. The presence of FtsH in the membrane may increase the likelihood of a membrane protein's being degraded even if it does not have a recognition motif for which FtsH has strong affinity. Other types of interactions, such as between proteases and chaperones or between proteases and DNA, may also compensate for low affinity in recognition motifs.

A. *E. coli* Degradation Signals

Several degradation signals that might function in *E. coli* have been described or proposed. They include the N-end degron (Tobias et al., 1991), ribosome/nascent polypeptide complexes stalled at the end of truncated RNA (Keiler et al., 1995), oxidatively damaged regions of proteins (Davies and Lin, 1988), and exposed hydrophobic bonding domains (Gottesman et al., 1997; Maurizi et al., 1997b). There have also been some attempts to identify features like destruction boxes in *E. coli* proteins (Parsell et al., 1990; Silber et al., 1992; Laachouch et al., 1996). Of these signals, the first two are the best defined, and the second is the most elegant and complete. However, it is likely that exposed hydrophobic domains, the least specifically defined motif, is the most common.

The N-end Degron

Proteins with bulky hydrophobic amino terminal residues are rapidly degraded in both eukaryotes and prokaryotes (Varshavsky, 1992). Moreover, the range of this degradation signal is expanded to include additional proteins by an enzymatic system that catalyzes the transfer of leucine or phenylalanine residues onto the alpha amino group of proteins with basic amino terminal residues. In *E. coli*, proteins with the N-end degron are degraded by ClpAP; mutations in either *clpP* or *clpA* increase the half-lives of substrates from 1–2 min to >30 minutes (Tobias et al., 1991). No other genetic loci have been found to affect this degradation, which suggests that the N-end degron also acts as a recognition motif for ClpAP. However, studies done both *in vivo* and *in vitro* clearly show that this motif is not obligatory in ClpAP substrates (Gottesman et al., 1990a; Tobias et al., 1991; Thompson and Maurizi, 1994), but in fact few *in vivo* protein substrates are known. A more serious problem with understanding the relevance of this degradation signal is that conditions that give rise to proteins bearing the N-end degron are not known.

Translation Complexes at the Ends of mRNA

By far the most general degradation signal described to date is that generated by a translating ribosome stalled at the end of an abnormally terminated mRNA

(Keiler et al., 1996). Proteins produced from abnormally terminated mRNA are by definition abnormal and defective, are potentially harmful, and may tie up the translation apparatus. They are thus universally appropriate targets for degradation, and the mechanism of signal transduction by which information about this biological problem is converted to a recognition motif for proteolytic systems is elegant and efficient. The nascent polypeptides in such complexes are tagged in a series of steps involving: 1) addition of an alanine residue to the carboxy terminus, 2) translocation of the ribosome-nascent polypeptide complex to a unique 10 S RNA (SsrA) that contains a 10-codon open reading frame following by a stop codon, 3) continuation of translation thereby adding the additional 10 unique amino acids to the carboxy terminus, and 4) release of the modified polypeptide (Keiler et al., 1996). The sequence added in this way is AADENYLALAA, which acts as a recognition motif for both cytoplasmic and periplasmic proteolytic systems. In the cytoplasm, proteins with the alanine-rich tail recognition motif are targeted by several proteases, including ClpXP and ClpAP, and FtsH (see below).

Because AADENYLALAA appears to be a recognition motif for proteases, that sequence would act as a degradation signal when it occurred naturally in a protein. Such a signal could be exposed in the native state of an intrinsically unstable protein or it could become exposed upon structural perturbation of a protein. Depending on the stringency of the recognition, variants of this sequence could also function as either weaker or stronger signals.

Oxidized Regions of Proteins and Exposed Hydrophobic Bonding Domains

Oxidatively damaged proteins are rapidly degraded in cells. However, enzymatic systems for marking such proteins or proteases that directly recognize them have not been demonstrated (Davies and Lin, 1988). It is possible that oxidative damage causes structural perturbations in the protein, similar to those caused by heat or chemical damage. Studies of fragments of the lambda repressor (Parsell and Sauer, 1989) and T4 lysozymes expressed in *E. coli* (Inoue and Rechsteiner, 1994) indicate that proteolytic susceptibility increases for structurally unstable proteins. Similarly, puromycyl peptides and other protein fragments which cannot fold into stable structures are rapidly degraded. The instability of poorly folded proteins suggests that features that are usually buried in proteins, such as hydrophobic faces of helices or series of hydrophobic amino acids, will act as degradation signals when exposed.

Proteins that form hetero-oligomeric complexes are usually rapidly degraded when they are expressed in the absence of the proteins with which they interact (Gottesman et al., 1997). An exposed bonding interface between subunits may act as a degradation signal. Such bonding sites generally have an amphipathic character, but otherwise do not have a defined structural or chemical composition for all proteins. Degradation signals of this nature may be found on normal proteins that are rapidly degraded, because essentially all of the proteins regulated by degradation *in vivo* function as parts of high molecular weight protein or protein-nucleic acid complexes (Gottesman and Maurizi, 1992). In most cases, these proteins are stabilized by interaction with their functional partners and degraded when they are free—thus SulA is stabilized by SulB (Lutkenhaus et al., 1986), RcsA by RcsB (Stout et al., 1991), CcdA by CcdB (Van Melderen et al., 1994), RepA by its DNA binding site (Wickner, S., personal communication), and λ O protein by the lambda replication complex (Szalewska et al., 1994). These proteins could be released from complexes during the normal cycle of their activity and be degraded before they can reassociate with their partners.

B. Recognition Motifs for ATP-Dependent Proteases

Proteases in *E. coli* probably directly interact with some of the proposed degradation signals, which thus serve as recognition motifs. The contribution of a recognition motif to proteolysis may be affected by other factors such as conformational flexibility or malleability of the substrate and colocalization of the protease and the protein. Thus the same motif may not produce the same effect on the degradation rates of two different proteins. Also, in the cases where proteins are tagged for degradation, it is important to distinguish between the tag's serving as a direct recognition motif and its affecting the protein to which it is attached to expose a degradation motif. For example, it is possible that ubiquitin serves to partially unfold the protein to which it is attached, a factor that might aid in targeting the protein to the proteasome.

Hydrophobic Peptides and Regions of Proteins

The discussion of recognition motifs should perhaps start where the previous section left off, with exposed hydrophobic domains of proteins. Exposed hydrophobic regions could be expected to interact with the chaperone components of ATP-dependent proteases. Such regions are also generally thought to be bound by molecular chaperones and tend to be the parts of unfolded proteins that interact with each other leading to aggregation. The interplay among these competing but also cooperating pathways is discussed below.

Both Lon and ClpP tend to cleave proteins after hydrophobic residues in amphipathic regions. Cleavage specificity, however, is unlikely to reflect the same specificity as interaction in the ATPase domain. ATPase activity of ClpA is activated by a number of peptides, the best one of which had several hydrophobic residues in the carboxy terminal regions (Thompson and Maurizi, 1994). No data is available on the composition of peptides that bind to Lon, but the correlation between denaturation of a protein and its ability to activate ATP hydrolysis by Lon (Waxman and Goldberg, 1986; Menon and Goldberg, 1987) suggests that hydrophobic regions might interact better with Lon.

Alanine-Rich Peptides

The alanine-rich tail, AADENYLALAA, leads to rapid degradation when added to the carboxy terminus of a protein (Keiler et al., 1996). When the last two alanines were replaced with aspartates, the peptide was no longer a recognition sequence and the proteins were stable. When the protein with the alanine-rich tail is expressed in the cytoplasm, its degradation is dependent on ClpXP (Gottesman et al., 1998). ClpAP also had an effect, although about 20–30% that of ClpXP. This alanine rich sequence is not found at the carboxy terminus of ClpXP substrates such as RpoS, Phd, and λ O protein. Thus, ClpXP must be able to recognize more than one motif for degradation.

Carboxy-Terminal Motifs and Other Destruction Boxes

Earlier studies had identified other hydrophobic carboxy-terminal sequences that caused rapid degradation *in vivo* (Parsell et al., 1990). Addition of five amino acids at the carboxy termini of fragments of the CI repressor led to their rapid degradation only when the amino acids were predominantly hydrophobic. In those studies, addition of more than five residues did not increase the degradation rate. Recently, it has been shown that degradation of some of these proteins *in vivo* requires FtsH and that ClpXP has a relatively minor effect (Herman, C., personal communication). These results suggest that the recognition motifs for the various proteases may have very subtle differences.

The carboxy terminal eight amino acids of MuA (QNRRKKAI) is important for recognition by ClpX in disassembly of the strand transfer complex. That sequence resembles a sequence at the carboxy terminus of the mutant Mu repressors, which are degraded by ClpXP. The carboxy termini of the Mu repressor mutants are FRNKVL and FRNHRR. Changing the N to D in both sequences destroys the apparent recognition motif. It has been suggested that sequences similar to these are present in other ClpXP substrates, sometimes internally (Levchenko et al., 1995), and recently ClpX has been shown to bind *in vitro* to proteins fused to the carboxy terminus of MuA (Levchenko et al., 1997).

Do Chaperones Target Proteins for Degradation?

Exposed hydrophobic regions are sites of interaction between unfolded proteins that lead to aggregation. Molecular chaperones are thought to bind to these regions, preventing aggregation and allowing the bound protein time to undergo some degree of remodeling before a new cycle of release, rebinding, and remodeling. The extent to which proteins are released from chaperones before achieving a state from which they can rapidly collapse into the native state is a matter of some debate and may vary with different chaperones and with different proteins. Proteins that cannot be refolded will eventually aggregate or be degraded. How chaperones effect the degradation of unfoldable proteins will depend on several factors (for a more complete discussion see Gottesman et al., 1997). For those proteins that remain bound to the chaperones, the chaperone could bind one region allowing another region with a recognition motif for the protease to become exposed. Alternatively, the protease could interact with the protein/chaperone complex itself, meaning that the chaperone itself becomes part of the recognition motif. Neither of these cases has been demonstrated experimentally.

If chaperones release the unfoldable protein, it can aggregate, interact with a protease, or bind again to the chaperone. By keeping the amount of free misfolded protein below the threshhold needed for aggregation, chaperones maintain a pool of soluble protein accessible to proteases and thereby contribute to protein degradation. If the recognition motif for the protease is the same one recognized by the chaperone, the chaperone and the protease will essentially compete for the substrate. This model may be similar to what the chaperone components of the proteases themselves do, and seems to provide the most efficient system that ensures against degradation of proteins that can be refolded. Chaperones in fact can protect some conformationally unstable proteins from degradation.

VII. RELATIONSHIP BETWEEN STRUCTURE AND MECHANISM

ATP-dependent proteases are bifunctional enzymes consisting of a chaperone linked to a protease. The structures of the ClpAP, ClpXP, and ClpYQ proteases suggest that they are designed to shield the proteolytic active sites from most soluble proteins in the cell. Restricting access to the active sites ensures that damage to functional proteins is avoided. According to the current model, potential substrates are first screened by the ATP-dependent chapterone activity, which functions to unfold or disassemble the bound protein (Gottesman et al., 1997). Discrimination may occur through kinetic partitioning between a refolding/release and a translocation/degradation pathway (Figure 2). The similarity in structure between the Clp proteases and the 26 S proteasome suggests that a similar mechanism applies for the latter.

The Clp ATPases can display chaperone activity on their own. Whether they are ever free in wild-type cells needs to be determined. Properties of the ClpAP complex suggest that ClpA should be bound to ClpP under most physiological conditions. Exchange studies indicated that the ClpA and ClpP oligomers interact very strongly and dissociate very slowly as long as ATP is in excess. Exchange of ClpP from the ClpAP complex has a $t_{1/2}$ ~8 minutes at 37 °C, which is enough time to degrade 40–50 subunits of casein (Singh, S.K. and Maurizi, M.R., unpublished). Thus processive degradation occurs to completion and fresh substrate binds through multiple rounds without dissociation of ClpAP. Similar studies have not been done with ClpXP or with ClpYQ, although the high affinity between ClpX and ClpP suggests they also should remain tightly bound through multiple cycles of protein turn-

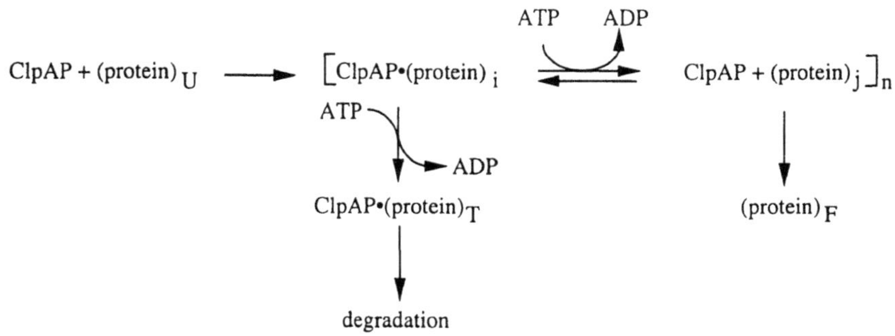

Figure 2. Screening of substrates by ATP-dependent proteases, which carry out kinetic partitioning of proteins between a folding pathway and degradation. The number of folding steps (n) will depend on how far from the native state the protein was when it bound to the chaperone component of the protease and on its degree of folding competence. The rate at which the complex is converted to one committed to degradation may depend on how long it takes to encounter a high-affinity, motif-signaling degradation or on the accumulation of a number of low-affinity contacts that trap the protein. The subscripts are U, unfolded; F, folded or folding competent; T, trapped or committed to degradation; i, j, different conformational states of protein bound to protease; n, number of cycles of conformational changes without dissociation of the

over. A major difference between ClpA and ClpX or ClpY, that could affect the partitioning of substrates between refolding and degradation is the presence of the second ATPase domain on the former.

In vivo, it is likely that proteins encounter the ClpAP complex rather than ClpA alone and the ClpAP complex has chaperone activity that can release an active form of the protein. The extent of that reaction will depend on the rate of refolding versus the rate of commitment to degradation, where commitment probably occurs following translocation of a portion of the protein to the vicinity of the proteolytic cavity. What determines these relative rates for a given substrate is the critical remaining question regarding ClpAP activity. It may be that, under some conditions, free ClpA or ClpX encounter substrates independently and recruit ClpP for degradation only after that need is evaluated.

The mechanism by which the protein, once committed to degradation, is retained by the protease and cleaved processively is also an intriguing issue. Several mechanisms of preventing release of the protein can be imaged and were recently proposed with ClpAP in mind (Maurizi et al., 1997b). For one, the mechanism could entail facilitated diffusion of the substrate through the channel in ClpA, similar to what has been proposed for transport of proteins through membrane channels. An end or loop of the substrate could be threaded into and anchored to either the proteolytic site or a binding site deep within the access channel. Alternatively, a major conformational change could open a door or hatch covering a wide channel with the proteolytic sites at the other end. The door could close on the protein after it has entered or there could be protrusions on the channel sides preventing backslipping. A third possibility is that the ATPase clamps a portion of the protein at the mouth of the channel and allows the remainder to find its way through the channel into the proteolytic sites. This model leaves the protein bound to the protease in a more exposed state. In a variation of this last model, the protein could be anchored by covalent attachment to a site on the ATPase. In this case, a mechanism must exist to remove the attached piece when degradation is completed. A final possibility is a corkscrewing mechanism obtained by rotation or ratcheting of the ATPase subunits with respect to the protease subunits. This model would require that the access channel have a polarity that could change or alternate during the cycles of ATP hydrolysis.

In the end, the protein substrate must effectively be fully extended one segment at a time until it is transported into the proteolytic cavity. The amount of protein that must be inserted into the active site cavity during a cycle could be rather large. For ClpP, there is presumably sufficient room for 196 amino acid residues because 14 14-amino acid propeptides appear to fill the cavity in the pro-form of the mutant, ClpP(S111A) (Kessel et al., 1995). In fact, the crystal structure of ClpP revealed a cavity with a capacity of more than 20 kDa of protein (Wang et al., 1997). If the cavity must be filled before cleavage can occur and assuming one cleavage per active site, then an average size of 14 amino acid residues for the products is expected, a size larger than that observed for degradation of most proteins. It is more likely that the protein is cleaved as it enters the aqueous cavity and does not accumulate there, because ClpP has the capacity to cleave peptide bonds at more than 10,000 min^{-1} (Thompson and Maurizi, 1994). Discovering how transfer of the protein is coupled to its cleavage will be a major step forward in our understanding of the mechanism of action of these amazing degradative machines.

REFERENCES

Aizenman, E., Engelberg-Kulka, H., & Glaser, G. (1996). An *Escherichia coli* chromosomal "addiction module" regulated by ppGpp: a model for programmed bacterial cell death. Proc. Natl. Acad. Sci. USA 93, 6059-6063.

Akiyama, Y., Kihara, A., & Ito, K. (1996a). Subunit a of proton ATPase F0 sector is a substrate of the FtsH protease in *Escherichia coli*. FEBS Lett. 399(1-2), 26-28.

Akiyama, Y., Kihara, A., Tokuda, H., & Ito, K. (1996b). FtsH (HflB) is an ATP-dependent protease selectively acting on SecY and some other membrane proteins. J. Biol. Chem. 271(49), 31196-31201.

Akiyama, Y., Ogura, T., & Ito, K. (1994a). Involvement of FtsH in protein assembly into and through the membrane. I. Mutations that reduce retention efficiency of a cytoplasmic reporter. J. Biol. Chem. 269(7), 5218-5224.

Akiyama, Y., Shirai, Y., & Ito, K. (1994b). Involvement of FtsH in protein assembly into and through the membrane. II. Dominant mutations affecting FtsH functions. J. Biol. Chem. 269(7), 5225-5229.

Akiyama, Y., Yoshihisa, T., & Ito, K. (1995). FtsH, a membrane-bound ATPase, forms a complex in the cytoplasmic membrane of *Escherichia coli*. J. Biol. Chem. 270(40), 23485-23490.

Arlt, H., Tauer, R., Feldmann, H., Neupert, W., & Langer, T. (1996). The YTA10-12 complex, an AAA protease with chaperone-like activity in the inner membrane of mitochondria. Cell 85(6), 875-885.

Bahl, H., Echols, H., Straus, D. B., Court, D., Crowl, R., & Georgopoulos, C. P. (1987). Induction of the heat shock response of *E. coli* through stabilization of sigma 32 by the phage lambda cIII protein. Genes Devel. 1, 57-64.

Baker, T. A., Grossman, A. D., & Gross, C. A. (1984). A gene regulating the heat shock response in *Escherichia coli* also affects proteolysis. Proc. Natl. Acad. Sci. USA 81, 6779-6783.

Banuett, F., & Herskowitz, I. (1987). Identification of polypeptides encoded by an *Escherichia coli* locus (*hflA*) that governs the lysis-lysogeny decision of bacteriophage l. J. Bacteriol. 169, 4076-4085.

Banuett, F., Hoyt, M. A., McFarlane, L., Echols, H., & Herskowitz, I. (1986). hflB, a new *Escherichia coli* locus regulating lysogeny and the level of bacteriophage lambda CII protein. J. Mol. Biol. 187, 213-224.

Bross, P., Andresen, B. S., Knudsen, I., Kruse, T. A., & Gregersen, N. (1995). Human ClpP protease: cDNA sequence, tissue-specific expression and chromosomal assignment of the gene. FEBS Lett. 377, 249-252.

Cheng, H. H., & Echols, H. (1987). A class of *Escherichia coli* proteins controlled by the *hflA* locus. J. Mol. Biol. 196, 737-740.

Cheng, H. H., Muhlrad, P. J., Hoyt, A., & Echols, H. (1988). Cleavage of the cII protein of phage lambda by purified HflA protease: control of the switch between lysis and lysogeny. Proc. Natl. Acad. Sci. USA 85, 7882-7886.

Chin, D. T., Goff, S. A., Webster, T., Smith, T., & Goldberg, A. L. (1988). Sequence of the *lon* gene in *Escherichia coli*: A heat-shock gene which encodes the ATP-dependent protease La. J. Biol. Chem. 263, 11718-11728.

Chuang, S.-E., Burland, V., III, G. P., Daniels, D. L., & Blattner, F. R. (1993). Sequence analysis of four new heat shock genes constituting the hslTS/ibpAB and hslVU operons in *Escherichia coli*. Gene 134, 1-6.

Chung, C. H., & Goldberg, A. L. (1982). DNA stimulates ATP-dependent proteolysis and protein-dependent ATPase activity of protease La from *Escherichia coli*. Proc. Natl. Acad. Sci. USA 79(3), 795-799.

Damerau, K., & St. John, A. C. (1993). Role of Clp protease subunits in degradation of carbon starvation proteins in *Escherichia coli*. J. Bacteriol. 175, 53-63.

Davies, K. J., & Lin, S. W. (1988). Oxidatively denatured proteins are degraded by an ATP-independent proteolytic pathway in *Escherichia coli*. Free Rad. Biol. Med. 5(4), 225-236.

de Feyter, R., Wallace, C., & Lane, D. (1989). Autoregulation of the ccd operon in the F plasmid. Mol. Gen. Genet. 218(3), 481-486.

Fischer, H., & Glockshuber, R. (1993). ATP hydrolysis is not stoichiometrically linked with proteolysis in the ATP-dependent protease La from *Escherichia coli*. J. Biol. Chem. 268(30), 22502-22507.

Fischer, H., & Glockshuber, R. (1994). A point mutation within the ATP-binding site inactivates both catalytic functions of the ATP-dependent protease La (Lon) from *Escherichia coli*. FEBS Lett. 356(1), 101-103.

Frank, E. G., Ennis, D. G., Gonzalez, M., Levine, A. S., & Woodgate, R. (1996). Regulation of SOS mutagenesis by proteolysis. Proc. Natl. Acad. Sci. USA 93(19), 10291-10296.

Fu, G. K., Smith, M. J., & Markovitz, D. M. (1997). Bacterial protease Lon is a site-specific DNA-binding protein. J. Biol. Chem. 272, 534-538.

Gill, R. E., Karlok, M., & Benton, D. (1993). Myxococcus xanthus encodes an ATP-dependent protease which is required for developmental gene transcription and intercellular signaling. J. Bacteriol. 175(14), 4538-4544.

Glotzer, M., Murray, A. W., & Kirschner, M. W. (1991). Cyclin is degraded by the ubiquitin pathway. Nature 349, 132-138.

Goff, S. A., Casson, L. P., & Goldberg, A. L. (1984). Heat shock regulatory gene *htpR* influences rates of protein degradation and expression of the *lon* gene in *Escherichia coli*. Proc. Natl. Acad. Sci. USA 81, 6647-6651.

Goldberg, A. L. (1992). The mechanism and functions of ATP-dependent proteases in bacterial and animal cells. Eur. J. Biochem. 203, 9-23.

Goldberg, A. L., Moerschell, R. P., Chung, C. H., & Maurizi, M. R. (1994). ATP-dependent protease La (Lon) from *Escherichia coli*. Methods Enzymol. 244, 350-375.

Goldberg, A. L., & Waxman, L. (1985). The role of ATP hydrolysis in the breakdown of proteins and peptides by protease La from *Escherichia coli*. J. Biol. Chem. 260, 12029-12034.

Gottesman, S. (1987). Regulation by proteolysis. In: *Escherichia coli and Salmonella typhimurium: Cellular and Molecular Biology* (Neidhardt, F. C., Ingraham, J. L., Low, K. B., Magasanik, B., Schaechter, M., & Umbarger, H. E., eds.), pp. 1308-1312. American Society for Microbiology, Washington, DC.

Gottesman, S. (1996). Proteases and their targets in *Escherichia coli*. Annu. Rev. Genet. 30, 465-506.

Gottesman, S., Clark, W. P., Crecy-Lagard, V. d., & Maurizi, M. R. (1993). ClpX, an alternative subunit for the ATP-dependent Clp protease of *Escherichia coli*: sequence and *in vivo* activities. J. Biol. Chem. 268, 22618-22626.

Gottesman, S., Clark, W. P., & Maurizi, M. R. (1990a). The ATP-dependent Clp protease of *Escherichia coli*: sequence of clpA and identification of a Clp-specific substrate. J. Biol. Chem. 265, 7886-7893.

Gottesman, S., Gottesman, M. E., Shaw, J. E., & Pearson, M. L. (1981). Protein degradation in *E. coli*: the *lon* mutation and bacteriophage lambda N and cII protein stability. Cell 24, 225-233.

Gottesman, S., & Maurizi, M. R. (1992). Regulation by proteolysis: energy-dependent proteases and their targets. Microbiol. Rev. 56, 592-621.

Gottesman, S., Roche, E., Zhou, Y.-N., & Sauer, R. T. (1998). The ClpXP and ClpAP proteases degrade proteins with C-terminal peptide tails added by the SsrA tagging system. Genes and Devel. (In press).

Gottesman, S., Squires, C., Pichersky, E., Carrington, M., Hobbs, M., Mattick, J. S., Dalrymple, B., Kuramitsu, H., Shiroza, T., Foster, T., Clark, W. P., Ross, B., Squires, C. L., & Maurizi, M. R. (1990b). Conservation of the regulatory subunit for the Clp ATP-dependent protease in prokaryotes and eukaryotes. Proc. Natl. Acad. Sci. USA 87, 3513-3517.

Gottesman, S., Wickner, S., Jubete, Y., Singh, S. K., Kessel, M., & Maurizi, M. R. (1995). Selective energy-dependent proteolysis in *Escherichia coli*. Cold Spring Harbor Sympos. Quant. Biol. Volume LX, 533-548.

Gottesman, S., Wickner, S., & Maurizi, M. R. (1997). Protein quality control: triage by chaperones and proteases. Genes and Devel. 11, 815-823.

Grimaud, R., Kessel, M., Beuron, F., Steven, A. C., & Maurizi, M. R. (1998). Enzymatic and structural similarities between the *E. coli* ATP-dependent proteases ClpXP and ClpAP. J. Biol. Chem. (In press).

Guelin, E., Rep, M., & Grivell, L. A. (1994). Sequence of the AFG3 gene encoding a new member of the FtsH/Yme1/Tma subfamily of the AAA-protein family. Yeast 10(10), 1389-1394.

Herman, C., Lecat, S., D'Ari, R., & Bouloc, P. (1995a). Regulation of the heat-shock response depends on divalent metal ions in an hflB mutant of *Escherichia coli*. Mol. Microbiol. 18(2), 247-255.

Herman, C., Ogura, T., Tomoyasu, T., Hiraga, S., Akiyama, Y., Ito, K., Thomas, R., D'Ari, R., & Bouloc, P. (1993). Cell growth and lambda phage development controlled by the same essential *Escherichia coli* gene, ftsH/hflB. Proc. Natl. Acad. Sci. USA 90(22), 10861-10865.

Herman, C., Thevenet, D., D'Ari, R., & Bouloc, P. (1995b). Degradation of sigma 32, the heat shock regulator in *Escherichia coli*, is governed by HflB. Proc. Natl. Acad. Sci. USA 92(8), 3516-3520.

Herman, C., Thevenet, D., D'Ari, R., & Bouloc, P. (1997). The HflB protease of *Escherichia coli* degrades its inhibitor lambda cIII. J. Bacteriol. 179(2), 358-363.

Hoyt, M. A., Knight, D. M., Das, A., Miller, H. I., & Echols, H. (1982). Control of phage lambda development by stability and synthesis of cII protein: Role of the viral cIII and host hflA, himA and himD genes. Cell 31, 565-573.

Hwang, B. J., Woo, K. M., Goldberg, A. L., & Chung, C. H. (1988). Protease Ti, a new ATP-dependent protease in *Escherichia coli* contains protein-activated ATPase and proteolytic functions in distinct subunits. J. Biol. Chem. 263, 8727-8734.

Inoue, I., & Rechsteiner, M. (1994). On the relationship between the metabolic and thermodynamic stabilities of T4 lysozymes; measurements in *Escherichia coli*. J. Biol. Chem. 269, 29241-29246.

Jensen, R. B., Grohmann, E., Schwab, H., Diaz-Orejas, R., & Gerdes, K. (1995). Comparison of ccd of F, parDE of RP4, and parD of R1 using a novel conditional replication control system of plasmid R1. Mol. Microbiol. 17(2), 211-220.

Jones, C., & Holland, I. B. (1985). Role of the SulB (FtsZ) protein in division inhibition during the SOS response in *Escherichia coli*: FtsZ stabilizes the inhibitor SulA in maxicells. Proc. Natl. Acad. Sci. USA 82, 6045-6049.

Jubete, Y., Maurizi, M. R., & Gottesman, S. (1996). Role of the heat shock protein DnaJ in the lon-dependent degradation of naturally unstable proteins. J. Biol. Chem. 271(48), 30798-30803.

Kandror, O., Busconi, L., Sherman, M., & Goldberg, A. L. (1994). Rapid degradation of an abnormal protein in *Escherichia coli* involves the chaperones GroEL and GroES. J. Biol. Chem. 269(38), 23575-23582.

Kandror, O., Sherman, M., Rhode, M., & Goldberg, A. L. (1995). Trigger factor is involved in GroEL-dependent protein degradation in *Escherichia coli* and promotes binding of GroEL to unfolded proteins. Embo. J. 14(23), 6021-6027.

Katayama, Y., Gottesman, S., Pumphrey, J., Rudikoff, S., Clark, W. P., & Maurizi, M. R. (1988). The two-component, ATP-dependent Clp protease of *Escherichia coli*. Purification, cloning, and mutational analysis of the ATP-binding component. J. Biol. Chem. 263(29), 15226-15236.

Katayama, Y., Kasahara, A., Kuraishi, H., & Amano, F. (1990). Regulation of activity of an ATP-dependent protease, Clp, by the amount of a subunit, ClpA, in the growth of *Escherichia coli* cells. J. Biochem. 108(1), 37-41.

Katayama-Fujimura, Y., Gottesman, S., & Maurizi, M. R. (1987). A multiple-component ATP-dependent protease from *Escherichia coli*. J. Biol. Chem. 262, 4477-4485.

Keiler, K. C., Waller, P. R., & Sauer, R. T. (1996). Role of a peptide tagging system in degradation of proteins synthesized from damaged messenger RNA. Science 271, 990-993.

Kessel, M., Maurizi, M. R., Kim, B., Trus, B. L., Kocsis, E., Singh, S. K., & Steven, A. C. (1995). Homology in structural organization between *E. coli* ClpAP protease and the eukaryotic 26 S proteasome. J. Mol. Biol. 250, 587-594.

Kessel, M., Wu, W.-F., Gottesman, S., Kocsis, E., Steven, A., & Maurizi, M. R. (1996). Six-fold rotational symmetry of ClpQ, the *E. coli* homolog of the 20 S proteasome, and its ATP-dependent activator, ClpY. FEBS Lett. 398, 274-278.

Kihara, A., Akiyama, Y., & Ito, K. (1995). FtsH is required for proteolytic elimination of uncomplexed forms of SecY, an essential protein translocase subunit. Proc. Natl. Acad. Sci. USA 92(10), 4532-4536.

Kihara, A., Akiyama, Y., & Ito, K. (1996). A protease complex in the *Escherichia coli* plasma membrane: HflKC (HflA) forms a complex with FtsH (HflB), regulating its proteolytic activity against SecY. Embo. J. 15(22), 6122-6131.

Kornitzer, D., Altuvia, S., & Oppenheim, A. B. (1991). The activity of the CIII regulator of lamboid bacteriophages resides within a 24-amino acid protein domain. Proc. Natl. Acad. Sci. USA 88, 5217-5221.

Kroh, H. E., & Simon, L. E. (1990). The ClpP component of Clp protease is the s^{32}-dependent heat shock protein F21.5. J. Bacteriol. 172, 6026-6034.

Kruklitis, R., Welty, D. J., & Nakai, H. (1996). ClpX protein of *Escherichia coli* activates bacteriophage Mu transposase in the strand transfer complex for initiation of Mu DNA synthesis. Embo. J. 15(4), 935-944.

Laachouch, J. E., Desmet, L., Geuskens, V., Grimaud, R., & Toussaint, A. (1996). Bacteriophage Mu reppressor as a target for the *Escherichia coli* ATP-dependent Clp Protease. Embo. J. 15(2), 437-444.

Langer, T., & Neupert, W. (1996). Regulated protein degradation in mitochondria. Experientia 52(12), 1069-1076.

Leffers, Jr., G. G., & Gottesman, S. (1998). λ Xis degradation *in vivo* by lon and FtsH. J. Bacteriol. 180 (In press).

Lehnherr, H., Maguin, E., Jafri, S., & Yarmolinsky, M. B. (1993). Plasmid addiction genes of bacteriophage P1: doc, which causes cell death on curing of prophage, and phd, which prevents host death when prophage is retained. J. Mol. Biol. 233(3), 414-428.

Lehnherr, H., & Yarmolinsky, M. B. (1995). Addiction protein Phd of plasmid prophage P1 is a substrate of the ClpXP serine protease of *Escherichia coli*. Proc. Natl. Acad. Sci. USA 92, 3274-3277.

Levchenko, I., Luo, L., & Baker, T. A. (1995). Disassembly of the Mu transposase tetramer by the ClpX chaperone. Genes Dev. 9(19), 2399-27408.

Levchenko, I., Yamauchi, M., & Baker, T. A. (1997). ClpX and MuB interact with overlapping regions of Mu transposase: Implications for control of the transposition pathway. Genes & Dev. 11, 1561-1572.

Lindahl, M., Tabak, S., Cseke, L., Pichersky, E., Andersson, B., & Adam, Z. (1996). Identification, characterization, and molecular cloning of a homologue of the bacterial FtsH protease in chloroplasts of higher plants. J. Biol. Chem. 271(46), 29329-293734.

Lowe, J., Stock, D., Jap, B., Zwickl, P., Baumeister, W., & Huber, R. (1995). Crystal structure of the 20 S proteasome from the Archaeon T. acidophilum at 3.4 A resolution. Science 268, 533-539.

Lutkenhaus, J., Sanjanwala, B., & Lowe, M. (1986). Overproduction of FtsZ suppresses sensitivity of lon mutants to division inhibition. J. Bacteriol. 166, 756-762.

Masuda, Y., Miyakawa, K., Nishimura, Y., & Ohtsubo, E. (1993). chpA and chpB, *Escherichia coli* chromosomal homologs of the pem locus responsible for stable maintenance of plasmid R100. J. Bacteriol. 175(21), 6850-6856.

Maurizi, M. R. (1987). Degradation in vitro of bacteriophage lambda N protein by Lon protease from *Escherichia coli*. J. Biol. Chem. 262, 2696-2703.

Maurizi, M. R. (1992). Proteases and protein degradation in *Escherichia coli*. Experientia 48, 178-201.

Maurizi, M. R., Clark, W. P., Katayama, Y., Rudikoff, S., Pumphrey, J., Bowers, B., & Gottesman, S. (1990a). Sequence and structure of Clp P, the proteolytic component of the ATP-dependent Clp protease of *Escherichia coli*. J. Biol. Chem. 265(21), 12536-12545.

Maurizi, M. R., Clark, W. P., Kim, S.-H., & Gottesman, S. (1990b). ClpP represents a unique family of serine proteases. J. Biol. Chem. 265, 12546-12552.

Maurizi, M. R., Clark, W. P., Kim, S. H., & Gottesman, S. (1990c). Clp P represents a unique family of serine proteases. J. Biol. Chem. 265(21), 12546-12552.

Maurizi, M. R., Singh, S. K., Thompson, M. W., & Ginsburg, A. (1997a). ATP-dependent association between ClpA and ClpP in the ClpAP protease from *Escherichia coli*. Biochemistry. In preparation.

Maurizi, M. R., Thompson, M. W., Singh, S. K., & Kim, S.-H. (1994). Endopeptidase Clp: the ATP-dependent Clp protease from *Escherichia coli*. Methods Enzymol. 244, 314-331.

Maurizi, M. R., Trisler, P., & Gottesman, S. (1985). Insertional mutagenesis of the *lon* gene in *Escherichia coli*: *lon* is dispensable. J. Bacteriol. 164, 1124-1135.

Maurizi, M. R., Wickner, S., & Gottesman, S. (1997b). Chaperones and charonins: protein unfolding enzymes and proteolysis. In: Molecular Biology of Chaperones and Folding Catalysis (Bukau, B., ed.). Harwood Academic Publishers, Amsterdam.

Menon, A. S., & Goldberg, A. L. (1987). Protein substrates activate the ATP-dependent protease La by promoting nucleotide binding and release of bound ADP. J. Biol. Chem. 262, 14929-14934.

Menon, A. S., Waxman, L., & Goldberg, A. L. (1987). The energy utilized in protein breakdown by the ATP-dependent protease La from *Escherichia coli*. J. Biol. Chem. 262, 722-726.

Mhammedi-Alaoui, A., Pato, M., Gama, M. J., & Toussaint, A. (1994). A new component of bacteriophage Mu replicative transposition machinery: the *Escherichia coli* ClpX protein. Mol. Microbiol. 11(6), 1109-1116.

Miller, C. G. (1996). Protein degradation and proteolytic modification. In: *Escherichia coli* and Salmonella: Cellular and Molecular Biology, 2nd edn. (Naedhardt, F. C., Curtiss III, R.,

Ingraham, J. L., Lin, E. C. C., Low, K. B., Magasanik, B., Reznikoff, W. S., Riley, M., Schaechter, M., & Umberger, H. E., eds.), pp. 938-954. ASM, Washington, DC.

Missiakas, D., Schwager, F., Betton, J. M., Georgopoulos, C., & Raina, S. (1996). Identification and characterization of HslV HslU (ClpQ ClpY) proteins involved in overall proteolysis of misfolded proteins in *Escherichia coli*. Embo. J. 15(24), 6899-6909.

Mizusawa, S., & Gottesman, S. (1983). Protein degradation in *Escherichia coli*: the *lon* gene controls the stability of the SulA protein. Proc. Natl. Acad. Sci. USA 80, 358-362.

Muffler, A., Trauslen, D. D., Altuvia, S., Storz, G., & Hengge-Aronis, R. (1996). Posttranscriptional osmotic regulation of the s^S subunit of RNA polymerase in *Escherichia coli*. J. Bacteriol. 178, 1607-1613.

Noble, J. A., Innis, M. A., Koonin, E. V., Rudd, K. E., Banuett, F., & Herskowitz, I. (1993). The *Escherichia coli* hflA locus encodes a putative GTP-binding protein and two membrane proteins, one of which contains a protease-like domain. Proc. Natl. Acad. Sci. USA 90(22), 10866-10870.

Nohmi, T., Battista, J. R., Dodson, L. A., & Walker, G. C. (1988). RecA-mediated cleavage activates UmuD for mutagenesis: Mechanistic relationship between transcriptional derepression and posttranslational activation. Proc. Natl. Acad. Sci. USA 85, 1816-1820.

Ostersetzer, O., Tabak, S., Yarden, O., Shapira, R., & Adam, Z. (1996). Immunological detection of proteins similar to bacterial proteases in higher plant chloroplasts. Eur. J. Biochem. 236(3), 932-936.

Parsell, D. A., & Sauer, R. T. (1989). The structural stability of a protein is an important determinant of its proteolytic susceptibility in *E. coli*. J. Biol. Chem. 264, 7590-7595.

Parsell, D. A., Silber, K. R., & Sauer, R. T. (1990). Carboxy-terminal determinants of intracellular protein degradation. Genes Dev. 4, 277-286.

Pratt, L. A., & Silhavy, T. J. (1996). The response regulator SprE controls the stability of RpoS. Proc. Natl. Acad. Sci. USA 93, 2488-2492.

Qu, J. N., Makino, S. I., Adachi, H., Koyama, Y., Akiyama, Y., Ito, K., Tomoyasu, T., Ogura, T., & Matsuzawa, H. (1996). The tolZ gene of *Escherichia coli* is identified as the ftsH gene. J. Bacteriol. 178(12), 3457-3461.

Rep, M., Nooy, J., Guelin, E., & Grivell, L. A. (1996a). Three genes for mitochondrial proteins suppress null-mutations in both Afg3 and Rca1 when over-expressed. Curr. Genet. 30(3), 206-2711.

Rep, M., van Dijl, J. M., Suda, K., Schatz, G., Grivell, L. A., & Suzuki, C. K. (1996b). Promotion of mitochondrial membrane complex assembly by a proteolytically inactive yeast Lon. Science 274(5284), 103-106.

Roberts, J. W., & Roberts, C. W. (1975). Proteolytic cleavage of bacteriophage lambda repressor in induction. Proc. Natl. Acad. Sci. USA 72, 147-151.

Rogers, S., Wells, R., & Rechsteiner, M. (1986). Amino acid sequences common to rapidly degraded proteins: the PEST hypothesis. Science 234, 364-368.

Rohrwild, M., Coux, O., Huang, H. C., Moerschell, R. P., Yoo, S. J., Seol, J. H., Chung, C. H., & Goldberg, A. L. (1996). HslV-HslU: a novel ATP-dependent protease complex in *Escherichia coli* related to the eukaryotic proteasome. Proc. Natl. Acad. Sci. USA 93, 5808-5813.

Rohrwild, M., Pfeifer, G., Santarius, U., Muller, S. A., Huang, H. C., Engel, A., Baumeister, W., & Goldberg, A. L. (1997). The ATP-dependent HslVU protease from *Escherichia coli* is a four-ring structure resembling the proteasome. Nature Stuct. Biol. 4, 133-139.

Schimke, R. T., & Doyle, D. (1970). Control of enzyme levels in animal tissues. Annu. Rev. Biochem. 39, 929-976.

Schirmer, E. C., Glover, J. R., Singer, M. A., & Lindquist, S. (1996). HSP100/Clp proteins: a common mechanism explains diverse functions. Trends Biochem. Sci. 21, 289-296.

Schmidt, R., Decatur, A. L., Rather, P. N., Moran, C., Jr., & Losick, R. (1994). Bacillus subtilis lon protease prevents inappropriate transcription of genes under the control of the sporulation transcription factor sigma G. J. Bacteriol. 176(21), 6528-6537.

Schweder, T., Lee, K. H., Lomovskaya, O., & Matin, A. (1996). Regulation of *Escherichia coli* starvation sigma factor (s^s) by ClpXP protease. J. Bacteriol. 178, 470-476.

Seol, J. H., Baek, S. H., Kang, M. S., Ha, D. B., & Chung, C. H. (1995). Distinctive roles of the two ATP-binding sites in ClpA, the ATPase component of protease Ti in *Escherichia coli*. J. Biol. Chem. 270(14), 8087-8092.

Shanklin, J., DeWitt, N. D., & Flanagan, J. M. (1995). The stroma of higher plant plastids contain ClpP and ClpC, functional homologs of *Escherichia coli* ClpP and ClpA: an archetypal two-component ATP-dependent protease. Plant Cell 7(10), 1713-1722.

Shin, D. H., Yoo, S. J., Shim, Y. K., Seol, J. H., Kang, M.-S., & Chung, C. H. (1996). Mutational analysis of the ATP-binding site in HslU, the ATPase component of the HslVU protease in *Escherichia coli*. FEBS Lett. 398, 151-154.

Shirai, Y., Akiyama, Y., & Ito, K. (1996). Suppression of ftsH mutant phenotypes by overproduction of molecular chaperones. J. Bacteriol. 178(4), 1141-1145.

Silber, K. R., Keiler, K. C., & Sauer, R. T. (1992). Tsp: A tailspecific protease that selectively degrades proteins with nonpolar C termini. Proc. Natl. Acad. Sci. USA 89, 295-299.

Singh, S. K., & Maurizi, M. R. (1994). Mutational analysis demonstrates different functional roles for the two ATP-binding sites in ClpAP protease from *Escherichia coli*. J. Biol. Chem. 269, 29537-29545.

Singh, S. K., & Maurizi, M. R. (1997). Kinetics of subunit exchange in the ClpAP *Escherichia coli* protease complex. J. Biol. Chem. In preparation.

Shotland, Y., Koby, S., Teff, D., Mansur, N., Oren, D. A., Tatematsu, K., Tomoyasu, T., Kessel, M., Bukau, B., Ogura, T., & Oppenheim, A. B. (1997). Proteolysis of the phage λ CII regulatory protein by FtsH (HflB) of *Escherichia coli*. Mol. Microbiol. 24, 1303-1310.

Sonezaki, S., Okita, K., Oba, T., Ishii, Y., Kondo, A., & Kato, Y. (1995). Protein substrates and heat shock reduce the DNA-binding ability of *Escherichia coli* Lon protease. Appl. Microbiol. Biotech. 44(3-4), 484-488.

Stewart, B. J., Enos-Berlage, J. L., & McCarter, L. L. (1997). The lonS gene regulates swarmer cell differentiation of Vibrio parahaemolyticus. J. Bacteriol. 179(1), 107-114.

Stout, V., Torres-Cabassa, A., Maurizi, M. R., Gutnick, D., & Gottesman, S. (1991). RcsA, an unstable positive regulator of capsular polysaccharide synthesis. J. Bacteriol. 173, 1738-1747.

Straus, D. B., Walter, W. A., & Gross, C. (1988). *Escherichia coli* heat shock gene mutants are defective in proteolysis. Genes Dev. 2, 1851-1858.

Suzuki, C. K., Suda, K., Wang, N., & Schatz, G. (1994). Requirement for the yeast gene LON in intramitochondrial proteolysis and maintenance of respiration [published erratum appears in Science 1994 May 13;264(5161):891]. Science 264(5156), 273-276.

Szalewska, A., Wegrzyn, G., & Taylor, K. (1994). Neither absence nor excess of lambda O initiator-digesting ClpXP protease affects lambda plasmid or phage replication in *Escherichia coli*. Mol. Microbiol. 13(3), 469-474.

Tauer, R., Mannhaupt, G., Schnall, R., Pajic, A., Langer, T., & Feldmann, H. (1994). Yta10p, a member of a novel ATPase family in yeast, is essential for mitochondrial function. FEBS Lett. 353(2), 197-200.

Teichmann, U., van Dyck, L., Guiard, B., Fischer, H., Glockshuber, R., Neupert, W., & Langer, T. (1996). Substitution of PIM1 protease in mitochondria by *Escherichia coli* Lon protease. J. Biol. Chem. 271(17), 10137-10142.

Thompson, M. W., & Maurizi, M. R. (1994). Activity and specificity of *Escherichia coli* ClpAP protease in cleaving model peptide substrates. J. Biol. Chem. 269, 18201-18208.

Thompson, M. W., Singh, S. K., & Maurizi, M. R. (1994). Processive degradation of proteins by the ATP-dependent Clp protease from *Escherichia coli*: requirement for the multiple array of active sites in ClpP but not ATP hydrolysis. J. Biol. Chem. 169, 18209-18215.

Tobias, J. W., Shrader, T. E., Rocap, G., & Varshavsky, A. (1991). The N-End rule in bacteria. Science 254, 1374-1376.

Tojo, N., Inouye, S., & Komano, T. (1993a). Cloning and nucleotide sequence of the Myxococcus xanthus lon gene: indispensability of lon for vegetative growth. J. Bacteriol. 175(8), 2271-2277.

Tojo, N., Inouye, S., & Komano, T. (1993b). The lonD gene is homologous to the lon gene encoding an ATP-dependent protease and is essential for the development of *Myxococcus xanthus*. J. Bacteriol. 175(14), 4545-4549.

Tomoyasu, T., Gamer, J., Bukau, B., Kanemori, M., Mori, H., Rutman, A. J., Oppenheim, A. B., Yura, T., Yamanaka, K., Niki, H. et al. (1995). *Escherichia coli* FtsH is a membrane-bound, ATP-dependent protease which degrades the heat-shock transcription factor sigma 32. Embo. J. 14(11), 2551-2560.

Tomoyasu, T., Yamanaka, K., Murata, K., Suzaki, T., Bouloc, P., Kato, A., Niki, H., Hiraga, S., & Ogura, T. (1993). Topology and subcellular localization of FtsH protein in *Escherichia coli*. J. Bacteriol. 175(5), 1352-1357.

Torres-Cabassa, A. S., & Gottesman, S. (1987). Capsule synthesis in *Escherichia coli* K-12 is regulated by proteolysis. J. Bacteriol. 169, 981-989.

Tsuchimoto, S., Nishimura, Y., & Ohtsubo, E. (1992). The stable maintenance system pem of plasmid R100: Degradation of PemI protein may allow PemK protein to inhibit cell growth. J. Bacteriol. 174, 4205-4211.

Van Dyck, L., Pearce, D. A., & Sherman, F. (1994). PIM1 encodes a mitochondrial ATP-dependent protease that is required for mitochondrial function in the yeast *Saccharomyces cerevisiae*. J. Biol. Chem. 269(1), 238-242.

Van Melderen, L., Bernard, P., & Couturier, M. (1994). Lon-dependent proteolysis of CcdA is the key control for activation of CcdB in plasmid-free segregant bacteria. Mol. Microbiol. 11(6), 1151-1157.

Van Melderen, L., Dao Thi, M. H., Lecchi, P., Gottesman, S., Couturier, M., & Maurizi, M. R. (1996). ATP-dependent degradation of CcdA by lon protease: effects of secondary structure and heterologous subunit interactions. J. Biol. Chem. 271, 27730-27738.

Varshavsky, A. (1992). The N-end Rule. Cell 69, 725-735.

Wagner, I., Arlt, H., van Dyck, L., Langer, T., & Neupert, W. (1994). Molecular chaperones cooperate with PIM1 protease in the degradation of misfolded proteins in mitochondria. Embo. J. 13(21), 5135-5145.

Wang, J., Hartling, J. A., & Flanagan, J. (1997). The structure of ClpP at 2.3 Å resolution suggests a model for ATP-dependent proteolysis. Cell 91, 447-456.

Wang, N., Gottesman, S., Willingham, M. C., Gottesman, M. M., & Maurizi, M. R. (1993). A human mitochondrial ATP-dependent protease that is highly homologous to bacterial Lon protease. Proc. Natl. Acad. Sci. USA 90, 11247-11251.

Wang, N., Huang, N.-n., Gottesman, M. M., & Maurizi, M. R. (1997). Human mitochondria contain a homolog of ClpP, the proteolytic component of the *E. coli* ATP-dependent ClpAP protease. In preparation.

Wawrzynow, A., Wojtkowiak, D., Marszalek, J., Banecki, B., Jonsen, M., Graves, B., Georgopoulos, C., & Zylicz, M. (1995). The ClpX heat-shock protein of *Escherichia coli*, the ATP-dependent substrate specificity component of the ClpP-ClpX protease, is a novel molecular chaperone. Embo. J. 14(9), 1867-1877.

Waxman, L., & Goldberg, A. L. (1982). Protease La from *Escherichia coli* hydrolyzes ATP and proteins in a linked fashion. Proc. Natl. Acad. Sci. USA 79, 4883-4887.

Waxman, L., & Goldberg, A. L. (1985). Protease La, the *lon* gene product, cleaves specific fluorogenic peptides in an ATP-dependent reaction. J. Biol. Chem. 260, 12022-12028.

Waxman, L., & Goldberg, A. L. (1986). Selectivity of intracellular proteolysis: protein substrates activate the ATP-dependent protease (La). Science 232, 500-503.

Weisberg, R. A., & Gottesman, M. E. (1971). The stability of Int and Xis functions. In: The Bacteriophage Lambda (Hershey, A. D., ed.), pp. 489-500. Cold Spring Harbor Laboratory, Cold Spring Harbor, NY.

Wickner, S., Gottesman, S., Skowyra, D., Hoskins, J., McKenney, K., & Maurizi, M. R. (1994). A molecular chaperone, ClpA, functions like DnaK and DnaJ. Proc. Natl. Acad. Sci. USA 91(25), 12218-12222.

Wickner, S., Hoskins, J., & McKenney, K. (1991). Monomerization of RepA dimers by heat shock proteins activates binding to DNA replication origin. Proc. Natl. Acad. Sci. USA 88, 7903-7907

Wojtkowiak, D., Georgopoulos, C., & Zylicz, M. (1993). Isolation and characterization of ClpX, a new ATP-dependent specificity component of the Clp protease of *Escherichia coli*. J. Biol. Chem. 268(30), 22609-22617.

Woo, K. M., Chung, W. J., Ha, D. B., Goldberg, A. L., & Chung, C. H. (1989). Protease Ti from *Escherichia coli* requires ATP hydrolysis for protein breakdown but not for hydrolysis of small peptides. J. Biol. Chem. 264, 2088-2091.

Wright, R., Stephens, C., Zweiger, G., Shapiro, L., & Alley, M. R. (1996). *Caulobacter* Lon protease has a critical role in cell-cycle control of DNA methylation. Genes Dev. 10(12), 1532-1542.

Yarmolinsky, M. B. (1989). Of genes and genesis [editorial]. New Biol. 1(2), 111-113.

Yoo, S. J., Seol, J. H., Shin, D. H., Rohrwild, M., Kang, M.-S., Tanaka, K., Goldberg, A. L., & Chung, C. H. (1996). Purification and characaterization of the heat shock propteins HslV and HslU that form a new ATP-dependent protease in *Escherichia coli*. J. Biol. Chem. 271, 14035-14040.

Zehnbauer, B. A., Foley, E. C., Henderson, G. W., & Markovitz, A. (1981). Identification and purification of the Lon+ (capR+) gene product, a DNA-binding protein. Proc. Natl. Acad. Sci. USA 78, 2043-2047.

FROM PROTEASOME TO LYSOSOME:
STUDIES ON YEAST DEMONSTRATE THE PRINCIPLES OF PROTEIN DEGRADATION IN THE EUKARYOTE CELL

Michael Thumm and Dieter H. Wolf

Advances in Molecular and Cell Biology
Volume 27, pages 43-70.

ISBN: 0-7623-0387-5

I. INTRODUCTION

Intracellular proteolysis is an essential tool to assure cellular life. Protein degradation is fundamentally involved in such different processes as adaptation to changes in environmental conditions, the targeting and sorting of proteins, cell differentiation, and the passage through the cell cycle. Here the principles of intracellular protein degradation in eukaryotic cells are discussed, using the yeast *Saccharomyces cerevisiae* as a simple, but powerful, model organism for higher cells. The structure and function of the cytoplasmic and nuclear proteasome, responsible for selective protein degradation in many life-promoting processes and its implication in endoplasmic reticulum associated protein degradation is outlined. The recent efforts to genetically dissect the unselective, starvation induced protein breakdown inside the vacuole via autophagocytosis are presented. Also the proteolytic systems found inside the mitochondrion, to some extent resembling procaryotic proteinases, are described.

II. YEAST AS A MODEL ORGANISM OF THE EUKARYOTE CELL

Since ancient times yeast is one of the most important organisms for man. It has been used as a technological tool for thousands of years: the Sumerians used yeast to produce beer and the Egyptians used the fermantation capacity of yeast to acidify bread. In modern times yeast has been used by man as a tool to uncover the basis of life. Work with yeast has led to breakthroughs in biochemistry, among which are the detection of specific catalysts of chemical reactions in organisms, the enzymes, by Berthelot in 1860 or the discovery by Buchner in 1897 showing that the activity of yeast enzymes is retained in the cell-free system of a press juice by this demonstrating that the reactions which occur in living cells can be simulated in the test tube. It is thus not surprising that the name enzyme, which is derived from the Greek, means "in yeast." The complementation of biochemical tools with the development of genetic and molecular biological tools for work with yeast in more recent years has made the organism a pacemaker in the study of fundamental biological problems. The belief that analysis of central biological problems with yeast as a model organism is directly relevant to the biology of more complex eukaryotes has become an irrefutable guiding principle over the last decade. The recent completion of the entire genome sequence of 12068 kB of the yeast *Saccharomyces cerevisiae* with its 5885 genes represents an important milestone in the biology of eukaryotes (Hieter et al., 1996; Goffeau et al., 1996). This resource is of tremendous value for all researchers working on higher organisms. A protein sequence from a multicellular eukaryote, once identified, can be used in a similarity search to find its most related counterpart in yeast and the biological information associated with the homologous yeast gene can immediately be linked to a similar function in the multicellular eukaryote. As yeast is an extremely powerful experimental organism for gaining an understanding of the function of genes in eukaryotic cell biology, an unknown function of any gene can be easily tested.

The elucidation of the function of proteolysis in cellular regulation of eukaryotic cells, the analysis of the enzymes involved and their mechanisms is exemplary for the unique value of the yeast *S. cerevisiae* in eukaryotic cell biology.

At first, intracellular proteolytic events were thought to be exclusively associated with the lysosomal (vacuolar) compartment of the cell. This compartment was considered to be the gut of the cell, only digesting cellular waste. The idea that destructive enzymes could not possibly be involved in cellular regulation had prevented research in this field for decades.

It was as early as 1898 that proteolytic activities were detected in yeast (Hahn, 1898). During the first three quarters of this century a variety of distinct proteolytic enzymes were identified and purified in yeast. As could be expected from the unspecific substrates used to test these enzymes, most of them were of vacuolar nature (Wolf, 1980). Biochemical studies tried to unravel their intracellular function, with little success (Wolf, 1982). The use of genetic tools opened a new era in the understanding of proteinase function (Wolf, 1982; Jones, 1984). They not only laid the basis for the elucidation of vacuolar proteinase function, but also led to the discovery of a variety of new proteolytic enzymes (Achstetter and Wolf, 1985; Wolf, 1986; Rendueles and Wolf, 1988; Hirsch et al., 1989).

It soon became clear that there were many proteins that had to undergo proteolytic maturation processes. Here one function of proteolysis rests in the activation of formerly inactive proteins at precise times and certain intracellular spaces. The peptides cleaved off such proteins in some cases serve additional functions such as targeting to localize a protein to a certain intracellular or extracellular compartment or act as intramolecular chaperones to allow proper folding. Examples of proteinases involved in targeting signal cleavage are the endoplasmic reticulum (ER) localized signal peptidase cleaving the signal peptides off proteins traversing the ER membrane (Dalbey and Von Heijne, 1992) or mitochondrial proteases located in the mitochondrial matrix (Hartl and Neupert, 1990) and intermembrane space (Nunnari et al., 1993) to cleave the targeting sequence off nuclear encoded proteins imported into the matrix or intermembrane space of mitochondria, respectively. Cleavage of propeptides from peptide hormones via highly specific proteases induces signaling cascades. A well-established example for such a proteinase is the yeast Kex2p (proteinase yscF) (Julius et al., 1984; Achstetter and Wolf, 1985) with its many related dibasic processing endoproteases in mammalian cells (Seidah and Chretien, 1994).

The cellular functions, loci and mechanisms of complete, unlimited proteolysis, however, remained an enigma. Only the lysosome was known to be a central organelle of cellular protein turnover. Studies on this organelle, however, had always left the question open about the selectivity of degradation and occurrence of different half-lives measured for a variety of cytoplasmic proteins (Bohley and Seglen, 1992). With the discovery of a very large and complex protease located in the cytoplasm and nucleus of all eukaryotic cells, the proteasome, a completely new understanding of protein degradation, its mechanisms and implication for cellular regulation was initiated (Rivett, 1990, 1993; Peters, 1994; Hochstrasser, 1995; Rubin and Finley, 1995; Coux et al., 1996; Hilt and Wolf, 1996).

The view emerges that the proteasome is centrally involved in the regulation of protein levels in the cytoplasm and nucleus which are critical for many cellular functions. Also improperly folded proteins of the cytoplasm and the ER seem to end up in proteasomal proteolysis.

In contrast, the lysosome (vacuole) seems to have its major role in rather slow general protein turnover of total cellular content which becomes important under nutritional starvation conditions. Interestingly, the degradation of plasma membrane proteins is a vacuolar task. The mitochondrion seems to have its own proteolytic system which controls proper protein stoichiometry in the organelle.

III. PROTEOLYSIS IN THE CYTOPLASM AND THE NUCLEUS; ENDOPLASMIC RETICULUM ASSOCIATED DEGRADATION

A. Structure of the Proteasome

In 1984 a high molecular mass multisubunit protease of about 700 kDa, called proteinase yscE, was isolated from *S. cerevisiae* (Achstetter et al., 1984). This proved to be the yeast homologue of the 20S multisubunit proteasome complex found in all eukaryotic cells (Kleinschmidt et al., 1988). As higher eukaryotic proteasomes, the enzyme complex of yeast is able to cleave chromo- and fluorogenic peptides at the carboxy terminus of hydrophobic, basic, and acidic amino acids (chymotrypsin-like, trypsin-like, and peptidyl glutamylpeptide hydrolyzing activity, respectively). The characteristic protein banding pattern of the complex in SDS polyacrylamide gel electrophoresis with molecular masses ranging from 20 to 35 kDa (Kleinschmidt et al., 1988) could be further separated by two-dimensional gel electrophoresis into 14 distinct protein spots (Heinemeyer et al., 1991). The discovery and structural analysis of the much simpler proteasome from the archaebacterium *Thermoplasma acidophilum* (Puhler et al., 1992; Zwickl et al., 1992; Löwe et al., 1995) laid the basis for a structural concept of all eukaryotic proteasomes including yeast: the 20S proteasome complex is a hollow cylinder of four stacks of proteins, each stack containing seven subunits. The only two subunits, named α and β, which build up the archaebacterial proteasome exhibit a distinct arrangement in the cylinder: the α-subunits are located in the two outer rings, the β-subunits in the two inner rings (Kopp et al., 1993). Sequencing of the genes of the yeast 20S proteasome subunits (Balzi et al., 1989; Fujiwara et al., 1990; Emori et al., 1991; Haffter and Fox, 1991; Heinemeyer et al., 1991, 1993; Basile et al., 1992; Friedman et al., 1992; Georgatsou et al., 1992, Lee et al., 1992; Hilt et al., 1993; Enenkel et al., 1994; Heinemeyer et al., 1994; Chen and Hochstrasser, 1995) uncovered a distinct sequence similarity of seven different yeast subunits with the *Thermoplasma* α-subunit and seven different yeast subunits with the *Thermoplasma* β-subunit (Heinemeyer et al., 1994). This led to the model of a cylindrical structure of the 20S proteasome, whereby seven different α-subunits each compose the two outer rings and seven different β-subunits each compose the two inner rings of the particle. The 14 proteasome subunits of yeast show strong structural relationships to the respective subunits of higher eukaryotes (Heinemeyer et al., 1994). Crosslinking studies in combination with immuno-electronmicroscopy of the human 20S proteasome has revealed the defined position of each subunit in the complex (Kopp et al., 1997). The evolutionarily highly conserved structure of all eukaryotic 20S proteasome complexes may lead to the assumption that the yeast subunits which are similar to the respective human subunits in sequence occupy the equivalent positions in the complex.

The location of the proteasome in the cytoplasm and the nucleus as discovered in higher eukaryotic cells (for review see Rivett and Knecht, 1993) and its high specificity led very early to the assumption that the active sites of of the complex must be located in the inner core of the proteasome α7β7β7α7-cylinder into which only unfolded proteins have access (Hilt and Wolf, 1992). Furthermore mutational analysis of the yeast proteasome yielded only inactive species against chromogenic substrates upon mutations in β-type subunits suggesting that this subunit type carried the active sites (Heinemeyer et al., 1991, 1993;), Hilt et al., 1993; Enenkel et al. 1994; for review see Hilt and Wolf, 1995). These hypotheses and findings were confirmed and combined when the X-ray structure of the archaebacterial proteasome became available (Löwe et al., 1995). It was shown that the active sites of the *Thermoplasma* particle resided in the β-subunits and were located in the inner core of the cylinder. The N-terminal threonine of the β-type subunit was identified as the active nucleophile classifying the archaebacterial proteasome as a threonine protease (Löwe et al., 1995; Seemuller et al., 1995). In addition a lysine residue at position 33 of the β-subunit was found to be crucial for catalytic activity of the archaebacterial enzyme (Löwe et al., 1995; Seemuller et al., 1995).

In yeast, as in all other eukaryotic proteasomes sequenced so far, only three of the seven β-type subunits—Pre2, Pre3 and Pup1—carry a threonine in a position homologous to β-Thr1 and a lysine homologous to βLys33 from *Thermoplasma*

Mutational analysis of the respective active site threonine residues in yeast showed that Pre2 is the active site subunit which confers the chymotrypsin-like activity to the particle (Chen and Hochstrasser, 1996; Heinemeyer et al., 1997), Pre3 is the active site subunit responsible for the peptidyl-glutamyl peptide hydrolyzing activity (Heinemeyer et al., 1997), and Pup1 is necessary for the trypsin-like activity of the enzyme (Heinemeyer et al., 1997). The recently analyzed X-ray structure of the yeast 20S proteasome nicely matches these genetic findings (Groll et al., 1997). In the case of the Pre2 subunit it was shown that processing of this subunit is autocatalytic and requires association of the two half ($\alpha7\beta7$) complexes to the $\alpha7\beta7\beta7\alpha7$ proteasome unit (Chen and Hochstrasser, 1996). Furthermore, the prosequence of Pre2 functions in trans, indicating a chaperonelike function (Chen and Hochstrasser, 1996). Processing of nonactive site β-subunits is dependent on the action of active β-subunits. This *trans*-processing must occur in a fully assembled 20S proteasome (Heinemeyer et al., 1997). The question of why three different proteolytically active subunits have evolved in yeast and higher eukaryotic cells may have the following answer: the three active site subunits complement each other with regard to their cleavage specificity, by improving and ensuring the processivity during substrate degradation. However, the differentiation of cleavage site preferences found for chromogenic peptides might not necessarily apply for natural substrates. For instance Pup1, which is correlated with the peptidyl glutamyl peptide hydrolyzing activity cleaves in the Pre4 propeptide region which does not contain any acidic residue (Heinemeyer, et al. 1997).

Twenty six S proteasomes had first been found in rabbit reticulocytes (Hough et al., 1986). In contrast to the 20S proteasome this 26S complex had been found to degrade ubiquitinated proteins *in vitro* in an ATP-dependent reaction. This 1,700 kDa complex contains the 20S proteasome as a core (for review see Peters, 1994). Using the cleavage of ubiquitinated proteins as a test, the 26S proteasome was also found in the yeast *S. cervisiae* (Fischer et al., 1994; Elias et al., 1995; Rubin et al. 1996).

In mammalian cells, the two 19S cap units docking symmetrically on the α-subunits of the 20S proteasome cylinder consist of at least 15 different subunits (for review see Tanaka 1995; Dubiel et al., 1995). Ten yeast 19S cap subunits homologous to mammalian 19S cap subunits have been cloned and sequenced. Some of these have been characterized as a new family of ATPases, called AAA-ATPases (Frohlich et al. 1991): *CIM3* is identical with the formerly cloned yeast gene *SUG1* (Swaffield et al., 1992; Ghislain et al., 1993; Rubin et al., 1996). The product of *CIM5/YTA3* exhibits 70% similarity with the human S7/MSS1 protein (Ghislain et al., 1993; Schnall et al. 1994). *YTA1*, *YTA2*, and *YTA5* are proposed to encode additional yeast 19S cap ATPase components (Schnall et al., 1994). Recently *SUG2* was found to also encode an ATPase component of the 19S cap of the 26S proteasome (Russell et al., 1996). ATPase subunits of the proteasome are thought to be responsible for the unfolding and the transport of substrate proteins into the 20S cylinder core. Of non-ATPase subunits Nin1, a homologue of p31 was discovered. The *SUN1* and *SUN2* (homologous to human p28 and p58, respectively) genes are multicopy suppressors of the temperature sensitivity of a *nin1-1* mutation (Dubiel et al., 1995; Tanaka 1995). Sen3 (homologous to human S1) was found as a yeast 19S cap protein which limits the expression of Sen1 (De Marini et al., 1995). Mcb1 was found as a homologue of the human multi-ubiquitin chain binding subunit S5a (van Nocker et al., 1996). Interestingly, chromosomal knock-out of the *MCB1* gene does not lead to cell death, indicating that additional factors may cooperate in substrate recognition (van Nocker et al., 1996). Two additional proteins, Nas1/Hrd2, and Nas3 are thought to be subunits of the 19S cap (Hochstrasser, 1996). Also the de-ubiquitinating enzyme Doa4 has been linked with the 26S proteasome (Papa and Hochstrasser, 1993).

B. General Proteasome Functions

Proteasomes are essential to life. With one exeption (Y13), deletion of individual chromosomal 20S proteasomal genes is lethal (Fujiwara et al., 1990; Heinemeyer et al., 1991, 1994; Hilt and Wolf, 1995). Chromosomal deletion of one of the 20S proteasome subunits most likely leads to a failure of proper assembly of the 20S proteasome. Also chromosomal deletion of genes of the 19S cap subunits of the 26S proteasome leads to cell death; *SUG1* (*CIM3*), *CIM5* (Ghislain et al., 1993), *YTA1*, *YTA2* (Schnall et al., 1994), *SUG2* (Russell et al., 1996), *SEN3* (De Marini et al., 1995), and *NIN1* (Kominami et al., 1995) are essential genes.

When the first nonlethal mutants of the proteasome became available (Heinemeyer et al., 1991) they opened the possibility to study the intracellular function of the enzyme complex. It could be shown that the proteasome is involved in turnover of short-lived proteins and abnormal proteins. It could be shown further that the enzyme complex functions in the degradation of ubiquitinated proteins *in vivo* (Heinemeyer et al., 1991). Ubiquitin, a 76 amino acid containing protein, had been considered for a long time to be a tool to tag proteins for degradation in the cytoplasm of eukaryotic cells (Hershko and Ciechanover, 1982). An array of enzymatic reactions—activation of ubiquitin via ATP and an activating enzyme E1, transfer of the activated ubiquitin to a conjugating enzyme E2, and linkage of ubiquitin with the help of a ligating enzyme E3 to a lysine residue of proteins with a concomitant build-up of a polyubiquitin chain (Chau et al., 1989; Finley and Chau, 1991; Hershko and Ciechanover, 1992) mediates rapid disappearance of the protein. The mutants defective in proteasome activity linked the ubiquitination event to subsequent degradation via the proteasome (Heinemeyer et al., 1991, 1993; Hilt et al., 1993). The fact that the proteasome is involved in the degradation of short-lived ubiquitinated proteins was further substantiated by following the degradation of substrates of the N-end rule pathway (Bachmair et al., 1986): β-galactosidase fusion proteins carrying a destabilizing amino acid and by this being a target for ubiquitination and rapid degradation, were stabilized in mutants (*pre1-1* and *pre2-2*) of the proteasome (Richter-Ruoff et al., 1992; Seufert and Jentsch, 1992).

In accordance with elevated levels of ubiquitinated proteins, yeast cells defective in the chymotrypsin-like activity of the proteasome exhibit increased sensitivity to stresses as are elevated temperature (38 °C) and application of the amino acid analogue canavanine. Even growth at 30 °C on complete and, more pronounced, poor media is affected by mutations in the Pre1 and Pre2 β-subunits of the enzyme (Heinemeyer et al., 1991, 1993). The cell differentiation process of sporulation is also defective in these mutants (Heinemeyer et al., 1991).

C. Specific Proteasome Functions

Cell Cycle

An intact proteasome is necessary for cell proliferation. Sporulation of heterozygous diploids carrying a wild-type proteasomal gene and a deletion of this gene (*PRE2/Δpre2* (Heinemeyer et al., 1993), *PRE4/Δpre4* (Hilt et al., 1993)) leads to asci of which only the two spores harboring a wild-type proteasome grow up to visible colonies. The *pre2* or *pre4* null mutant spores are capable of germinating, however, they arrest after two or three cycles of cell division (Heinemeyer et al., 1993; Hilt et al., 1993). The cell division cycle of yeast as in all eukaryotic cells is controlled through oscillations in the activity of cyclin dependent kinases (CDKs). This control is brought about by synthesis and degradation of kinase activators, the cyclins, and inhibitors (King et al., 1996). All data which are available up to now point to the proteasome as the essential catalyst for the proteolytic control. The G1 cyclins Cln1, Cln2, and Cln3 drive cells through G1 by activating the Cdc28 kinase. Following multiubituitination by the ubiquitin conjugating enzyme Cdc34 (Ubc3), Cln2, and most probably Cln1 are degraded in the ensuing S-phase. Also Cln3, which does not exhibit periodic expression, is degraded via the Cdc34 mediated proteasomal pathway (Deshaies et al., 1995; Yaglom et al., 1995; Willems et al., 1996). Carboxy terminal PEST sequences are required for proteolysis of these proteins. Phosphorylation by Cdc28 is a necessary prelude to ubiquitination and degradation (Deshaies et al., 1995; Yaglom et al., 1995). Crucial for G1-S transition is the degradation of an inhibitor of CDK, Sic1 (Mendenhall, 1993; Nugroho and Mendenhall, 1994). Additional genetic evidence for the proteasome being the proteolytic catalyst comes from *nin1-1* mutants conditionally defective in a 19S cap subunit of the enzyme: they are blocked in G1-S transition at nonpermissive temperature (Kominami et al., 1995).

B-type cyclins regulate the kinase activity during S, G2, and M-phases of the cell cycle. Their rapid degradation is dependent on a conserved nine amino acid motif, the destruction box, within the molecule, which is necessary for ubiquitination and degradation (King et al., 1996). Several observations point to the proteasome as being the proteolytic regulator also in these phases of the cell cycle. Mutant cells defective in the 19S cap subunit Nin1 of the proteasome are blocked not only in the G1-S transition but also in the G2-M transition (Kominami et al., 1995). Mutations in two ATPase subunits of the 26S proteasome, Cim3 (Sug1) and Cim5 confer a temperature-sensitive cell cycle phenotype to the respective strains. At the nonpermissive temperature the mutants stop growth with characteristics of cells arrested in the G2-metaphase (Ghislain et al., 1993). Direct involvement of the proteasome in the degradation of mitotic cyclins is indicated by overexpression experiments. Overexpression of Clb2 (Richter-Ruoff and Wolf, 1993) or Clb5 (Seufert et al., 1995) is not tolerated by mutants defective in the Pre1 subunit of the proteasome. Pds1, an inhibitor of anaphase, which also contains a destruction box—and probably functions as a chromosomal gene product holding chromosomes together—is rapidly degraded when anaphase should proceed (Yamamoto et al., 1996). The proteasome is the most likely catalyst. With its function in the cell cycle the proteasome occupies a prominent position in cellular control.

Cell Differentiation

The enzyme complex is furthermore involved in cell differentiation through regulation of the mating type of yeast cells. *S. cerevisiae* can exist in three cell types, MATα cells, MATa cells, and MATα/MATa diploid cells. The MATα2 repressor is required for the regulation of expression of the different cell types. It represses MATa-specific genes in haploid MATα cells and haploid specific genes in diploid cells (Herskowitz 1988). The MATα2 repressor is very short-lived. It is degraded in an ubiquitinated fashion (Chen et al., 1993). In mutants defective in Pre1 and Pre2 subunits of the proteasome MATα2 is stabilized demonstrating the involvement of the protease in the degradation of this transcriptional regulator (Richter-Ruoff et al., 1994). Also the Gα protein, a G-protein which is part of the cell differentiation cascade in response to mating type pheromones, is assumed to be a target of the proteasome. It is the first known *in vivo* substrate of the N-end rule pathway (Madura and Varshavsky, 1994).

Metabolic Adaptation

Two metabolic enzymes, fructose-1,6-bisphosphatase and ornithine decarboxylase, were uncovered to be targets of proteasome dependent proteolysis, demonstrating that proteasomes are involved in regulation of metabolism and adaptation of the cell to environmental changes. Fructose-1,6-bisphosphatase is a key enzyme in gluconeogenesis. It is responsible for hydrolyzing fructose-1,6-bisphosphate to yield fructose-6-phosphate. When glycolysis can proceed at the same time a futile cycle of ATP wasting would be initiated if the enzyme were not inactivated.

In *S. cerevisiae* fructose-1,6-bisphosphatase is subject to catabolite inactivation (Holzer, 1976). Upon addition of glucose to yeast cells growing on a nonfermentable carbon source fructose-1,6-bisphosphatase is inactivated by phosphorylation and subsequent degradation. Under inactivation conditions the enzyme is multiubiquitinated via the ubiquitin conjugating enzymes Ubc1, Ubc 4, and Ubc 5 (Schork et al. 1995). In mutants defective in subunits of the 20S proteasome core (*pre1-1*, *pre1-1 pre2-1* (Schork et al., 1994a,b) or of the 19S cap (*cim3-1*) (Schork et al., 1995)) fructose-1,6-bisphosphatase is considerably stabilized indicating the involvement of the proteasome in its degradation.

Ornithine decarboxylase is a key enzyme in polyamine synthesis. In mammalian cells the enzyme is one of the most rapidly turned over proteins. Mammalian ornithine decarboxylase is degraded via the 26S proteasome by a ubiquitin independent mechanism, which uses antizyme, a protein induced by polyamines as an accelerator of proteolysis. Yeast mutants defective in the Pre1 subunit of the proteasome are able to stabilize mouse ornithine decarboxylase. Also the yeast enzyme is clearly stabilized in *pre1-1* mutants of the proteasome indicating a similar inactivation mechanism for yeast ornithine decarboxylase (Mamroud-Kidron et al., 1994).

Besides degrading metabolic enzymes directly, proteasomes also function in metabolic adaptation by modulating the intracellular levels of metabolic regulator proteins. This has been demonstrated for the transcriptional activator Gcn4 (Kornitzer et al., 1994). This regulatory protein is required for high level expression of genes involved in biosynthesis of amino acids and purines. Starvation for these compounds induces Gcn4 on the level of transcription. Gcn4 contains several PEST-sequences and is a proteolytically extremely unstable protein. It is ubiquitinated via the ubiquitin conjugating enzymes Ubc3 (Cdc34) and Ubc2 (Rad6). In a mutant carrying the *cim5-1* mutation in the 19S cap of the proteasome Gcn4 is significantly stabilized, demonstrating that the proteasome is involved in its degradation. Upon amino acid starvation the rate of Gcn4 degradation is decreased. Thus besides enhanced synthesis downregulation of proteolysis is also used to increase cellular levels of this protein under these conditions (Kornitzer et al., 1994).

Stress Response and Degradation of Abnormal Proteins

The proteasome functions in stress dependent proteolytic pathways. Mutants defective in Pre1, Pre2, and Pre4 subunits of the proteasome are sensitive to heat stress and to the arginine analogue canavanine (Heinemeyer et al., 1991, 1993; Hilt et al., 1993). Under such stress conditions the mutants accumulate ubiquitinated proteins. Heat and canavanine stress are suggested to induce the formation of large numbers of abnormal proteins which are recognized and modified by the ubiquitin system. They are stabilized in the proteasome mutants due to defective proteolysis. *In vivo* degradation by the proteasome of a protein which is abnormal because it lacks its association partner has been demonstrated for the fatty acid synthase α-subunit: Fatty acid synthase is a cytoplasmically localized complex consisting of six α and six β-subunits. The complex is a long-lived protein which is degraded in the vacuole (Egner et al., 1993). However, if due to deletion of the *FAS1* gene no β-subunits are present in the cell, the unassembled, free α-subunits are rapidly degraded via the proteasome (Egner et al., 1993). This indicates that the proteasome may be generally responsible for balancing the subunit concentration of multisubunit complexes in the cell.

ER Degradation

The ER is the site of entry of proteins into the secretory pathway. Such proteins are translocated in an unfolded state through the membrane of this organelle, and, during the subsequent folding process N- and O-linked glycosylation occurs and disulfide linkages are formed. The large number of unfolded proteins that enter the lumen of the ER must be protected from aggregation and be maintained in a folding-competent state (Pryer et al., 1992). Thus the ER contains a high concentration of molecular chaperones that promote protein translocation, folding, and oligomerization (Rothman, 1989). Inefficient folding, unbalanced subunit synthesis or mutations in secretory proteins result in translocated polypeptides which fail to assume their correct conformation (Bonifacio and Klausner, 1994). The ER can eliminate these proteins through selective degradation. It was proposed that the degradation process is associated with the ER or a pre-Golgi compartment (Bonifacio and Klausner 1994). By using a mutated form of the vacuolar carboxypeptidase yscY (CPY*) (Finger et al., 1993) it was shown that this protein variant is translocated into the ER, glycosylated and thereafter rapidly degraded. Surprisingly, however, this degradation was found to be a ubiquitin mediated process, dependent on the ubiquitin conjugating enzymes Ubc6 and Ubc7, and finally, executed by the 26S proteasome (Hiller et al., 1996). Degradation by the ubiquitin-proteasome system was shown to be a cytoplasmic event indicating retrograde transport of CPY* out of the ER back into the cytoplasm (Hiller et al., 1996). Retrograde transport of the CPY* was found to be dependent on the translocon components Sec61p and Sec63-p_1 as well as the ER-lumenal chaperone Kar2p (Bip) (Plemper et al., 1997). ER-associated degradation of misfolded proteins on the cytoplasmic side of the ER is also supported by the observation that the secretory protein pro-α-factor mutated in such a manner as to be incapable of undergoing glycosylation required cytosol, ATP (McCracken and Brodsky, 1996), and the proteasome *in vitro* (Werner et al., 1996). Also here, *in vitro* studies point to the channel protein Sec61p as being involved in retrograde ER transport of the mutated α-factor (Pilon et al., 1997). Using a heterologous protein, a mutant form of human α-1-proteinase inhibitor, which is translocated into the ER of *S. cerevisiae* and glycosylated, degradation was found to be ubiquitin-independent but relied on the proteasome as well (Werner et al., 1996). When the fate of a mutated, multispanning ER-membrane protein, Sec61p, was followed, its degradation was found to be connected to ubiquitin conjugation via the conjugating enzymes Ubc6 and Ubc7 (Biederer et al., 1996). Degradation of the mutated Sec61 protein was severely retarded in a mutant defective in the Pre1 subunit of the proteasome indicating this enzyme complex to be the responsible proteinase (Biederer et al., 1996). The involvement of the proteasome was also shown in the regulated degradation of the ER-membrane located 3-hydroxy-3-

methylglutaryl-CoA (HMG-CoA) reductase (Hampton et al., 1996). The available examples may suggest that the proteasome is responsible for most ER-associated protein degradation.

IV. PROTEOLYSIS IN THE VACUOLE (LYSOSOME)

A. The Proteolytic Machinery, its Import Into the Organelle and Activation

The vacuole of the yeast *S. cerevisiae* is the functional equivalent of the mammalian lysosome. This acidic compartment plays an important role in the degradation of cellular proteins. It is furthermore involved in the storage of ions, amino acids and in pH and osmoregulation (for reviews see Achstetter and Wolf, 1985; Rendueles and Wolf, 1988; Hirsch et al., 1989; Klionsky et al., 1990; Jones, 1991; Thumm, 1993; van den Hazel et al., 1996). A large number of rather unspecific endo- and exoproteinases have been localized to the vacuole.

B. The Main Proteinases, Their Transport, and Function

Proteinase yscA (PrA) is one of the two most dominant vacuolar endoproteinases. The aspartic acid proteinase is encoded by the *PRA1/PEP4* gene (Mechler and Wolf, 1981; Jones et al., 1982). It shares significant homologies to human pepsin and the human lysosomal cathepsin D (Ammerer et al., 1986; Woolford et al., 1986).

Proteinase yscA as a powerful, highly unspecific proteinase may be harmful to the cell in the case of vacuolar leakage. This is illustrated by the existence of the strong cytoplasmic inhibitor I_{A_3} (Saheki et al., 1974). Strains deleted for the respective *pai3* gene exhibited a 2–3-fold increased protein degradation rate during the first six hours after shifting from rich to starvation medium (Schu and Wolf, 1991). The subtilisin related serine endoproteinase yscB (PrB, gene *PRB1*) is the second vacuolar endoproteinase with a very broad substrate specificity (Wolf and Ehmann, 1979; Zubenko et al., 1979; Moehle et al., 1987). Also for this enzyme a cytoplasmic inhibitor I_2^B (*PBI2* gene) was detected. Under starvation conditions Δ*pbi2* strains had a 50% increased protein degradation rate (Schu et al., 1991). Carboxypeptidase yscY (CPY, gene *PRC1*), is a rather unspecific serine exopeptidase. The gene encodes a preproenzyme which is matured to a protein of a molecular mass of 61 kDa (Wolf and Fink, 1975; Wolf and Weiser, 1977; Hasilik and Tanner, 1978; Stevens et al., 1982, 1986; Klionsky et al., 1990; Sorensen et al., 1994). A cytoplasmic inhibitor I^C of carboxypeptidase yscY with a molecular mass of 25 kDa was found (Lenney, 1975).

Carboxypeptidase yscY, like most of the soluble vacuolar proteinases, is synthesized as a enzymatically inactive precursor which is proteolytically matured and glycosylated on its transit to the vacuole via the ER and the Golgi apparatus. This enzyme has been used extensively as a marker protein to characterize the vacuolar protein sorting pathway (Stack and Emr, 1993; Conibear and Stevens, 1995). The correct sorting of soluble vacuolar proteinases requires a vacuolar targeting signal within the proteins. In the propeptide of carboxypeptidase yscY a four amino acid motif, QRPL, which is essential for vacuolar targeting has been identified (Johnson et al., 1987; Valls et al., 1990). Overexpression of this soluble proteinase leads to mislocalization to the extracellular medium (Stevens et al., 1986) suggesting a saturable receptor protein, which has been identified as the Vps10p (Marcusson et al., 1994; Cereghino et al., 1995).

C. Maturation of Vacuolar Proteinases

Proteinase yscA deficient mutant cells exhibit significantly reduced levels of active proteinase yscB, carboxypeptidase yscY, aminopeptidase I and several other vacuolar hydrolases (Hemmings et al., 1981; Jones et al., 1982; Klionsky et al., 1990). Proteinase yscA is synthesized as an inactive pre-proenzyme which is proteolytically matured and glycosylated to the 42 kDa mature vacuolar form with a half-time of about six minutes (Mechler et al., 1982; Ammerer et al., 1986; Woolford et al., 1986; Klionsky et al., 1988; Mechler et al., 1988). Proteinase yscB is initially transcribed as a pre-super-proenzyme (Moehle et al., 1987; Mechler et al., 1988). The aminoterminal presequences of these two endoproteinases are proteolytically removed by the signal peptidase upon arrival in the ER (Klionsky et al., 1988; Mechler et al., 1988; Moehle et al., 1989; Yadeau et al., 1991). The generated pro-proteinase yscA transits to the vacuole. The superpeptide of the super-pro form of proteinase yscB is autocatalytically removed inside the ER, but most likely remains attached to the pro-proteinase yscB molecule on the way to the vacuole (Mechler et al., 1988; Moehle et al., 1989; Nebes and Jones, 1991; Hirsch et al., 1992a). Pro-proteinase yscA can initate its own maturation in the vacuole or a prevacuolar compartement yielding a 43 kDa active species (PrA^a) which is further processed by proteinase yscB to the 42 kDa mature form (Rupp et al., 1991; Hirsch et al., 1992b; van den Hazel et al., 1992; Woolford et al., 1993; Rupp and Wolf, 1995). Active proteinase yscA triggers the maturation of proteinase yscB, followed by a self-processing reaction leading to the mature 33 kDa proteinase yscB species (Mechler et al. 1988), (Hirsch et al. 1992a). At this point the superpeptide of proteinase yscB is also degraded (Hirsch et al., 1992a,b). *In vivo* experiments suggested that after formation of small amounts of active proteinase yscB by the action of proteinase yscA, proteinase yscB plays the major part in maturing itself and the other vacuolar hydrolases (Hirsch et al., 1992a,b).

D. Membrane Bound Vacuolar Proteinases and Their Sorting

Mature, active carboxypeptidase yscS (CPS, encoded by the *CPS1* gene (Spormann et al., 1991; Bordallo et al., 1991)) is a soluble vacuolar proteinase, which reaches the vacuole as a membrane associated precursor (Spormann et al., 1992). Pro-carboxypeptidase yscS is matured with a half-time of 20 minutes. In the absence of carboxypeptidase yscY, carboxypeptidase yscS is essential for growth of a leucine auxotrophic strain on benzyloxycarbonyl-glycyl-leucine as a sole leucine source.

The membrane bound dipeptidylaminopeptidase B (dipeptidylaminopeptidase yscV) encoded by the *DAP2* gene (Suarez Rendueles and Wolf, 1987) is not matured inside the vacuole (Roberts et al., 1989).

The sorting of membrane proteins to the vacuole follows the same route as outlined for the soluble proteinases. Detailed studies using dipeptidylaminopeptidase B, however, demonstrated the lack of a specific vacuolar targeting signal. For membrane proteins the vacuole seems to be the default sorting compartment. The overexpression of membrane proteins results in mislocalization to the vacuole (Roberts et al., 1992; Nothwehr and Stevens, 1994).

E. Autophagocytosis as a Route for the Uptake of Proteins Destined for Proteolysis Into the Vacuole

The levels of vacuolar proteinases raises severalfold, when cells enter the stationary growth phase or are subject to nutrient limitation (Achstetter and Wolf, 1985; Rendueles and Wolf, 1988; Hirsch et al. 1989; Klionsky et al., 1990; van den Hazel et al., 1996). This suggests an essential role for vacuolar proteolysis under these conditions. Indeed, the two vacuolar proteinases yscA and yscB are responsible for more than 80% of total protein breakdown when cells were starved for nitrogen, while in rich media only 30 to 40% of the total protein breakdown is due to both endopeptidases (Teichert et al., 1989). Under starvation conditions about 45% of all cellular proteins are degraded inside the vacuole during a 24 hour period.

A very interesting question is how cytosolic proteins that are to be degraded in the vacuole reach this organelle. Three different mechanisms have been reported so far. (i) A selective pathway based on a KFERQ related pentapeptide motif has been found in mammalian cells (Dice, 1990; Cuervo et al., 1995). This pathway involves the action of a 73 kDa protein, which is a member of the 70 kDa heat shock protein family (Chiang et al., 1989). Most recently the lysosomal membrane glycoprotein LGP96 was identified as a receptor molecule in this pathway (Cuervo and Dice, 1996). (ii) Microautophagocytosis, a process involving invaginations of the vacuolar membrane and subsequent pinching off of vesicles. These microautophagic vesicles are surrounded by parts of the vacuolar membrane itself (Dunn, 1994). In *Pichia pastoris* the microautophagic uptake of peroxisomes into the vacuole after a shift of the cells from methanol to glucose containing media was reported (Tuttle and Dunn, 1995). (iii) The best documented mechanism for protein uptake into the vacuole, especially in mammalian cells, is autophagocytosis (Seglen and Bohley, 1992; Dunn, 1994). Hereby double-layered nascent autophagic vesicles or early autophagosomes containing parts of the cytoplasm are formed by invaginations of parts of the rough ER (Dunn, 1990a; Ueno, et al. 1991). These vesicles are maturing into late autophagosomes and autolysosomes by acquiring vacuolar proteins and vacuolar proteinases (Dunn, 1990b; Tooze et al., 1990; Aplin et al., 1992; Gordon et al., 1992; Lawrence and Brown, 1992; Dunn, 1994).

Most recently, in *S. cerevisiae* evidence for the occurence of autophagocytosis appeared by visualizing autophagic vesicles inside the vacuole of proteinase yscB deficient cells starved for nitrogen. Inside these vesicles cytoplasmic proteins, ribosomes, and whole mitochondria could be detected (Takeshige et al., 1992; Thumm et al., 1994). Also in wild-type cells the accumulation of these monolayered autophagic vesicles in the vacuole could be induced by applying the proteinase yscB inhibitor phenylmethylsulfonylfluoride (PMSF) to the cells. Detailed electron microscopic studies further demonstrated the occurence of double-layered autophagosome-like structures in the cytosol. In some cases the outer membrane of these vesicles had been found in continuity with the vacuolar membrane, thereby suggesting a mechanism for the entrance of autophagic vesicles into the vacuole (Baba et al., 1994, 1995).

Following the concomitant uptake of several cytoplasmic proteins during starvation into the vacuole of proteinase yscA deficient cells, the unselective nature of this process has been demonstrated. Interestingly the enzymes accumulated in active form inside the vacuole. Only long-lived proteins were found in significant amounts inside the vacuole, because short-lived proteins like the α-subunit of fatty acid synthase are rapidly degraded in the cytoplasm via the proteasome (Egner et al., 1993). This is in agreement with findings in mammalian cells, which unraveled autophagy as an unspecific bulk flow process (Kopitz et al., 1990).

To genetically dissect the autophagic pathway, two sets of mutant strains (*aut*- and *apg*-) have been isolated. The *aut*-strains were isolated for their inability to degrade a cytosolic marker protein, fatty acid synthase, whose degradation has been shown to depend mainly on the vacuole (Thumm et al., 1994). The *apg*-strains were isolated for their reduced survival rate during starvation (Tsukada and Ohsumi, 1993). Both sets of autophagocytosis mutants were further screened for their inability to accumulate autophagic vesicles in the vacuole during a starvation period for nitrogen in the presence of phenylmethylsulfonylfluoride.

The autophagocytosis mutant strains share several phenotypes with proteinase yscA deficient strains. Their survival rate upon nutrient limitation is significantly reduced and homozygous diploids fail to form asci (Tsukada and Ohsumi, 1993; Thumm et al., 1994; Schlumpberger et al., 1997).

Several *AUT* genes were isolated (Schlumpberger et al., 1997). Chromosomal *aut* null mutant strains were fully viable on rich media. The morphology of their vacuole was unchanged compared to a wild-type strain. Vacuolar acidification was normal. A block of the autophagic pathway has no significant influence on the vacuolar protein sorting (Harding et al., 1996). Also fluid phase endocytosis or secretion was not influenced, suggesting that autophagocytosis may not be an essential prerequisite for these processes (Schlumpberger et al., 1997). Various *aut* deleted strains showed a reduction in the starvation induced protein breakdown rate of about 80% (Schlumpberger et al., 1997). A *pra1 Δaut1* double mutant strain showed no further reduction in the protein breakdown rate. This confirms the vacuolar localization of the autophagocytosis dependent protein breakdown. *Aut3* mutants defective in a protein kinase were identified. They showed the same phenotype with respect to defects in protein degradation (Straub et al., 1997). Most recently, $APG1_1$, a gene that is identical to *AUT3*, has been identified (Matsuura et al., 1997). Two new complementation groups, *aut4* and *aut5,* have been isolated, which exhibit a defect in lysing autophagic vesicles inside the vacuole (Thumm et al., unpublished observations).

F. The Cytoplasm to Vacuole Targeting of Aminopeptidase I and the Autophagic Process

Aminopeptidase I (gene *APEI*) is a soluble vacuolar enzyme (Frey and Röhm, 1978; Chang and Smith, 1989; Cueva et al., 1989; Bordallo and Suarez-Rendueles, 1995). After uptake of the cytosolic 514 amino acid precursor molecule into the vacuole, the amino terminal 45 amino acids are split off in a proteinase yscB dependent manner. The existence of a 55 kDa intermediate form of aminopeptidase I (iAPI) due to only partial processing by proteinase yscA has been reported (Segui et al., 1995). Several data clearly indicate that aminopeptidase I is delivered to the vacuole independently of the early secretory pathway (Klionsky et al., 1992). The half-time of the maturation process with 45 minutes is significantly longer than the six minutes found for other soluble vacuolar proteinases which transit through the secretory pathway. Mature aminopeptidase I is not glycosylated. Overexpression of pro-aminopeptidase I resulted in the accumulation of unprocessed precursor in the cytosol. This suggests the idea of a saturable component, maybe a receptor molecule, involved in the aminopeptidase I import from the cytoplasm into the vacuole (Klionsky et al., 1992).

Most of the autophagocytosis mutant cells are unable to transport precursor aminopeptidase I to the vacuole (Harding et al., 1996; Scott et al., 1996). *Cvt*-mutants with a defect in the cytoplasm to vacuole targeting of aminopeptidase I have been isolated (Harding et al., 1995). Besides the phenotypical overlap between autophagocytosis and aminopeptidase I targeting, a genetic overlap between the *aut*- and *cvt*-mutant strains has also been detected (Harding et al., 1996).

G. Degradation of Membrane Proteins in the Vacuole

As outlined, autophagocytosis is responsible for the transport of large amounts of soluble cytosolic proteins and whole organelles (as are mitochondria or ribosomes) to the vacuole (Thumm et al., 1994; Takeshige et al., 1992). For the degradation of several individual membrane proteins an endocytosis dependent mechanism carrying these proteins to the vacuole was reported.

The plasma membrane Pdr5 ABC transporter is a short-lived protein with a half-life of about 60–90 minutes, which accumulates and is significantly stabilized in vacuoles of proteinase yscA deficient cells (Egner et al., 1995). A block of endocytosis in *end4* mutant cells prevented the transport of this protein to the vacuole and subsequent degradation. Furthermore ubiquitination of Pdr5p *in vivo* could be shown (Egner and Kuchler, 1996). No involvement of the cytoplasmic proteasome in the degradation of the Pdr5p was found (Egner et al., 1995).

Also the plasma membrane localized α-factor receptor Ste2p was found to be degraded in the vacuole. Ligand induced ubiquitination as a prerequisite for its endocytic internalization was shown. The proteasome was not involved in the degradation process (Hicke and Riezman, 1996).

Kölling and colleagues (1994) reported the accumulation of an ABC-transporter involved in a-factor secretion, Ste6p, at the plasma membrane in ubiquitinated form in mutant cells defective in endocytosis. The very unstable Ste6p with a half-life of 13 minutes was stabilized approximately threefold in a *ubc4ubc5* mutant strain. In a proteinase yscA deficient strain, the Ste6p is strongly stabilized (half-time 2 hours) indicating the vacuole as the locus of proteolysis (Kölling and Hollenberg, 1994). Similar results were obtained for the degradation of the uracil permease Fur4p (Volland et al., 1994). The degradation process of Fur4p depends on a functional *NPL1/RSP5* gene, which encodes a ubiquitin-protein ligase (E3 enzyme) (Hein et al., 1995). Also for this protein the vacuole is the degrading organelle. Two other plasma membrane proteins involved in sugar transport, the maltose transporter (Riballo et al., 1995) and the galactose transporter Gal2 (Horak and Wolf, 1997), were shown to be endocytosed and degraded upon glucose treatment of cells via a vacuolar proteinase dependent mechanism. Also for Gal2 ubiquitination prior to degradation was shown (Horak and Wolf, 1997).

Taken together these results suggest ubiquitination as a signal for the endocytic delivery of membrane proteins to the vacuole where subsequent degradation of these proteins takes place. This suggests a role for ubiquitination as a degradation signal for membrane proteins without the involvement of the proteasome. Glucose induced degradation of the soluble cytoplasmic enzyme fructose-1,6-bisphosphatase had also been claimed to be of vacuolar nature (Chiang and Schekman, 1991). However, detailed experiments indicate the ubiquitin-proteasome pathway responsible for catabolite inactivation of this enzyme (Schork et al., 1994a,b 1995).

V. PROTEOLYSIS IN THE IMMIGRANT, THE MITOCHONDRION

Due to its evolutionary history as an immigrant in the form of endosymbiotic bacteria, the mitochondrion is unique among the organelles of the eukaryote cell. Not only is it surrounded by two membranes, it has also retained an independent genetic apparatus and protein machinery which encode and supply part of the protein potential to exert mitochondrial function. From this point of view protein regulation via proteolysis in mitochondria independent from other cellular proteolytic systems could be expected. However, until recently only complete engulfment via autophagocytosis of mitochondria in the vacuole had been visualized (Takeshige et al., 1992; Thumm et al., 1994) which most likely constitutes a mechanism to regulate the intracellular concentration of mitochondria. Only very recently proteases, which could fulfil the specific functions of regulating protein levels in the mitochondrion individually, were uncovered. Two proteolytic systems were found, one in the mitochondrial matrix (Kutejova et al., 1993; Suzuki et al., 1994; Van Dyck et al., 1994) and one in the mitochondrial inner membrane (Arlt et al., 1996; Leonhard et al., 1996). The matrix enzyme, Pim1 (Van Dyck et al., 1994) or Lon (Suzuki et al., 1994) is an ATP-dependent protease of 1,133 amino acids, which exhibits 30% identity with the ATP-dependent protease La from *Escherichia coli* and Lon from *Bacillus brevis*. The enzyme contains two highly conserved regions, a putative ATP binding motif and a region around a serine residue at the putative catalytic site (Suzuki et al., 1994; Van Dyck et al. 1994). Cells defective in the proteolytic function of Pim1 (Lon) are respiratory deficient and contain a nonfunctional mitochondrial genome (Suzuki et al., 1994; Van Dyck et al., 1994). The enzyme is responsible for selective mitochondrial protein turnover and proteolytic breakdown of misfolded proteins in the matrix space site (Suzuki et al., 1994; Van Dyck et al., 1994; Wagner et al., 1994). Interestingly, the Lon protease of *E. coli* can substitute for Pim functions to a large extent in a *PIM1* deleted yeast strain (Teichmann et al., 1996). ATP-dependent degradation mediated by the Pim1 protease depends on the mitochondrial chaperone proteins mt-hsp70 and Mdj1p, which maintain misfolded proteins in a soluble state (Wagner et al., 1994).

A second proteolytic system was found in the inner mitochondrial membrane. It consists of highly conserved members of the AAA-family of proteins. Two different protease complexes are present: one complex contains multiple copies of the ATP-dependent metallopeptidases Yta10p and Yta12p (Schnall et al., 1994; Arlt et al., 1996; Leonhard et al., 1996). The other complex is thought to be built up of several units of Yme1 (Thorsness et al., 1993) (Yta11) (Schnall et al., 1994) yielding a complex of about 850 kDa (Leonhard et al., 1996), also exhibiting metallopeptidase activity. Yme1, Yta10, and Yta12 exhibit high sequence homology to the *E. coli* FtsH, the prokaryotic AAA protease. Both protease complexes are embedded into the inner mitochondrial membrane in opposite directions. According to the opposite topology of the two protease complexes the Yme1 derived complex, the active site of which faces the intermembrane space is called i-AAA protease, the Yta10-Yata12 containing complex, which with its active site faces the mitochondrial matrix is named m-AAA protease (Leonhard et al., 1996). Both protease complexes are thought to degrade the hydrophilic segments of inner membrane proteins on the respective side of the inner membrane. The cooperation of both proteinases in the degradation of inner mitochondrial membrane proteins is an open question. Interestingly, the m-AAA protease (Yta10-Yta12) has chaperonelike activity. By this the enzyme complex mediates assembly and degradation processes of membrane protein complexes and may be a key player in the maintenance of inner mitochondrial membrane integrity (Arlt et al., 1996).

REFERENCES

Achstetter, T., Ehmann, C., Osaki, A., & Wolf, D. H. (1984). Proteolysis in eukaryotic cells. Proteinase yscE, a new yeast peptidase. J. Biol. Chem. 259, 13344-13348.

Achstetter, T., & Wolf, D. H. (1985). Hormone processing and membrane-bound proteinases in yeast. Embo J. 4, 173-177.

Achstetter, T., & Wolf, D. H. (1985). Proteinases, proteolysis and biological control in the yeast *Saccharomyces cerevisiae*. Yeast 1, 139-157.

Ammerer, G., Hunter, C. P., Rothman, J. H., Saari, G. C., Valls, L. A., & Stevens, T. H. (1986). PEP4 gene of *Saccharomyces cerevisiae* encodes proteinase A, a vacuolar enzyme required for processing of vacuolar precursors. Mol Cell Biol 6, 2490-2499.

Aplin, A., Jasionowski, T., Tuttle, D. L., Lenk, S. E., & Dunn, W. J. (1992). Cytoskeletal elements are required for the formation and maturation of autophagic vacuoles. J. Cell Physiol. 152, 458-466.

Arlt, H., Tauer, R., Feldmann, H., Neupert, W., & Langer, T. (1996). The YTA10-12 complex, an AAA protease with chaperone-like activity in the inner membrane of mitochondria. Cell 85, 875-885.

Baba, M., Osumi, M., & Ohsumi, Y. (1995). Analysis of the membrane structures involved in autophagy in yeast by freeze-replica method. Cell Struct. Func. 20, 465-471.

Baba, M., Takeshige, K., Baba, N., & Ohsumi, Y. (1994). Ultrastructural analysis of the autophagic process in yeast: detection of autophagosomes and their characterization. J. Cell Biol. 124, 903-913.

Bachmair, A., Finley, D., & Varshavsky, A. (1986). In vivo half-life of a protein is a function of its amino-terminal residue. Science 234, 179-86.

Balzi, E., Chen, W. N., Capieaux, E., McCusker, J. H., Haber, J. E., & Goffeau, A. (1989). The suppressor gene scl1+ of Saccharomyces cerevisiae is essential for growth [published erratum appears in Gene 1990 Apr 30; 89(1):151]. Gene 83, 271-279.

Basile, G., Aker, M., & Mortimer, R. K. (1992). Nucleotide sequence and transcriptional regulation of the yeast recombinational repair gene RAD51. Mol. Cell. Biol. 12, 3235-3246.

Biederer, T., Volkwein, C., & Sommer, T. (1996). Degradation of subunits of the Sec61p complex, an integral component of the ER membrane, by the ubiquitin-proteasome pathway. EMBO J. 15, 2069-2076.

Bohley, P., & Seglen, P. O. (1992). Proteases and proteolysis in the lysosome. Experientia 48, 151-157.

Bonifacio, J. S., & Klausner, R. D. (1994). Degradation of proteins retained in the endoplasmic reticulum. In: Cellular Proteolytic Systems (Ciechanover, A., & Schwartz, A.L., eds.) Vol. 15, pp. 137-160. Wiley-Liss, New York.

Bordallo, J., Bordallo, C., Gascon, S., & Suarez-Rendueles, P. (1991). Molecular cloning and sequencing of genomic DNA encoding yeast vacuolar carboxypeptidase yscS. FEBS Lett. 283, 27-32.

Bordallo, J., & Suarez-Rendueles, P. (1995). Identification of regulatory proteins that might be involved in carbon catabolite repression of the aminopeptidase I gene of the yeast Saccharomyces cerevisiae. FEBS Lett. 376, 120-124.

Cereghino, J. L., Marcusson, E. G., and Emr, S. D. (1995). The cytoplasmic tail domain of the vacuolar protein-sorting receptor Vps10p and a subset of VPS gene products regulate receptor stability, function, and localization. Mol. Biol. Cell 6, 1089-1102.

Chang, Y. H., & Smith, J. A. (1989). Molecular cloning and sequencing of genomic DNA encoding aminopeptidase I from *Saccharomyces cerevisiae*. J. Biol. Chem. 264, 6979-6983.

Chau, V., Tobias, J. W., Bachmair, A., Marriott, D., Ecker, D. J., Gonda, D. K., & Varshavsky, A. (1989). A multiubiquitin chain is confined to specific lysine in a targeted short-lived protein. Science 243, 1576-1583.

Chen, P., & Hochstrasser, M. (1995). Biogenesis, structure and function of the yeast 20S proteasome. EMBO J. 14, 2620-2630.

Chen, P., & Hochstrasser, M. (1996). Autocatalytic subunit processing couples active site formation in the 20S proteasome to completion of assembly. Cell 86, 961-972.

Chen, P., Johnson, P., Sommer, T., Jentsch, S., & Hochstrasser, M. (1993). Multiple ubiquitin-conjugating enzymes participate in the in vivo degradation of the yeast MAT alpha 2 repressor. Cell 74, 357-369.

Chiang, H. L., & Schekman, R. (1991). Regulated import and degradation of a cytosolic protein in the yeast vacuole. Nature 350, 313-318.

Chiang, H. L., Terlecky, S. R., Plant, C. P., & Dice, J. F. (1989). A role for a 70-kilodalton heat shock protein in lysosomal degradation of intracellular proteins. Science 246, 382-385.

Conibear, E., & Stevens, T. H. (1995). Vacuolar biogenesis in yeast: sorting out the sorting proteins. Cell 83, 513-516.

Coux, O., Tanaka, K., & Goldberg, A. L. (1996). Structure and functions of the 20S and 26S proteasomes. Annu. Rev. Biochem. 65, 801-847.

Cuervo, A. M., & Dice, J. F. (1996). A receptor for the selective uptake and degradation of proteins by lysosomes. Science 273, 501-503.

Cuervo, A. M., Knecht, E., Terlecky, S. R., & Dice, J. F. (1995). Activation of a selective pathway of lysosomal proteolysis in rat liver by prolonged starvation. Am. J. Physiol. 269, C1200-C1208.

Cueva, R., Garcia, A. N., & Suarez, R. P. (1989). Yeast vacuolar aminopeptidase yscI. Isolation and regulation of the APE1 (LAP4) structural gene. FEBS Lett. 259, 125-129.

Dalbey, R. E., & Von Heijne, G. (1992). Signal peptidases in prokaryotes and eukaryotes—a new protease family [published erratum appears in Trends Biochem Sci 1993 Jan; 18(1):25]. Trends Biochem. Sci. 17, 474-478.

De Marini, D. J., Papa, F. R., Swaminathan, S., Ursic, D., Rasmussen, T. P., Culbertson, M. R., & Hochstrasser, M. (1995). The yeast SEN3 gene encodes a regulatory subunit of the 26S proteasome complex required for ubiquitin-dependent protein degradation in vivo. Mol. Cell. Biol. 15, 6311-6321.

Deshaies, R. J., Chau, V., & Kirschner, M. (1995). Ubiquitination of the G1 cyclin Cln2p by a Cdc34p-dependent pathway. EMBO J. 14, 303-312.

Dice, J. F. (1990). Peptide sequences that target cytosolic proteins for lysosomal proteolysis. Trends Biochem. Sci. 15, 305-309.

Dubiel, W., Ferrell, K., & Rechsteiner, M. (1995). Subunits of the regulatory complex of the 26S protease. Mol. Biol. Rep. 21, 27-34.

Dunn, W. J. (1990a). Studies on the mechanisms of autophagy: formation of the autophagic vacuole. J. Cell Biol. 110, 1923-1933.

Dunn, W. J. (1990b). Studies on the mechanisms of autophagy: maturation of the autophagic vacuole. J. Cell Biol. 110, 1935-1945.

Dunn, W. J. (1994). Autophagy and related mechanisms of lysosome-mediated protein degradation. Trends Cell Biol. 4, 139-143.

Egner, R., & Kuchler, K. (1996). The yeast multidrug transporter Pdr5 of the plasma membrane is ubiquitinated prior to endocytosis and degradation in the vacuole. FEBS Lett. 378, 177-181.

Egner, R., Mahe, Y., Pandjaitan, R., & Kuchler, K. (1995). Endocytosis and vacuolar degradation of the plasma membrane-localized Pdr5 ATP-binding cassette multidrug transporter in *Saccharomyces cerevisiae*. Mol. Cell Biol. 15, 5879-5887.

Egner, R., Thumm, M., Straub, M., Simeon, A., Schüller, H. J., & Wolf, D. H. (1993). Tracing intracellular proteolytic pathways. Proteolysis of fatty acid synthase and other cytoplasmic proteins in the yeast *Saccharomyces cerevisiae*. J. Biol. Chem. 268, 27269-27276.

Elias, S., Bercovich, B., Kahana, C., Coffino, P., Fischer, M., Hilt, W., Wolf, D. H., & Ciechanover, A. (1995). Degradation of ornithine decarboxylase by the mammalian and yeast 26S proteasome complexes requires all the components of the protease. Eur. J. Biochem. 229, 276-283.

Emori, Y., Tsukahara, T., Kawasaki, H., Ishiura, S., Sugita, H., & Suzuki, K., (1991). Molecular cloning and functional analysis of three subunits of yeast proteasome. Mol. Cell Biol. 11, 344-353.

Enenkel, C., Lehmann, H., Kipper, J., Guckel, R., Hilt, W., & Wolf, D. H. (1994). PRE3, highly homologous to the human major histocompatibility complex- linked LMP2 (RING12) gene, codes for a yeast proteasome subunit necessary for the peptidylglutamyl-peptide hydrolyzing activity. FEBS Lett. 341, 193-196.

Finger, A., Knop, M., & Wolf, D. H. (1993). Analysis of two mutated vacuolar proteins reveals a degradation pathway in the endoplasmic reticulum or a related compartment of yeast. Eur. J. Biochem. 218, 565-574.

Finley, D., & Chau, V. (1991). Ubiquitination. Ann. Rev. Cell Biol. 7, 25-69.

Fischer, M., Hilt, W., Richter, R. B., Gonen, H., Ciechanover, A., & Wolf, D. H. (1994). The 26S proteasome of the yeast *Saccharomyces cerevisiae.* FEBS Lett. 355, 69-75.

Frey, J., & Röhm, K. H. (1978). Subcellular localization and levels of aminopeptidases and dipeptidase in *Saccharomyces cerevisiae.* Biochim. Biophys. Acta 527, 31-41.

Friedman, H., Goebel, M., & Snyder, M. (1992). A homolog of the proteasome-related RING10 gene is essential for yeast cell growth. Gene 122, 203-206.

Frohlich, K. U., Fries, H. W., Rudiger, M., Erdmann, R., Botstein, D., & Mecke, D. (1991). Yeast cell cycle protein CDC48p shows full-length homology to the mammalian protein VCP and is a member of a protein family involved in secretion, peroxisome formation, and gene expression. J. Cell Biol. 114, 443-53.

Fujiwara, T., Tanaka, K., Orino, E., Yoshimura, T., Kumatori, A., Tamura, T., Chung, C. H., Nakai, T., Yamaguchi, K., Shin, S. et al. (1990). Proteasomes are essential for yeast proliferation. cDNA cloning and gene disruption of two major subunits. J. Biol. Chem. 265, 16604-16613.

Georgatsou, E., Georgakopoulos, T., & Thireos, G. (1992). Molecular cloning of an essential yeast gene encoding a proteasomal subunit. FEBS Lett. 299, 39-43.

Ghislain, M., Udvardy, A., & Mann, C. (1993). *S. cerevisiae* 26S protease mutants arrest cell division in G2/metaphase. Nature 366, 358-362.

Goffeau, A., Barrell, B. G., Bussey, H., Davis, R. W., Dujon, B., Feldmann, H. F., Galibert, Hoheisel, J. D., Jacq, C., Johnston, M., Louis, E. J., Mewes, H. W., Murakami, Y., Philippsen, P., Tettelin, H., & Oliver, S. G. (1996). Life with 6000 genes. Science 274, 546.

Gordon, P. B., Hoyvik, H., & Seglen, P. O. (1992). Prelysosomal and lysosomal connections between autophagy and endocytosis. Biochem. J. 283, 361-369.

Groll, M., Ditzel, L., Lowe, J., Stock, D., Bochtler, M., Bartunik, H. D., and Huber, R. (1997). Structure of 20S proteasome from yeast at 2.4 A resolution. Nature 386, 463-471.

Haffter, P., & Fox, T. D. (1991). Nucleotide sequence of PUP1 encoding a putative proteasome subunit in *Saccharomyces cerevisiae.* Nucleic Acids Res. 19, 5075.

Hahn, M. (1898). Das proteolytische Enzym des Hefepreßsaftes. Ber. Dt. Chem. Ges. 31, 200.

Hampton, R. Y., Gardner, R. G., and Rine, J. (1996). Role of 26S proteasome and HRD genes in the degradation of 3-hydroxy-3- methylglutaryl-CoA reductase, an integral endoplasmic reticulum membrane protein. Mol. Biol. Cell 7, 2029-2044.

Harding, T. M., Hefner-Gravink, A., Thumm, M., & Klionsky, D. J. (1996). Genetic and phenotypic overlap between autophagy and the cytoplasm to vacuole targeting pathway. J. Biol. Chem. 271, 17621-17624.

Harding, T. M., Morano, K., A. Scott, S. V., & Klionsky, D. J. (1995). Isolation and characterization of yeast mutants in the cytoplasm to vacuole protein targeting pathway. J. Cell Biol. 131, 591-602.

Hartl, F. U., & Neupert, W. (1990). Protein sorting to mitochondria: evolutionary conservations of folding and assembly. Science 247, 930-938.

Hasilik, A., & Tanner, W. (1978). Biosynthesis of the vacuolar yeast glycoprotein carboxypeptidase Y. Conversion of precursor into the enzyme. Eur. J. Biochem. 85, 599-608.

Hein, C., Springael, J. Y., Volland, C., Haguenauer, T. R., & Andre, B. (1995). NPI1, an essential yeast gene involved in induced degradation of Gap1 and Fur4 permeases, encodes the Rsp5 ubiquitin-protein ligase. Mol. Microbiol. 18, 77-87.

Heinemeyer, W., Fischer, M., Krimmer, T., Stachon, U., & Wolf, D. H. (1997). The active sites of the eukaryotic 20S proteasome and their involvement in the subunit precursor processing. J. Biol. Chem. 272, 25200-25209.

Heinemeyer, W., Gruhler, A., Mohrle, V., Mahe, Y., & Wolf, D. H. (1993). PRE2, highly homologous to the human major histocompatibility complex-linked RING10 gene, codes for a yeast proteasome subunit necessary for chrymotryptic activity and degradation of ubiquitinated proteins. J. Biol. Chem. 268, 5115-5120.

Heinemeyer, W., Kleinschmidt, J. A., Saidowsky, J., Escher, C., & Wolf, D. H. (1991). Proteinase yscE, the yeast proteasome/multicatalytic-multifunctional proteinase: mutants unravel its function in stress induced proteolysis and uncover its necessity for cell survival. EMBO J. 10, 555-562.

Heinemeyer, W., Trondle, N., Albrecht, G., & Wolf, D. H. (1994). PRE5 and PRE6, the last missing genes encoding 20S proteasome subunits from yeast? Indication for a set of 14 different subunits in the eukaryotic proteasome core. Biochemistry 33, 12229-12237.

Hemmings, B. A., Zubenko, G. S., Hasilik, A., & Jones, E. W. (1981). Mutant defective in processing of an enzyme located in the lysosome-like vacuole of *Saccharomyces cerevisiae*. Proc. Natl. Acad. Sci. USA 78, 435-439.

Hershko, A., & Ciechanover, A. (1982). Mechanisms of intracellular protein breakdown. Ann. Rev. Biochem. 51, 335-364.

Hershko, A., & Ciechanover, A. (1992). The ubiquitin system for protein degradation. Ann. Rev. Biochem. 61, 761-807.

Herskowitz, I. (1988). Life cycle of the budding yeast *Saccharomyces cerevisiae*. Microbiol. Rev. 52, 536-553.

Hicke, L., & Riezman, H. (1996). Ubiquitination of a yeast plasma membrane receptor signals its ligand-stimulated endocytosis. Cell 84, 277-287.

Hieter, P., Bassett, Jr., D. E., & Valle, D. (1996). The yeast genome—a common currency. Nat. Genet. 13, 253-255.

Hiller, M. M., Finger, A., Schweiger, M., & Wolf, D. H. (1996). ER degradation of a misfolded luminal protein by the cytosolic ubiquitin-proteasome pathway. Science 273, 1725-1728.

Hilt, W., Enenkel, C., Gruhler, A., Singer, T., & Wolf, D. H. (1993). The PRE4 gene codes for a subunit of the yeast proteasome necessary for peptidylglutamyl-peptide-hydrolyzing activity. Mutations link the proteasome to stress- and ubiquitin-dependent proteolysis. J. Biol. Chem. 268, 3479-3486.

Hilt, W., & Wolf, D. H. (1992). Stress-induced proteolysis in yeast. Mol. Microbiol. 6, 2437-2442.

Hilt, W., & Wolf, D. H. (1995). Proteasomes of the yeast S. cerevisiae: genes, structure and functions. Mol. Biol. Rep. 21, 3-10.

Hilt, W., & Wolf, D. H. (1996). Proteasomes: destruction as a programme. Trend. Biochem. Sci. 21, 96-101.

Hirsch, H. H., Schiffer, H. H., Muller, H., & Wolf, D. H. (1992a). Biogenesis of the yeast vacuole (lysosome). Mutation in the active site of the vacuolar serine proteinase yscB abolishes proteolytic maturation of its 73-kDa precursor to the 41.5-kDa pro-enzyme and a newly detected 41-kDa peptide [published erratum appears in Eur J Biochem 1992 May 1; 205(3):1217]. Eur. J. Biochem. 203, 641-653.

Hirsch, H. H., Schiffer, H. H., & Wolf, D. H. (1992b). Biogenesis of the yeast vacuole (lysosome). Proteinase yscB contributes molecularly and kinetically to vacuolar hydrolase-precursor maturation. Eur. J. Biochem. 207, 867-876.

Hirsch, H. H., Suarez-Rendueles, P., & Wolf, D. H. (1989). Yeast (*Saccharomyces cerevisiae*) proteinases: structure, characteristics, function. In: Molecular and Cell Biology of Yeasts (Walton, E.F., & Yarranton, G.T., eds.), pp. 134-200. Blackie, Glasgow and Van Nostrand Reinhold, New York.

Hochstrasser, M. (1995). Ubiquitin, proteasomes, and the regulation of intracellular protein degradation. Curr. Opin. Cell. Biol. 7, 215-223.

Hochstrasser, M. (1996). Ubiquitin-dependent protein degradation. Ann. Rev. Genet. 30, 405-439.

Holzer, H. (1976). Catabolite inactivation in yeast. Trends Biochem. Sci. 1, 178-181.

Horak, J., & Wolf, D. H. (1997). Catabolite Inactivation of the galactose transporter in the yeast *Saccharomyces cerevisiae*: Ubiquitination, endocytosis, and degradation in the vacuole. J. Bacteriol. 179, 1541-1549.

Hough, R., Pratt, G., & Rechsteiner, M. (1986). Ubiquitin-lysozyme conjugates. Identification and characterization of an ATP-dependent protease from rabbit reticulocyte lysates. J. Biol. Chem. 261, 2400-2408.

Johnson, L. M., Bankaitis, V. A., & Emr, S. D. (1987). Distinct sequence determinants direct intracellular sorting and modification of a yeast vacuolar protease. Cell 48, 875-885.

Jones, E. W. (1984). The synthesis and function of proteases in Saccharomyces: genetic approaches. Ann. Rev. Genet. 18, 233-270.

Jones, E. W. (1991). Three proteolytic systems in the yeast *Saccharomyces cerevisiae*. J. Biol. Chem. 266, 7963-7966.

Jones, E. W., Zubenko, G. S., & Parker, R. R. (1982). PEP4 gene function is required for expression of several vacuolar hydrolases in *Saccharomyces cerevisiae*. Genetics 102, 665-677.

Julius, D., Brake, A., Blair, L., Kunisawa, R., & Thorner, J. (1984). Isolation of the putative structural gene for the lysine-arginine- cleaving endopeptidase required for processing of yeast prepro-alpha- factor. Cell 37, 1075-1089.

King, R. W., Deshaies, R. J., Peters, J. M., & Kirschner, M. W. (1996). How proteolysis drives the cell cycle. Science 274, 1652-1659.

Kleinschmidt, J. A., Escher, C., & Wolf, D. H. (1988). Proteinase yscE of yeast shows homology with the 20 S cylinder particles of *Xenopus laevis*. FEBS Lett. 239, 35-40.

Klionsky, D. J., Banta, L. M., & Emr, S. D. (1988). Intracellular sorting and processing of a yeast vacuolar hydrolase: proteinase A propeptide contains vacuolar targeting information. Mol. Cell Biol. 8, 2105-2116.

Klionsky, D. J., Cueva, R., & Yaver, D. S. (1992). Aminopeptidase I of *Saccharomyces cerevisiae* is localized to the vacuole independent of the secretory pathway. J. Cell Biol. 119, 287-299.

Klionsky, D. J., Herman, P. K., & Emr, S. D. (1990). The fungal vacuole: composition, function and biogenesis. Microbiol. Rev. 54, 266-292.

Kölling, R., & Hollenberg, C. P. (1994). The ABC-transporter Ste6 accumulates in the plasma membrane in a ubiquitinated form in endocytosis mutants. EMBO J. 13, 3261-3271.

Kominami, K., DeMartino, G. N., Moomaw, C. R., Slaughter, C. A., Shimbara, N., Fujimuro, M., Yokosawa, H., Hisamatsu, H., Tanahashi, N., Shimizu, Y., et al. (1995). Nin1p, a regulatory subunit of the 26S proteasome, is necessary for activation of Cdc28p kinase of *Saccharomyces cerevisiae*. EMBO J. 14, 3105-3115.

Kopitz, J., Kisen, G. O., Gordon, P. B., Bohley, P., & Seglen, P. O. (1990). Nonselective autophagy of cytosolic enzymes by isolated rat hepatocytes. J. Cell Biol. 111, 941-953.

Kopp, F., Dahlmann, B., & Hendil, K. B. (1993). Evidence indicating that the human proteasome is a complex dimer. J. Mol. Biol. 229, 14-19.

Kopp, F., Hendil, K. B., Dahlmann, B., Kristensen, P., Sobek, A., & Uerkvitz, W. (1997). Subunit arrangement in the human 20S proteasome. Proc. Nat. Acad. Sci. USA. 94, 2939-2944.

Kornitzer, D., Raboy, B., Kulka, R. G., & Fink, G. R. (1994). Regulated degradation of the transcription factor Gcn4. EMBO J. 13, 6021-6030.

Kutejova, E., Durcova, G., Surovkova, E., & Kuzela, S. (1993). Yeast mitochondrial ATP-dependent protease: purification and comparison with the homologous rat enzyme and the bacterial ATP-dependent protease La. FEBS Lett. 329, 47-50.

Lawrence, B. P., & Brown, W. J. (1992). Autophagic vacuoles rapidly fuse with pre-existing lysosomes in cultured hepatocytes. J. Cell Sci. 102, 515-526.

Lee, D. H., Tanaka, K., Tamura, T., Chung, C. H., & Ichihara, A. (1992). PRS3 encoding an essential subunit of yeast proteasomes homologous to mammalian proteasome subunit C5. Biochem. Biophys. Res. Commun. 182, 452-460.

Lenney, J. F. (1975). Three yeast proteins that specifically inhibit yeast proteases A, B, and C. J. Bacteriol. 122, 1265-1273.

Leonhard, K., Herrmann, J. M., Stuart, R. A., Mannhaupt, G., Neupert, W., & Langer, T. (1996). AAA proteases with catalytic sites on opposite membrane surfaces comprise a proteolytic system for the ATP-dependent degradation of inner membrane proteins in mitochondria. EMBO J. 15, 4218-4229.

Löwe, J., Stock, D., Jap, B., Zwickl, P., Baumeister, W., & Huber, R. (1995). Crystal structure of the 20S proteasome from the archaeon T. acidophilum at 3.4 A resolution [see comments]. Science 268, 533-539.

Madura, K., & Varshavsky, A. (1994). Degradation of G alpha by the N-end rule pathway. Science 265, 1454-1458.

Mamroud-Kidron, E., Rosenberg-Hasson, Y., Rom, E., & Kahana, C. (1994). The 20S proteasome mediates the degradation of mouse and yeast ornithine decarboxylase in yeast cells. FEBS Lett. 337, 239-242.

Marcusson, E. G., Horazdovsky, B. F., Cereghino, J. L., Gharakhanian, E., & Emr, S. D. (1994). The sorting receptor for yeast vacuolar carboxypeptidase Y is encoded by the *VPS10* gene. Cell 77, 579-586.

Matsuura, A., Tsukada, M., Wada, Y., & Ohsumi, Y. (1997). Apg1p, a novel protein kinase required for the autophagic process in *Saccharomyces cerevisiae* Gene 192, 245-50.

McCracken, A. A., & Brodsky, J. L. (1996). Assembly of ER-associated protein degradation in vitro: dependence on cytosol, calnexin, and ATP. J. Cell Biol. 132, 291-298.

Mechler, B., Hirsch, H. H., Muller, H., & Wolf, D. H. (1988). Biogenesis of the yeast lysosome (vacuole): biosynthesis and maturation of proteinase yscB. EMBO J. 7, 1705-1710.

Mechler, B., Muller, M., Muller, H., Meussdoerffer, F., & Wolf, D. H. (1982). In vivo biosynthesis of the vacuolar proteinases A and B in the yeast *Saccharomyces cerevisiae*. J. Biol. Chem. 257, 11203-11206.

Mechler, B., & Wolf, D. H. (1981). Analysis of proteinase A function in yeast. Eur. J. Biochem. 121, 47-52.

Mendenhall, M. D. (1993). An inhibitor of p34CDC28 protein kinase activity from Saccharomyces cerevisiae. Science 259, 216-219.

Moehle, C. M., Dixon, C. K., & Jones, E. W. (1989). Processing pathway for protease B of *Saccharomyces cerevisiae*. J. Cell Biol. 108, 309-325.

Moehle, C. M., Tizard, R., Lemmon, S. K., Smart, J., & Jones, E. W. (1987). Protease B of the lysosomelike vacuole of the yeast Saccharomyces cerevisiae is homologous to the subtilisin family of serine proteases. Mol. Cell Biol. 7, 4390-4399.

Nebes, V. L., & Jones, E. W. (1991). Activation of the proteinase B precursor of the yeast *Saccharomyces cerevisiae* by autocatalysis and by an internal sequence. J. Biol. Chem. 266, 22851-22857.

Nothwehr, S. F., & Stevens, T. H. (1994). Sorting of membrane proteins in the yeast secretory pathway. J. Biol. Chem. 269, 10185-10188.

Nugroho, T. T., & Mendenhall, M. D. (1994). An inhibitor of yeast cyclin-dependent protein kinase plays an important role in ensuring the genomic integrity of daughter cells. Mol. Cell Biol. 14, 3320-3328.

Nunnari, J., Fox T. D., & Walter, P. (1993). A mitochondrial protease with two catalytic subunits of nonoverlapping specificities. Science 262, 1997-2004.

Papa, F. R., & Hochstrasser, M. (1993). The yeast DOA4 gene encodes a deubiquitinating enzyme related to a product of the human tre-2 oncogene. Nature 366, 313-319.

Peters, J. M. (1994). Proteasomes: protein degradation machines of the cell. Trends Biochem. Sci. 19, 377-382.

Pilon, M., Schekman, R., & Romisch, K. (1997). Sec61p mediates export of a misfolded secretory protein from the endoplasmic reticulum to the cytosol for degradation. Emb. J. 16, 4540-4548.

Plemper, R. K., Bohmler, S., Bordallo, J., Sommer, T., & Wolf, D. H. (1997). Mutant analysis links the translocon and BiP to retrograde protein transport for ER degradation. Nature 388, 891-895.

Pryer, N. K., Wuestehube, L. J., & Schekman, R. (1992). Vesicle-mediated protein sorting. Ann. Rev. Biochem. 61, 471-516.

Puhler, G., Weinkauf, S., Bachmann, L., Muller, S., Engel, A., Hegerl, R., & Baumeister, W. (1992). Subunit stoichiometry and three-dimensional arrangement in proteasomes from *Thermoplasma acidophilum*. EMBO J. 11, 1607-1616.

Rendueles, P. S., & Wolf, D. H. (1988). Proteinase function in yeast: biochemical and genetic approaches to a central mechanism of post-translational control in the eukaryote cell. Fems Microbiol. Rev. 4, 17-45.

Riballo, E., Herweijer, M., Wolf, D. H., & Lagunas, R. (1995). Catabolite inactivation of the yeast maltose transporter occurs in the vacuole after internalization by endocytosis. J. Bacteriol. 177, 5622-5627.

Richter-Ruoff, B., Heinemeyer, W., & Wolf, D. H. (1992). The proteasome/multicatalytic-multifunctional proteinase. In vivo function in the ubiquitin-dependent N-end rule pathway of protein degradation in eukaryotes. FEBS Lett. 302, 192-196.

Richter-Ruoff, B., & Wolf, D. H. (1993). Proteasome and cell cycle. Evidence for a regulatory role of the protease on mitotic cyclins in yeast. FEBS Lett. 336, 34-36.

Richter-Ruoff, B., Wolf, D. H., & Hochstrasser, M. (1994). Degradation of the yeast MAT alpha 2 transcriptional regulator is mediated by the proteasome [published erratum appears in FEBS Lett 1995 Jan 16; 358(1):104]. FEBS Lett. 354, 50-52.

Rivett, A. J. (1990). Eukaryotic protein degradation. Curr. Opin. Cell Biol. 2, 1143-1149.

Rivett, A. J. (1993). Proteasomes: multicatalytic proteinase complexes. Biochem. J. 291, 1-10.

Rivett, A. J., &Knecht, E. (1993). Proteasome location. Curr. Biol. 3, 127-129.

Roberts, C. J., Nothwehr, S. F., & Stevėns, T. H. (1992). Membrane protein sorting in the yeast secretory pathway: evidence that the vacuole may be the default compartment. J. Cell Biol. 119, 69-83

Roberts, C. J., Pohlig, G., Rothman, J. H., & Stevens, T. H. (1989). Structure, biosynthesis, and localization of dipeptidylaminopeptidase B, an integral membrane glycoprotein of the yeast vacuole. J. Cell Biol. 108, 1363-1373.

Rothman, J. E. (1989). Polypeptide chain binding proteins: catalysts of protein folding and related processes in cells. Cell 59, 591-601.

Rubin, D. M., Coux, O., Wefes, I., Hengartner, C., Young, R. A., Goldberg, A. L., & Finley, D. (1996). Identification of the gal4 suppressor Sug1 as a subunit of the yeast 26S proteasome. Nature 379, 655-657.

Rubin, D. M., & Finley, D. (1995). Proteolysis. The proteasome: a protein-degrading organelle? Curr. Biol. 5, 854-858.

Rupp, S., Hirsch, H. H., & Wolf, D. H. (1991). Biogenesis of the yeast vacuole (lysosome). Active site mutation in the vacuolar aspartate proteinase yscA blocks maturation of vacuolar proteinases. FEBS Lett. 293, 62-66.

Rupp, S., & Wolf, D. H. (1995). Biogenesis of the yeast vacuole (lysosome). The use of active-site mutants of proteinase yscA to determine the necessity of the enzyme for vacuolar proteinase maturation and proteinase yscB stability. Eur. J. Biochem. 231, 115-125.

Russell, S. J., Sathyanarayana, U. G., & Johnston, S. A. (1996). Isolation and characterization of SUG2. A novel ATPase family component of the yeast 26 S proteasome. J. Biol. Chem. 271, 32810-32817.

Saheki, T., Matsuda, Y., & Holzer, H. (1974). Purification and characterization of macromolecular inhibitors of proteinase A from yeast. Eur. J. Biochem. 47, 325-332.

Schlumpberger, M., Schaeffeler, E., Straub, M., Bredschneider, M., Wolf, D. H., & Thumm, M. (1997). *AUT1*, a gene essential for autophagocytosis in the yeast *Saccharomyces cerevisiae*. J. Bacteriol. 179, 1068-1076.

Schnall, R., Mannhaupt, G., Stucka, R., Tauer, R., Ehnle, S., Schwarzlose, C., Vetter, I., & Feldmann, H. (1994). Identification of a set of yeast genes coding for a novel family of putative ATPases with high similarity to constituents of the 26S protease complex. Yeast 10, 1141-1155.

Schork, S., Bee, G., Thumm, M., & Wolf, D. H. (1994). Catabolite inactivation of fructose-1,6-bisphosphatase in yeast is mediated by the proteasome. FEBS Lett. 349, 270-274.

Schork, S., Bee, G., Thumm, M., & Wolf, D. H. (1994). Site of catabolite inactivation. Nature 369, 283-284.

Schork, S. M., Thumm, M., & Wolf, D. H. (1995). Catabolite inactivation of fructose-1,6-bisphosphatase of *Saccharomyces cerevisiae*. Degradation occurs via the ubiquitin pathway. J. Biol. Chem. 270, 26446-26450.

Schu, P., & Wolf, D. H. (1991). The proteinase yscA-inhibitor, IA3, gene. Studies of cytoplasmic proteinase inhibitor deficiency on yeast physiology. FEBS Lett. 283, 78-84.

Schu, P., Suarez Rendueles, P., & Wolf, D. H. (1991). The proteinase yscB inhibitor (PB12) gene of yeast and studies on the function of its protein product. Eur. J. Biochem. 197, 1-7.

Scott, S. V., Hefner-Gravink, A., Morano, K. A., Noda, T., Ohsumi, Y., & Klionsky, D. J. (1996). Cytoplasm-to-vacuole targeting and autophagy employ the same machinery to deliver proteins to the yeast vacuole. Proc. Natl. Acad. Sci. USA 93, 12304-12308.

Seemuller, E., Lupas, A., Stock, D., Lowe, J., Huber, R., & Baumeister, W. (1995). Proteasome from Thermoplasma acidophilum: a threonine protease. Science 268, 579-582.

Seglen, P. O., & Bohley, P. (1992). Autophagy and other vacuolar protein degradation mechanisms. Experientia 48, 158-172.

Segui, R. B., Martinez, M., & Sandoval, I. V. (1995). Yeast aminopeptidase I is post-translationally sorted from the cytosol to the vacuole by a mechanism mediated by its bipartite N-terminal extension. EMBO J. 14, 5476-5484.

Seidah, N. G., & Chretien, M. (1994). Pro-protein convertases of subtilisin/kexin family. Methods Enzymol. 244, 175-188.

Seufert, W., Futcher, B., & Jentsch, S. (1995). Role of a ubiquitin-conjugating enzyme in degradation of S- and M-phase cyclins. Nature 373, 78-81.

Seufert, W., & Jentsch, S. (1992). *In vivo* function of the proteasome in the ubiquitin pathway. EMBO J. 11, 3077-3080.

Sorensen, S. O., van den Hazel, H. B., Kielland-Brandt, M. C., & Winther, J. R. (1994). pH-dependent processing of yeast procarboxypeptidase Y by proteinase A in vivo and in vitro. Eur. J. Biochem. 220, 19-27.

Spormann, D. O., Heim, J., & Wolf, D. H. (1991). Carboxypeptidase yscS: gene structure and function of the vacuolar enzyme. Eur. J. Biochem. 197, 399-405.

Spormann, D. O., Heim, J., & Wolf, D. H. (1992). Biogenesis of the yeast vacuole (lysosome). The precursor forms of the soluble hydrolase carboxypeptidase yscS are associated with the vacuolar membrane. J. Biol. Chem 267, 8021-8029.

Stack, J. H., & Emr, S. D. (1993). Genetic and biochemical studies of protein sorting to the yeast vacuole. Curr. Opin. Cell Biol. 5, 641-646.

Stevens, T., Esmon, B., & Schekman, R. (1982). Early stages in the yeast secretory pathway are required for transport of carboxypeptidase Y to the vacuole. Cell 30, 439-448.

Stevens, T. H., Rothman, J. H., Payne, G. S., & Schekman, R. (1986). Gene dosage-dependent secretion of yeast vacuolar carboxypeptidase Y. J. Cell. Biol. 102, 1551-1557.

Straub, M., Bredschneider, M., & Thumm, M. (1997). *AUT3*, a serine/threonine kinase gene, is essential for autophagocytosis in *Saccharomyces cerevisiae*. J. Bacteriol. 179, 3875-3883.

Suarez-Rendueles, P., & Wolf, D. (1987). Identification of the structural gene for dipeptidylaminopeptidase yscV (DAP2) of *Saccharomyces cerevisiae*. J. Bacteriol. 169, 4041-4048.

Suzuki, C. K., Suda, K., Wang, N., & Schatz, G. (1994). Requirement for the yeast gene LON in intramitochondrial proteolysis and maintenance of respiration [published erratum appears in Science 1994 May 13; 264(5161):891]. Science 264, 273-276.

Swaffield, J. C., Bromberg, J. F., & Johnston, S. A. (1992). Alterations in a yeast protein resembling HIV Tat-binding protein relieve requirement for an acidic activation domain in GAL4 [published erratum appears in Nature 1992 Dec 24-31; 360(6406):768]. Nature 357, 698-700.

Takeshige, K., Baba, M., Tsuboi, S., Noda, T., & Ohsumi, Y. (1992). Autophagy in yeast demonstrated with proteinase-deficient mutants and conditions for its induction. J. Cell Biol. 119, 301-311.

Tanaka, K. (1995). Molecular biology of proteasomes. Mol. Biol. Rep. 21, 21-26.

Teichert, U., Mechler, B., Muller, H., & Wolf, D. H. (1989). Lysosomal (vacuolar) proteinases of yeast are essential catalysts for protein degradation, differentiation, and cell survival. J. Biol. Chem. 264, 16037-16045.

Teichmann, U., van Dyck, L., Guiard, B., Fischer, H., Glockshuber, R., Neupert, W., & Langer, T. (1996). Substitution of PIM1 protease in mitochondria by *Escherichia coli* Lon protease. J. Biol. Chem. 271, 10137-10142.

Thorsness, P. E., White, K. H., & Fox, T. D. (1993). Inactivation of YME1, a member of the ftsH-SEC18-PAS1-CDC48 family of putative ATPase-encoding genes, causes increased escape of DNA from mitochondria in *Saccharomyces cerevisiae*. Mol. Cell Biol. 13, 5418-5426.

Thumm, M. (1993). Biological control through proteolysis. Yeast as a model of the eukaryote cell. In: Innovations on Proteases and their Inhibitors (Aviles, F.X., ed.), pp. 63-80. Walter de Gruyter, Berlin.

Thumm, M., Egner, R., Koch, B., Schlumpberger, M., Straub, M., Veenhuis, M., & Wolf, D. H. (1994). Isolation of autophagocytosis mutants of *Saccharomyces cerevisiae*. FEBS Lett. 349, 275-280.

Tooze, J., Hollinshead, M., Ludwig, T., Howell, K., Hoflack, B., & Kern, H. (1990). In exocrine pancreas, the basolateral endocytic pathway converges with the autophagic pathway immediately after the early endosome. J. Cell Biol. 111, 329-345.

Tsukada, M., & Ohsumi, Y. (1993). Isolation and characterization of autophagy-defective mutants of *Saccharomyces cerevisiae*. FEBS Lett. 333, 169-174.

Tuttle, D. L., & Dunn, W. J. (1995). Divergent modes of autophagy in the methylotrophic yeast Pichia pastoris. J. Cell Sci. 108, 25-35.

Ueno, T., Muno, D., & Kominami, E. (1991). Membrane markers of endoplasmic reticulum preserved in autophagic vacuolar membranes isolated from leupeptin-administered rat liver. J. Biol. Chem. 266, 18995-18999.

Valls, L. A.,Winther, J. R., & Stevens, T. H. (1990). Yeast carboxypeptidase Y vacuolar targeting signal is defined by four propeptide amino acids. J. Cell Biol. 111, 361-368.

van den Hazel, H. B., Kielland, B. M., & Winther, J. R. (1992). Autoactivation of proteinase A initiates activation of yeast vacuolar zymogens. Eur. J. Biochem. 207, 277-283.

van den Hazel, H. B., Kielland-Brandt, M. C., & Winther, J. R. (1996). Review: Biosynthesis and function of yeast vacuolar proteases. Yeast 12, 1-16.

Van Dyck, L., Pearce, D. A., & Sherman, F. (1994). PIM1 encodes a mitochondrial ATP-dependent protease that is required for mitochondrial function in the yeast *Saccharomyces cerevisiae*. J. Biol. Chem. 269, 238-242.

van Nocker, S., Sadis, S., Rubin, D. M., Glickman, M., Fu, H., Coux, O., Wefes, I., Finley, D., & Vierstra, R. D. (1996). The multiubiquitin-chain-binding protein Mcb1 is a component of the 26S proteasome in *Saccharomyces cerevisiae* and plays a nonessential, substrate-specific role in protein turnover. Mol. Cell Biol. 16, 6020-6028.

Volland, C., Urban, G. D., Geraud, G., & Haguenauer, T. R. (1994). Endocytosis and degradation of the yeast uracil permease under adverse conditions. J. Biol. Chem. 269, 9833-9841.

Wagner, I., Arlt, H., van Dyck, L., Langer, T., & Neupert, W. (1994). Molecular chaperones cooperate with PIM1 protease in the degradation of misfolded proteins in mitochondria. EMBO J. 13, 5135-5145.

Werner, E. D., Brodsky, J. L., & McCracken, A. A. (1996). Proteasome-dependent endoplasmic reticulum-associated protein degradation: an unconventional route to a familiar fate. Proc. Natl. Acad. Sci. USA 93, 13797-13801.

Willems, A. R., Lanker, S., Patton, E. E., Craig, K. L., Nason, T. F., Mathias, N. R., Kobayashi, Wittenberg, C., & Tyers, M. (1996). Cdc53 targets phosphorylated G1 cyclins for degradation by the ubiquitin proteolytic pathway. Cell 86, 453-463.

Wolf, D. H. (1980). Control of metabolism in yeast and other lower eukaryotes through action of proteinases. In: Advances in Microbial Physiology (Rose, A.H.& Gareth Morris, J. eds.), Vol. 21, pp. 267-338.Academic Press, London, New York.

Wolf, D. H. (1982). Proteinase action *in vitro* versus proteinase function *in vivo*: mutants shed light on intracellular proteolysis in yeast. Trends Biochem. Sci. 7, 35-37.

Wolf, D. H. (1986). Cellular control in the eukaryotic cell through action of proteinases: the yeast *Saccharomyces cerevisiae* as a model organism. Microbiol. Sci. 3, 107-111.

Wolf, D. H., & Ehmann, C. (1979). Studies on a proteinase B mutant of yeast. Eur. J. Biochem. 98, 375-384.

Wolf, D. H., & Fink, G. R. (1975). Proteinase C (carboxypeptidase Y) mutant of yeast. J. Bacteriol. 123, 1150-1156.

Wolf, D. H., & Weiser, U. (1977). Studies on a carboxypeptidase Y mutant of yeast and evidence for a second carboxypeptidase Activity. Eur. J. Biochem. 73, 553-6.

Woolford, C. A., Daniels, L. B., Park, F. J., Jones, E. W., Van, A. J., & Innis, M. A. (1986). The PEP4 gene encodes an aspartyl protease implicated in the posttranslational regulation of Saccharomyces cerevisiae vacuolar hydrolases. Mol. Cell Biol. 6, 2500-2510.

Woolford, C. A., Noble, J. A., Garman, J. D., Tam, M. F., Innis, M. A., & Jones, E. W. (1993). Phenotypic analysis of proteinase A mutants. Implications for autoactivation and the maturation pathway of the vacuolar hydrolases of Saccharomyces cerevisiae. J. Biol. Chem. 268, 8990-8998.

Yadeau, J. T., Klein, C., & Blobel, G. (1991). Yeast signal peptidase contains a glycoprotein and the Sec11 gene product. Proc. Natl. Acad. Sci. USA 88, 517-521.

Yaglom, J., Linskens, M. H., Sadis, S., Rubin, D. M., Futcher, B., & Finley, D. (1995). p34Cdc28-mediated control of Cln3 cyclin degradation. Mol. Cell Biol. 15, 731-741.

Yamamoto, A., Guacci, V., & Koshland, D. (1996). Pds1p, an inhibitor of anaphase in budding yeast, plays a critical role in the APC and checkpoint pathway(s). J. Cell Biol. 133, 99-110.

Zubenko, G. S., Mitchell, A. P., & Jones, E. W. (1979). Septum formation, cell division, and sporulation in mutants of yeast deficient in proteinase B. Proc. Natl. Acad. Sci. USA 76, 2395-2399.

Zwickl, P., Grziwa, A., Puhler, G., Dahlmann, B., Lottspeich, F., & Baumeister, W. (1992). Primary structure of the Thermoplasma proteasome and its implications for the structure, function, and evolution of the multicatalytic proteinase. Biochemistry 31, 964-72.

CELLULAR REGULATION BY UBIQUITIN-DEPENDENT PROCESSES

Keith D. Wilkinson

Advances in Molecular and Cell Biology
Volume 27, pages 71-104.

ISBN: 0-7623-0387-5

I. INTRODUCTION

This contribution will update the reader on the plethora of processes regulated by the covalent attachment of ubiquitin to other proteins. In general, these can be attributed to targeting; i.e., attachment of ubiquitin to a protein alters it's structure, localization, activity, or metabolism. Many excellent reviews have detailed the enzymology of ubiquitination and deubiquitination (Hershko and Ciechanover, 1992; Rivett, 1993; Attaix et al., 1994; Ciechanover and Schwartz, 1994; Driscoll, 1994; Hochstrasser, 1995; Rubin and Finley, 1995; Wilkinson, 1995) and so these aspects will be only briefly discussed here. I will concentrate on recent findings, especially as they relate to regulatory proteins that are substrates for the system, and the consequences of these ubiquitination events on biological function.

II. UBIQUITIN REGULATES MANY CELLULAR PROCESSES

Many of the biological effects of ubiquitination can be attributed to the targeting of ubiquitinated proteins to proteolytic processes. It has long been known that eukaryotic cells possess both lysosomal and ATP-dependent "soluble" proteolytic systems (Goldberg and Dice, 1974). Among the roles of these systems is degradation of abnormal or damaged proteins. Considerable effort has been expended to identify and characterize the systems responsible for this type of stress response (see reviews above). It was not until the late 1970s however, that these questions became approachable. Surprisingly, the same systems that participate in the stress response and degradation of abnormal proteins are also important in the degradation of short-lived regulatory proteins (Ciechanover et al., 1984). In the past 20 years, it has become apparent that this regulated proteolysis of proteins serves many functions in addition to removing damaged or denatured proteins.

As we have learned more about the molecular details of the cellular degradative machinery, the question of how the cell achieves specificity has remained in the forefront. The short answer appears to be that specificity is attained by sequestering the proteolytic machinery and controlling access to potential substrates. Proteases can be enclosed in an organelle such as the lysosome, encased in a multienzyme complex like the proteasome, or like calpain and caspase proteases, kept in an inactive state until needed. The cell then utilizes the energy of ATP to activate the enzyme or bring substrate and enzyme into proximity. The calpains (Suzuki et al.,

1992; Sorimachi et al., 1994) are activated by release of calcium and/or translocation to the membrane; the proteases of apoptosis, digestion, and clotting are themselves proteolytically processed to become activated; the lysosome can obtain its substrates by microautophagy, by endocytosis, or by the action of a specific transporter hsc73 (Dice, 1992) (see Knecht et al., chapter 7); while the muticatalytic protease of eukaryotic cells, called the proteasome (Tanaka et al., 1992; Rivett, 1993; Driscoll, 1994), binds and degrades proteins that have been covalently marked by the addition of a polyubiquitin chain (see Dubiel and Rechisteiner, chapter 5).

In addition to targeting proteins for proteolysis, the ubiquitin domain may play roles in regulating protein sorting and function. Polyubiquitin chains are the signal for many of these events. These are formed by first attaching the carboxyl-terminus of ubiquitin to a lysine on the target protein, and subsequently attaching additional ubiquitins by isopeptide bonds between a lysine on the proximal ubiquitin and the C-terminus of the next (Ciechanover and Schwartz, 1994; Wilkinson, 1995). Several different types of polyubiquitin chains can be formed by utilizing different ubiquitin lysines, and these variants may participate in targeting proteins to alternate fates (Arnason and Ellison, 1994; Spence et al., 1995; Baboshina and Haas, 1996). In addition, it appears that there are other ubiquitin-like proteins which can be conjugated to cellular proteins (Lowe et al., 1995; Haas et al., 1996; Matunis et al., 1996; Narasimhan et al., 1996) expanding the possible roles of the ubiquitin domain.

Even more intriguing than the mechanistic details, are the cellular consequences of such posttranslational modifications. A host of cellular responses is intimately regulated by ubiquitin-dependent processes (see Figure 1). It is the aim of this chapter to update the reader on some of these phenomena, and to point out how the covalent attachment of ubiquitin (or ubiquitin-like proteins) to proteins is involved.

A. Chromosome Structure and Genomic Stability

Chromatin Structure

The first identified role for ubiquitin involved the covalent attachment of ubiquitin to histone H2A. The two proteins are linked by an isopeptide bond between the C-terminus of ubiquitin and the ε-amino group of K115 of histone H2A (Busch, 1984). Approximately 10% of the core octomers contain this modified histone (referred to as uH2A) and its abundance varies in different locations on the chromosome and with the phase of the cell cycle (Bradbury, 1992). Levels of uH2A are highly upregulated in SV40 transformed cells (Vassilev et al., 1995). In contrast to many other ubiquitinated proteins, uH2A is stable and the ubiquitin is removed by the action of a protease which hydrolyzes the isopeptide bond (Andersen et al., 1981; Matsui et al., 1982). Current models suggest that this modification is involved in chromatin condensation or in DNA synthesis (Bradbury, 1992).

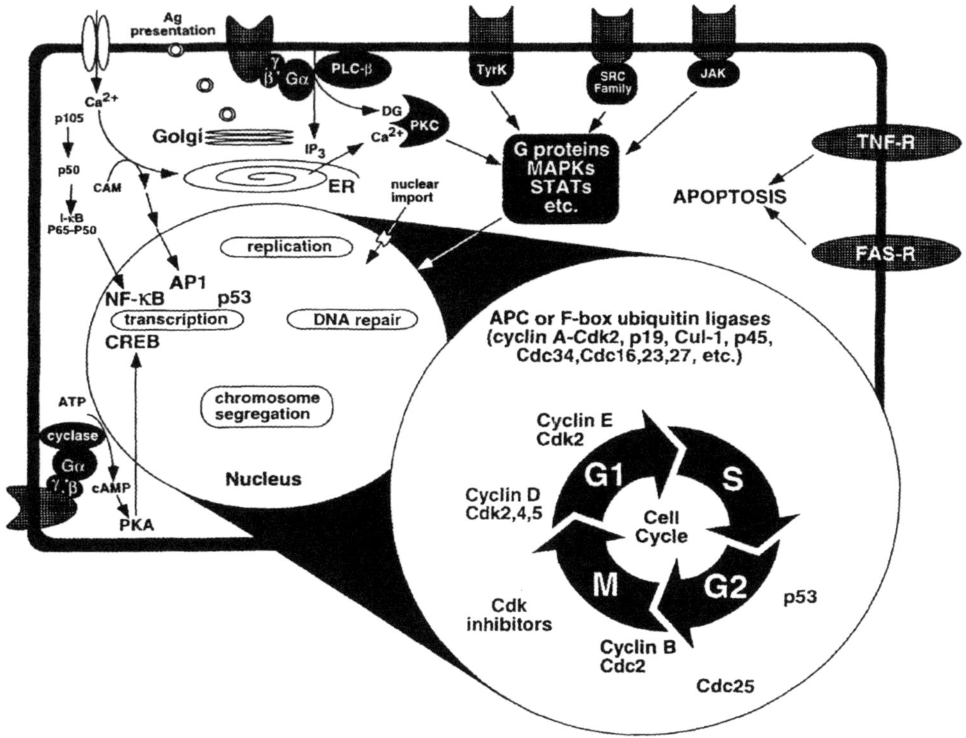

Figure 1. Role of ubiquitin in signal transduction and cellular growth. Ubiquitin-dependent processes are involved in virtually all these pathways, particularly the highlighted ones. Details are given in the text. This figure is adapted from a poster distributed by MBL International Corporation, Watertown, MA.

DNA Repair

The ubiquitin system is also involved in all three classical types of DNA repair; postreplicative repair, nucleotide excision repair, and recombination repair. The methylation repair enzyme, O6-methylguanine-DNA methyltransferase is an enzyme which protects from the mutagenic effects of nitroso compounds by transferring the methyl group of O6-methylguanine to a cysteine of the protein. Once alkylated, this protein is inactive and is degraded by a ubiquitin-dependent system (Srivenugopal et al., 1996). RAD6 is a DNA repair mutant in *Saccharomyces cerevisiae* which is involved in post-replicative DNA repair. The RAD6 mutation is a mutation in the structural gene for one of the ubiquitin conjugating enzymes (Jentsch et al., 1987) (see below). This highly-conserved protein also plays a role in damage-induced mutagenesis, sporulation, and protein degradation (Watkins et al., 1993b). RAD6 has been noted to associate with several other proteins, most notably the UBR1 ubiquitin protein ligase and RAD18, a damage-specific DNA binding protein (Bailly et al., 1994).

Cockayne's Syndrome is caused by a defect in transcription-coupled nucleotide excision repair, i.e., repair of transcriptionally active genes. A recent report (Bregman et al., 1996) suggests that ultraviolet exposure induces the ubiquitination of RNA polymerase II, and that this induction is not observed in mutants from either complementation group of the syndrome. Expression of the wild-type gene in these fibroblasts restores the ubiquitination, implying a link between effective transcription-coupled repair and ubiquitination of PolII. Interestingly, a gene immediately adjacent to RAD26, the *S. cerevisiae* homologue of human ERCC6 (a helicase defective in complimentation group B of Cockayne's), codes for an E3 ubiquitin protein ligase (Huang et al., 1995). If this juxtaposition is also present in humans, it is possible that part of the ubiquitination defect could be due to altered structure or expression of the adjacent E3 ubiquitin protein ligase.

The above examples (and those discussed below in the section on ubiquitin-like proteins) demonstrate that enzymes of the ubiquitin system and ubiquitin-like proteins can associate with specific DNA binding proteins. This may in part explain how the ubiquitin system can target proteins which must be degraded in order to accomplish DNA repair.

Chromosome Segregation

Possibly related to these structural functions, the transition from metaphase to anaphase requires the separation of sister chromatids. This separation requires ubiquitin-dependent processes, *in vitro* (Holloway, 1995) and *in vivo* (Glotzer, 1995). This requirement probably reflects the need to degrade one or more proteins at the time of separation. Ubiquitin-like proteins have also been implicated in spindle pole body duplication (see below).

B. Cell Cycle Progression

The above examples demonstrate that the ubiquitin domain has been used in a number of ways in modulating the synthesis, structure, and segregation of genetic material in higher eukaryotes. It has been argued that most mutations affecting chromosome structure also affect the cell cycle (Bradbury, 1992). However, there are additional points at which ubiquitin impacts the cell cycle, and thus the normal progression of growth and the regulation of differentiation.

Cyclins

One of the most common phenotypes of mutations in the ubiquitin pathway is cell cycle defects. In addition to its role in maintaining chromosome structure, ubiquitin also participates in the degradation of both mitotic (Glotzer et al., 1991; Murray, 1995) and G1 cyclins (Deshaies et al., 1995), the p27 cyclin-dependent kinase inhibitor (Pagano et al., 1995), and *Schizosaccharomyces pombe* cdc25 which is responsible for dephosphorylating and activating the cyclin-dependent kinase (Nefsky and Beach, 1996). The ubiquitination and degradation of cyclin B and cdc25 phosphatase is necessary for exit from mitosis. Efficient targeting of cyclin A and B requires a specific amino acid sequence, the "destruction box" (Klotzbucher et al., 1996), which also directs ubiquitination of uracil permease (Galan et al., 1996). An anaphase-promoting complex (APC) which assists in the ubiquitination of cyclin B has been characterized and shown to consist of homologues of *S. cerevisiae* CDC27p and CDC16p (King et al., 1995; Sudakin et al., 1995). This complex has the properties expected of a cyclin-specific E3. In addition, the G1 cylins CLN1 and CLN2 are phosphorylated and rapidly degraded at the "start," apparently in a ubiquitin-dependent pathway (Barral et al., 1995; Deshaies et al., 1995; Yaglom et al., 1995) utilizing CDC34p, a *S. cerevisiae* ubiquitin conjugating enzyme. Interestingly, mutation of GRR1, a gene involved in gene regulation during nutrient deprivation, also stabilizes the G1 cyclins (Barral et al., 1995). Finally, the degradation of p27, a cdk inhibitor involved in regulating the activity of cdks is also degraded by a ubiquitin-dependent pathway (Pagano et al., 1995).

p53

The tumor suppressor protein p53 mediates a G1 arrest that may be an important regulator of apoptosis and transformation. For example, cells exposed to ionizing radiation respond with an increased level of p53 which binds to the promotor for the cyclin-dependent kinase inhibitor p21 and stimulates transcription from that locus (Labrecque and Matlashewski, 1995). The resultant inhibition of cyclin-dependent kinase prevents progression through the G1 arrest. Mutation or disruption of p53 is one of the most common mutations in a variety of cellular carcinomas. It has been suggested, but not universally accepted (Murnane, 1995), that the G1 block allows damaged DNA to be repaired and that a failure in this block allows the fixation of mutations in subsequent generations. In addition to acting as a sequence specific transcriptional activator, p53 is thought to be a repressor of the transcription of other genes. As such, a variety of oncogenic virus mechanisms target this important protein. In particular, the human papillomavirus (HPV) encodes a viral protein (E6) which binds to and enhances the degradation of p53 by a ubiquitin-dependent mechanism (Scheffner et al., 1990). The HPV E6 protein and a cellular protein called E6AP form a complex which acts as an E3 (Scheffner et al., 1993) and uses the UbcH5 ubiquitin conjugating enzyme to ubiquitinate p53 (Scheffner et al., 1994). Ubiquitinated p53 is then degraded by the proteasome.

C. Signal Transduction

Receptor Tyrosine Kinases

A large number of receptor tyrosine kinases are ubiquitinated and degraded upon receptor engagement. These include the receptors for epidermal growth factor (EGF) (Galcheva-Gargova et al., 1995), platelet-derived growth factor (PDGF) (Yarden et al., 1986), mast cell growth factor (Miyazawa et al., 1994), fibroblast growth factor (FGF), and colony-stimulating factor (CSF-1) (Mori et al., 1995). Herbimycin A, an antibiotic which has been used as a general inhibitor of tyrosine kinases and the associated signal transduction pathways acts, at least in part, by inducing ubiquitin-dependent and proteasome-dependent degradation of this class of receptors (Sepp-Lorenzino et al., 1995).

In some cases, there also appears to be a downstream effect of ubiquitin on these pathways. The c-cbl proto-oncoprotein is normally found in the cytoplasm bound to growth factor receptor–bound protein 2 (GRB2), an adapter protein involved in signal transduction. Upon stimulation with EGF, CSF-1, erythropoietin (EPO) or cell-surface antigens, several cell types phosphorylate and ubiquitinate c-cbl (Wang et al., 1996). The modified c-cbl next appears in a complex with the receptor (via GRB2 recruitment (Meisner and Czech, 1995)), tyrosine-phosphorylated SHC (another adapter) and the 85 kDa inhibitory subunit of phosphoinositide 3 kinase (Panchamoorthy et al., 1996; Soltoff and Cantley, 1996). This complex is rapidly assembled at the membrane (Wang et al., 1996). It appears that c-cbl is a major player in coupling tyrosine kinase signaling to the phosphoinositide pathway. While the role of ubiquitination is unknown, it seems likely to be involved with downregulating this pathway by dissociating and degrading the components.

G-Protein Coupled Receptors

Upon binding the *S. cerevisiae* α-factor pheromone, the STE2p receptor triggers a MAP kinase cascade which results in transcriptional activation of the genes required for mating. This also results in the phosphorylation and ubiquitination of the C-terminus of the receptor. The modified receptor is internalized by endocytosis and degraded in the *S. cerevisiae* vacuole (Hicke and Riezman, 1996). Also, GPA1p, the α subunit of a heterotrimeric G protein involved in the mating pheromone response is degraded by the ubiquitin system (Madura and Varshavsky, 1994).

ABC Transporters

A similar degradative pathway is followed for STE6p, a member of the ABC family of transporters which is responsible for export of the a-factor pheromone. Normally this transporter is concentrated at the *trans*-Golgi, presumably waiting to export the pheromone. Endocytosis mutants accumulate the receptor at the plasma membrane and mutations in the ubiquitin system and the vacuolar proteases both stabilize the transporter (Kolling and Hollenberg, 1994). The Pdr5 multidrug resistance transporter undergoes a similar ubiquitin-dependent endocytosis step, followed by degradation in the vacuole (Egner and Kuchler, 1996).

Mutations in the cystic fibrosis transmembrane conductance regulator (CFTR) cause the most common autosomal recessive defect in Caucasians. This ABC transporter is responsible for ion transport, and is particularly important in the lung. Much of the degradation of wild type protein, and all the degradation of mutant protein occurs in the endoplasmic reticulum (ER), even before its appearance on the cell surface. A significant portion of its degradation requires ubiquitination and degradation by the proteasome (Jensen et al., 1995; Ward et al., 1995; see McCracken et al., chapter 6).

The degradation of *S. cerevisiae* Gap1p (the general amino acid transporter) and Fur4p (uracil permease) both require a specific E3 ligase encoded by the Npi1/Rsp5 locus. This suggests that these membrane transporters are also modified by the ubiquitin system, while the exact role is not certain (Hein et al., 1995)

Cytokine Receptors

The growth hormone receptor is monomeric and dimerizes upon binding its ligand. This receptor (Spencer et al., 1988) and at least one tumor necrosis factor receptor (Loetscher et al., 1990) contain ubiquitin as determined by sequencing of the purified receptors.

Protein Kinases

Protein kinases are important downstream mediators of signal transduction. Control of amount, localization and activity are important regulatory processes. The activity of protein kinase C (PKC), protein kinase A (PKA), and MAP kinases (indirectly) are all affected by ubiquitin metabolism.

PKC has been implicated in the signaling pathway for numerous responses. It is normally soluble and inactive until bound by diacylglycerol and translocated to the plasma membrane. Activation requires a short-lived phosphorylation of the protein. Chronic activation of PKC by bryostatin 1 or phorbol myristate results in the subsequent proteolytic degradation of PKC by the ubiquitin- and proteasome-dependent pathways (Lee et al., 1996; Smith et al., 1996). Interestingly, it is the dephospho-PKC which is ubiquitinated and degraded. Clearly, the desensitization event must involve reduced phosphorylation, but it is unclear if degradation occurs at the membrane or in the cytoplasm.

Under some circumstances, cAMP dependent protein kinase is also subject to regulation by ubiquitination. The regulatory subunit of PKA is selectively degraded during long-term potentiation in Aplysia (Hegde et al., 1993). It is thought that dissociation of the C and R subunits, triggered by cAMP binding, is required for ubiquitination and subsequent proteolysis by the proteasome

Cell Surface Molecules

Perhaps the most widely known role of ubiquitin involving cell surface molecules is the Class I major histocompatability complex (MHC) response, or antigen presentation. The Class 1 system responds to the presence of foreign, viral, or damaged proteins by degrading the protein to peptides which are transported into the ER and assembled in the peptide binding groove on Class I MHC molecules. Appearance of these loaded Class I molecules at the cell surface recruits specific cytotoxic T lymphocytes which recognize the subtype of MHC and the nature of the bound peptide, mounting a cellular response which results in the death of the presenting cell. Much evidence now implicates the ubiquitin- and proteasome-dependent system as the one responsible for producing the peptides for transport and presentation (Michalek et al., 1993; Driscoll, 1994; Tanaka, 1994). Interferons, cytokines produced in response to viral infections, induce both ubiquitin-like proteins and specific subunits of the proteasome (see Tanaka and Tamahaslin, Chapter 4). Thus, the ubiquitin system degrades many cellular proteins, and releases peptides which appear on the surface of the cell as a sampling of the internal protein metabolism. The immune system can then respond to significant changes in the protein metabolism of individual cells.

Additionally, a variety of cell surface molecules involved in adhesion or cell-cell interactions are ubiquitinated (Ciechanover, 1994), including the lymphocyte homing receptor (Siegelman et al., 1986), the IgE receptor (Paolini and Kinet, 1993), the T cell antigen receptor (Cenciarelli et al., 1992), and the gap junction protein connexin43 (Laing and Beyer, 1995). In addition, a number of cell-surface proteins react with anti-ubiquitin antibodies (Siegelman et al., 1986; Serdaroglu et al., 1992). These may be related to the examples described above, or may be molecules involved in still unidentified phenomena. For instance, a ubiquitin binding site has been reported on a hematopoietic progenitor cell (Parakh and Kannan, 1993), and antibodies to ubiquitin block murine T cell homing (Siegelman et al., 1986), and neurotransmitter uptake (Meyer et al., 1986, 1987). FUBI (also called Fau, see below), a protein with a ubiquitin-like domain fused to the S30 ribosomal protein is secreted by a murine T cell hybridoma and can act as a nonspecific suppressor of B cell immune responses (Nakamura et al., 1996). The ubiquitin-like domain is sufficient, and this effect can be antagonized by ubiquitin suggesting a ubiquitin binding site on the surface of B cells. Expression of the antisense strand of the FUBI/Fau gene slightly increases the transformation frequency in cultured cells (Michiels et al., 1993). Fau is the cellular counterpart of the fox sequence (in the reverse orientation) of the Finkel-Biskis-Reilly murine sarcoma virus (Michiels et al., 1993).

D. Regulation of Gene Expression

In addition to the effects on receptor structure and signaling described above, the half-life of many transcriptional regulators is also regulated by ubiquitination.

NFκB/IκBα

NFκB is heterodimeric transcription factor involved in the cellular response to a variety of extracellular signals, including viral proteins, cytokines, and T-cell receptor engagement. One subunit is synthesized as a 105 kDa precursor which is processed to the mature 50 kDa form by proteolysis of the carboxyl-terminal domain in a reaction requiring ubiquitination and degradation by the proteasome (Palombella et al., 1994). The N-terminal portion, p50, associates with p65/RelA to form the active transcription factor, but it is normally sequestered in the cytoplasm bound to the inhibitor IκBα. Upon stimulation by extracellular signals, IκBα is rapidly phosphorylated on serines 32 and 36 (Chen et al., 1995) and ubiquitinated on lysines 21 and/or 22 (Scherer et al., 1995; Baldi et al., 1996). The modified inhibitory subunit is then degraded by the proteasome. NFκB is then imported into the nucleus where it activates transcription of many genes. Unexpectedly, the phosphorylation requires a large molecular weight complex referred to as IκBα kinase and this kinase is activated by ubiquitination (Chen et al., 1996).

Fos/Jun

Fos is a short-lived member of a large family of related transcription factors which function as heterodimers. v-Fos is a truncated version missing the carboxyl-terminal PEST domain, is found in some viruses, is long-lived and is an oncoprotein. Association of c-Fos with c-Jun yields the heterodimer AP1. Phosphorylation of AP1 leads to ubiquitination and degradation of c-Fos by the proteasome (Ciechanover et al., 1991a; Tsurumi et al., 1995). v-Fos is not phosphorylated or ubiquitinated, and is thus stable, leading to transformation. In a recombinant system, both c-Fos and c-Jun homodimers are ubiquitinated, but the heterodimer is a poorer substrate (Hermida-Matsumoto et al., 1996). Another report with a different mixture of partially purified components suggests that heterodimerization induces c-Fos degradation (Stancovski et al., 1995). While the *in vitro* results are not definitive, it appears that phosphorylation may be required to dissociate the heterodimer and allow efficient ubiquitination.

p53

As discussed above, p53 is a transcriptional regulator involved in G1 arrest. It is important in preventing DNA replication in the presence of DNA damage. Wild-type p53 is normally short-lived, however it accumulates transiently in the presence of DNA damage. In turn, the putative DNA binding results in the upregulation of p21 and the G1 arrest. In the absence of functional p53, cells are prone to apoptosis and/or transformation because of the replication of accumulated damage. The human papillomavirus E6 binds p53 and participates in its ubiquitination. The DNA binding and E6 binding are mutually exclusive and the strength of binding by different E6 sequences correlates with the oncogenic potential of various isolates (Lechner and Laimins, 1994; Molinari and Milner, 1995).

GCN4

GCN4 is a transcriptional regulator of the basic leucine zipper family which is involved in the regulation of the biosynthesis of amino acids and purines. The protein is normally short-lived, but is stabilized during starvation for amino acids. This transcription factor is degraded by the ubiquitin proteasome pathway (Kornitzer et al., 1994).

MATα2

The *S. cerevisiae* mating pheromone, MATα2 is a transcriptional repressor involved in mating-type switching. It is rapidly degraded in a ubiquitin-dependent process (Hochstrasser et al., 1991). At least two pathways of ubiquitination involving different complementation groups of E2 conjugating enzymes are involved (Chen et al., 1993), one of which recognizes a specific peptide sequence of MATα2. Degradation also requires a deubiquitinating enzyme, DOA4p, which plays a role in removing the polyubiquitin chain from the residual peptide produced by degradation by the proteasome (Papa and Hochstrasser, 1993).

Estrogen Receptor

The estrogen receptor is a soluble estrogen binding protein that functions by binding its ligand, and subsequently binding to specific DNA sequences as a transcriptional regulator. It is ubiquitinated both *in vivo* and *in vitro* and the level of ubiquitination is directly proportional to the levels of estrogen (Nirmala and Thampan, 1995).

E. Regulation of Protein Localization

Internalization of Membrane Proteins

As noted above, it appears that one mechanism of receptor downregulation involves ligand-induced ubiquitination of the receptor, followed by endocytosis and proteolysis by tne endocytic machinery and/or the proteasome. It should be noted that these may not be mutually exclusive. Ubiquitin system enzymes, ubiquitin immunoreactivity and, as will be detailed elsewhere in this volume (see chapters 4 and 5), proteasome proteins have been noted in various membrane bound compartments (Magnani et al., 1991; Ciechanover et al., 1991b; Rivett et al., 1992; Schwartz et al., 1992; Gray et al., 1995; Low et al., 1995).

Lysosome and Peroxisome Biogenesis

Mutations in various ubiquitin system enzymes have been reported to interfere with organelle biogenesis. Mutations in the ubiquitin activating enzyme appear to prevent the maturation of autophagic vesicles (Schwartz et al., 1992), possibly due to the failure to degrade internalized receptors and cell surface proteins which are normally ubiquitinated and degraded by this system (see above). A mutation in one of the ubiquitin conjugating enzymes (PAS2 in *S. cerevisiae*, PAS4 in *Pichia pastoris*) blocks peroxisome assembly (Wiebel and Kunau, 1992; Crane et al., 1994). This conjugating enzyme is localized on the cytoplasmic face of the peroxisomal membrane.

Mitochondrial Proteins

The import into the mitochondria of both forms of monoamine oxidase, a protein important for the synthesis of neurotransmitters, is blocked by antibodies directed to ubiquitin (Zhaung and McCauley, 1989; Zhuang et al., 1992). Hexokinase bound to the outer membrane of mitochondria is not a substrate for ubiquitination, while the soluble enzyme is. Additionally, ubiquitin immunoreactivity and enzymes have been noted in mitochondria (Magnani et al., 1994).

ER Proteolysis

A significant fraction of many membrane or secreted proteins is degraded before they exit the secretory pathway. It is thought that this is a "proofreading" function that prevents inaccurately processed proteins from exiting the ER. The mutant cystic fibrosis transmembrane conductance regulator is a case in point. A significant fraction of wild type protein, and all of the mutant protein is degraded in the ER in a process that requires both ubiquitin and the proteasome (Jensen et al., 1995; Ward et al., 1995; McGee et al., 1996; Qu et al., 1996; Hiller et al., 1996). One of the *S. cere-*

visiae E2 conjugating enzymes is localized in the ER with its active site facing the cytoplasm (Sommer and Jentsch, 1993). This enzyme may participate in degradation of ER proteins, especially SEC61p and other components involved in protein import (Biederer et al., 1996).

A similar transport process, peptide uptake by *S. cerevisiae* is disrupted by ubiquitin system enzyme mutations. Mutations of the PTR1 locus of *S. cerevisiae* were selected on the basis of defects in peptide uptake, but when the gene was sequenced it was found to be identical to the gene for UBR1, an E3 ubiquitin protein ligase (Alagramam et al., 1995).

Nuclear Proteolysis

Many of the regulatory proteins mentioned above are degraded in the nucleus. In addition, a large amount of remodeling must occur with each cell division, including processes where proteolysis is important, i.e., chromatin condensation, spindle duplication and disassembly, laminin breakdown, and nuclear membrane dissolution. All of the enzymes of the ubiquitin system are represented in the nucleus, although the specific isozymes present and their amounts seem to vary with the cell cycle and other factors (Rivett, 1993; Watkins et al., 1993b; Goebl et al., 1994; Grenfell et al., 1994; Gray et al., 1995). Nuclear enzymes of the ubiquitin system include E1, E2, E3, deubiquitinating enzymes and the proteasome. The process of enzyme (and substrate) import and export have not been addressed, but are clearly potential regulatory processes which could determine the types of nuclear proteins degraded at any time.

F. Antigen Presentation

A variety of lines of evidence have demonstrated that the Class I MHC response (self vs. nonself) is accomplished by ubiquitin- and proteasome-dependent degradation of cellular proteins (reviewed in (Ciechanover & Schwartz, 1994; Lehner and Cresswell, 1996; Groettrup et al., 1996; York and Rock, 1996). Both genetic and biochemical studies have shown that significant peptides can be released from the proteasome during degradation. These peptides are transported into the ER by the TAP transporters where they assemble with nascent Class I MHC molecules and are subsequently displayed on the surface of the cell. In this way, a sampling of the internal protein metabolism can be displayed on the surface of the cell. Cells which display peptides from damaged or exogenous proteins can be recognized by the immune system and targeted for killing by specific cytotoxic T lymphocytes. Self peptides presumably do not elicit such a response due to the selection against self-reactive T cells early in development. At least part of the autoimmune diseases such as lupus and arthritis may be due to abberations in this system. It has been observed that interferon treatment results in proteasomes containing a different population of subunits. It has been speculated that these replacement subunits exhibit different proteolytic specificities and produce peptides which are more efficiently bound to and presented by the Class I MHC molecules.

III. STRUCTURAL BIOLOGY OF UBIQUITIN

A. Ubiquitin and Ubiquitinlike Proteins

Ubiquitin Gene Products

Ubiquitin is always synthesized as a fusion protein. For simplicity I will use the *S. cerevisiae* gene nomenclature, however, the basic gene structures are conserved throughout the eukaryotic kingdom. The C-terminal extensions are either additional copies of ubiquitin (UBI4) or ribosomal proteins (UBI1, UBI2, and UBI3) (Ozkaynak et al., 1987). The UBI4 gene encodes a proubiquitin consisting of several repeats of ubiquitin, is induced during the stress response, and is a major transcript in quiescent cells. In rapidly growing cells, UBI1–3 are actively transcribed to produce two fusion proteins; ubiquitin fused to the small ribosomal subunit protein S27a and ubiquitin fused to the large ribosomal subunit protein L40 (Finley et al., 1989; Chan et al., 1995). These are processed by proteolytic cleavage to yield ubiquitin and the ribosomal proteins (Finley et al., 1989).

Ubiquitinlike Proteins

In addition to the above genes in all eukaryotes, several additional proteins contain ubiquitin-like domains, some of which undergo processing similar to the ubiquitin fusions and some that are stable and not processed (see Table 1).

Viral Ubiquitin. Several virus strains have been found to contain ubiquitin-like sequences. These include *Autographa californica* nuclear polyhedrosis virus (baculovirus), as well as an RNA virus (bovine diarrhea virus, BDV) (Guarino, 1990; Tautz et al., 1993). The presence of ubiquitin in BVD correlates with pathogenicity (Tautz et al., 1993), perhaps because cellular ubiquitin processing enzymes cleave the viral protein and generate a new 80 kDa protein which confers cytopathogenicity. In *A. californica*, the viral ubiquitin is required for high yields of viral particles but not for replication (Reilly and Guarino, 1996). The *A. californica* ubiquitin can be conjugated to cellular proteins, but at a lower efficiency at K48 than at alternate linkage sites (Haas et al., 1996). This suggests that viral ubiquitin may specifically interfere with proteolysis requiring K48-linked chains but not processes requiring other types of linkages.

Ubiquitinlike Proteins Conjugated to Other Cellular Proteins. Treatment of mammalian cells with interferons induces a variety of cellular responses, many of

Table 1. Ubiquitin and Ubiquitin-Like Gene Products*

Protein	*Organism*[1]	*C-Terminus*	*Role*	*Accession*[2]
UBIQUITIN		PROCESSED TO GIVE		
Proubiquitin	Human	Ubiquitin	Ub precursor	GB:M26880
UbCEP52	Human	Ribosomal L40	Ub and L40 precursor	GB:M62406
UbCEP80	Human	Ribosomal S27a	Ub and S27a precursor	SW:P14798
UBIQUITIN-LIKE		PROBABLY PROCESSED TO GIVE		
FUBI	*Human*	Ribosomal S30	Ub-like (nonspecific immuno-supressor) and S30 precursor	SW:P35544
SMT3a	*Human*	VPESSLAGHSF	Similar to suppressor of mitosis fidelity mutations in yeast	EMBL:X99584
SMT3b	*Human*	VY	Similar to suppressor of mitosis fidelity mutations in yeast	EMBL:X99585
SMT3c (GMP1)	*Human*	HSTV	Conjugated to RanGAP1 protein	EMBL:X99586
ISG15	*Mouse*	GTEPGGRS	Conjugated to cellular proteins	GB:M13755
NEDD8	*Mouse*	GGLGQ	Induced in development	SW:P29595
UBL	*Caenorrhabitis elegans*	Ribosomal S27a	Ub-like (unknown function) and S27a precursor	SW:Q07371
Viral Ub	*Bacculovirus*	NGLRKGKRRCLSLLQFI or Y	Inhibitor of K48-linked poly-ubiquitin chain synthesis	SW:Q05120
UBIQUITIN-LIKE				
elongin B	Human	Probably not Processed	Elongation factor SIII subunit	GB:L42856
GDX	Human	Probably not Processed	Unknown	SW:P11441
BAT3	Human	Probably not Processed	Unknown	SW:P46379
hHR23a	Human	Probably not Processed	DNA repair in group C Xeroderma pigmentosum	SW:P54725
hHR23b	Human	Probably not Processed	DNA repair in group C Xeroderma pigmentosum	SW:P54727
AN1a	*X. laevis*	Probably not Processed	Unknown	GB:L08474
AN1b	*X. laevis*	Probably not Processed	Unknown	GB:L08475
DSK2p	Yeast	Probably not Processed	Spindle pole body duplication	SW:P48510

Notes: * Only one example of each type is given. [1]In many cases, homologous or analagous proteins are known in several organisms. [2]Accession number prefixes indicate the database: GB, GenBank; SW, SwissProt; and European Molecular Biology Laboratory (EMBL). See text for individual references.

which seem to be part of the organisms' response to viral infection. One such protein induced by interferon-β is detected by anti-ubiquitin antibodies and has been termed ubiquitin cross-reacting protein (UCRP) (Haas et al., 1987). This protein consists of two ubiquitin-like domains, with the C-terminal domain exhibiting more similarity to ubiquitin than the N-terminal domain. The protein is synthesized as a 17 kDa precursor and is processed to the 15 kDa active species by proteolysis at a position homologous with the C-terminus of free ubiquitin. The resultant protein can be either secreted, or activated and conjugated to a variety of intracellular proteins by enzymes which appear to be similar to, but distinct from, those that conjugate ubiquitin (Narasimhan et al., 1996). UCRP-protein conjugates selectively associate with intermediate filaments (Loeb and Haas, 1994). It is postulated that conjugation of UCRP to proteins may be a mechanism to localize those proteins to the intermediate filaments, although little more is known.

Interferon-γ induces another ubiquitin-like gene product (Nakamura et al., 1996), FUBI (also called Fau), is a protein with an N-terminal ubiquitin-like domain (ubi-L) fused to the ribosomal protein S30 (Olvera and Wool, 1993). FUBI can be processed to give S30 and ubi-L. It has been suggested that ubi-L can be conjugated to cellular proteins in a manner analogous to ubiquitin, although there is no direct evidence of this. The ubi-L domain is also a potent nonspecific immunosuppressor, although the immunoreactive band in active preparations migrates on SDS-PAGE with an apparent molecular weight of 70 kDa (Nakamura et al., 1995). It may be that a ubi-L protein conjugate can be released from cells in culture or that the active 70 kDa protein is a related protein with a ubi-L domain. The FUBI protein is also a target of a virus. The fox sequence of the Finkel-Biskis-Reilly murine sarcoma virus (Michiels et al., 1993) codes for an RNA which is the antisense strand of the FUBI message.

Finally, RanGAP1 (a GTPase activating protein specific for the Ran GTPase protein) has been shown to be covalently modified by the attachment of a small ubiquitin-like protein (Matunis et al., 1996). Ran has been shown to be involved in nuclear import and is thought to mediate the binding and release of the substrate/karyopherin complex to the docking sites in the central channel of the nuclear pore complex. The free RanGAP1 has an apparent molecular weight of 70 kDa and is found in the cytoplasm. Upon covalent ligation of the GAP protein and a ubiquitin-like protein (GMP1), the RanGAP1 is found in a 90 kDa form and is associated with the cytoplasmic fibers of the nuclear pore complex. A deubiquitinating enzyme activity has also been found in these preparations and this suggests that reversible modification of this protein by GMP1 may be involved in cycling this G protein required for nuclear transport. A system requiring ubiquitination of one side of the membrane and deubiquitination on the other could easily lead to regulation of vectoral transport,

Ubiquitin-Like Proteins Involved in DNA Structure and Function. The human homologue of *S. cerevisiae* RAD23p (HHR23B) is part of a heterodimeric complex which participates in nucleotide excision repair. This complex is involved in the general overall DNA repair pathway which is defective in complementation group C of xeroderma pigmentosum (XPC) (Masutani et al., 1994). The evidence suggests that the XPC protein is the other subunit of this heterodimer. HHR23B is a stable protein which has an N-terminal domain with significant similarity to ubiquitin. In *S. cerevisiae* the N-terminal domain of RAD23 is required for function but can be replaced by a ubiquitin domain (Watkins et al., 1993a).

In addition to its function in DNA repair, RAD23p and another protein with an N-terminal ubiquitin-like domain, DSK2p (Biggins et al., 1996), are required for spindle pole body duplication. The spindle pole body is similar to the centrosome in higher organisms and is a microtubule organizing center required for proper chromosome segregation. Both of these ubiquitin-like proteins apparently interact with KAR1p, a core component and CDC31p, a calcium binding protein.

Recently, it has been reported that a human protein, ubL1, forms a tight complex with RAD51 and RAD52 (Shen et al., 1996), proteins involved in recombination and repair of double strand breaks. RAD51 also binds Ubc9, a human ubiquitin conjugating enzyme (Kovalenko et al., 1996). hRAP12 has similarity to ubiquitin and to Smt3p, a suppressor of mutations in the MIF2p *S. cerevisiae* mitosis fidelity protein. MIF2p, and its mammalian counterpart CENP-C, functions in mitotic spindle formation and chromosome separation.

Finally, SIII is a mammalian general transcription factor required for efficient transcription by RNA polymerase II. It is a heterotrimer composed of p110, p18, and p15. The p18 component is a 118 residue protein required for optimal activity and contains an 84 residue N-terminal domain with significant homology to ubiquitin (Garrett et al., 1995). The role of this domain is unknown.

Ubiquitin-Like Proteins of Unknown Function. In addition to the above cases, where some information about function is known, there are several other genes which encode proteins containing ubiquitin like sequences and whose functions are unknown. These include: mouse NEDD8, a transcript which is downregulated during neural development (Kumar et al., 1993); human GDX, an X-linked gene near G6PDH (Toniolo et al., 1988); *X. laevis* AN1a and AN1b (Linnen et al., 1993), first noted as maternal transcripts localized in the animal hemisphere of unfertilized eggs; and human BAT3 (Banerji et al., 1990), a gene from the MHC region which codes for a proline-rich protein with a ubiquitin-like domain and multiple repeats of several other domains. In addition, another human protein similar to Smt3p and GMP1 has been detected (Matunis et al., 1996).

B. Structure of Ubiquitin

Ubiquitin is a 76 amino acid protein that is highly conserved and universally distributed in the eukaryotic kingdom. The X-ray crystal structure of human, *S. cerevisiae,* and oat ubiquitin has been determined and all three structures are very similar

(Vijay-Kumar et al., 1987). The protein folds into a compact, stable structure with a central five-stranded β sheet wrapped with one α helix and one short 3–10 helix. It exhibits three distinct "faces," a basic face, an acidic face ,and an hydrophobic face (Wilkinson, 1988). Well over 20 examples of this ubiquitin-like domain are known, and the sequence differences cluster on the face and ends of the αhelix. Ubiquitin itself is stable to proteases, with the exception of a trypsin catalyzed hydrolysis of the C-terminal dipeptide. The C-terminus projects from the body of the protein, and is the site of its covalent attachment to other proteins. Ubiquitin is conjugated to a large variety of proteins by formation of an isopeptide bond between the C-terminal glycine and a lysine on the target protein. The chemistry and biology of ubiquitin is dominated by this conjugation reaction.

C. Structure of Polymeric Ubiquitin

There are at least three types of polymeric ubiquitin. The primary gene products described above consist of ubiquitin (or a ubiquitin-like domain) fused to a C-terminal extension via a peptide bond. These single chain polymers are usually processed by the action of one or more ubiquitin carboxyl-terminal hydrolases (Wilkinson et al., 1992). No crystal structures or physical studies on these polymers have been reported.

Proteins are targeted for proteolysis by the attachment of a polyubiquitin chain. This polyubiquitin degradation signal is a branched polymer formed by ligating the C-terminus of ubiquitin to K48 of another ubiquitin. This chain has one free C-terminus (the proximal end) where it is attached to the target. Chains consisting of four or more ubiquitins are thought to be tightly bound by the S5a subunit of the proteasome and the attached protein is degraded. The structure of K48-linked tetraubiquitin has been reported (Cook et al., 1994). The chain is slightly elliptical in cross-section, resembling a flattened tube. There is a pseudo twofold symmetry with even numbered subunits having one orientation and the position of odd numbered subunits is derived by a 180° rotation about the screw axis and a translation down that axis. A prominent feature of the chain is a hydrophobic stripe extending along both flattened faces. Part of this stripe is formed by the exposed side chains of L8, I44, and V70. These residues have been shown to be important for both binding of the polyubiquitin chains and the subsequent degradation of the protein attached (Beal et al., 1996). The current evidence is that these chains are synthesized by specific ubiquitin conjugating enzymes, the UBC2 (RAD6) and UBC3 (CDC34) gene products in *S. cerevisiae* and their homologues in other organisms. The specificity of other E2s is less clear (see below) but others may also synthesize this linkage.

Alternate linkages can also be formed. Genetic evidence suggests that the stress response (Arnason and Ellison, 1994) and some component of the RAD6 pathway of DNA repair (Spence et al., 1995) requires UBC4 or UBC5. These appear to be involved in forming K29 and/or K63-linked polyubiquitin chains. UBC6 and UBC7 form a complex and participate (in addition to UBC4 and UBC5) in degradation of

MATα2 (Chen et al., 1993) and the Sec61p complex (Biederer et al., 1996). The precise nature of the linkages formed in these cases is not known. Finally, several of the ubiquitin conjugating enzymes autoubiquitinate. A novel keratinocyte E2, E2EPF, forms a K-11 polyubiquitin chain attached to itself, while recombinant RAD6 carries out a similar reaction making K6-linked polyubiquitin chains (Baboshina and Haas, 1996). It should be noted that there is no evidence for physiological K6 linkages as RAD6 (in crude extracts or *in vivo*) forms K48-linked chains.

IV. ENYZMES OF THE UBIQUITIN PATHWAY

We have already seen that many of the functions of ubiquitination require protein-protein associations, especially for the enzymes of the system. There are two putative protein interaction domains present on a number of the enzymes and ubiquitin-like proteins and which are thought to assist in this localization. The WW domain (see below) is present on several of the E3 enzymes and is thought to be important for binding to an SH3-like sequences (Sudol et al., 1995). The UBA domain is a motif found in some of the deubiquitinating enzymes, the E2 conjugating enzymes, ubiquitin-like proteins, protein kinases, and the c-cbl proto-oncogene (Hofmann and Bucher, 1996). Finally, the ubiquitin-like motif must also direct specific protein binding.

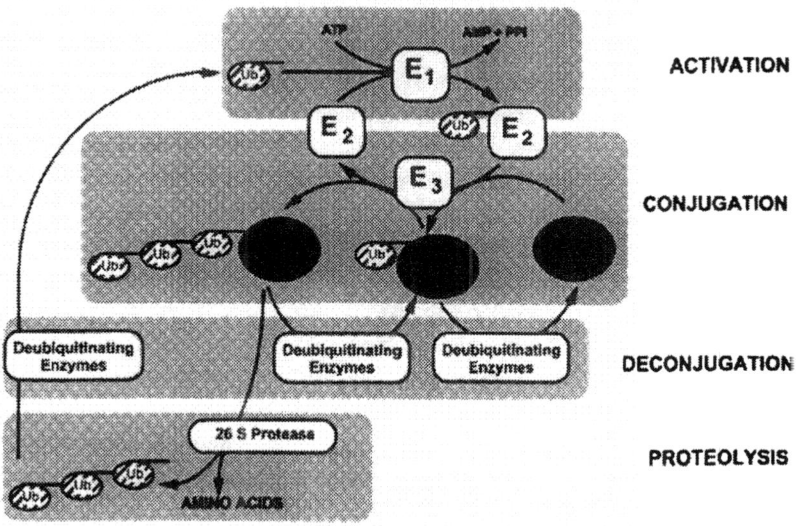

Figure 2. Ubiquitin-dependent proteolysis. The enzymes of the pathway are indicated. Four separate processes are indicated: activation, conjugation, deconjugation, and proteolysis. Polymeric ubiquitin chains are linked by the K48 isopeptide bond. Details are given in the text.

A. Formation of Ubiquitinated Proteins

The targeting functions of ubiquitin require its association with, or covalent attachment to, other proteins. The specificity of this involves modulation of the structure of the substrate, temporal and spatial regulation of the ubiquitinating and deubiquitinating enzymes of the system, and the identity and localization of ubiquitin binding proteins. One of the central themes of this regulation is the specific localization or recruitment of ubiquitin conjugating enzymes to the site(s) of substrate localization. Figure 2 details the enzymatic conversions required in ubiquitin-dependent proteolysis.

E1 Activating Enzymes

The first step of ubiquitination is the activation of the C-terminus of ubiquitin as a thiol-ester with a cysteine residue on the enzyme. There are only one to three different genes for the ubiquitin activating enzyme in most organisms (Cook and Chock, 1992; Hatfield and Vierstra, 1992; Dohmen et al., 1995). The enzyme is found both in the cytoplasm and nucleus, depending on the cell cycle (Grenfell et al., 1994). Several temperature sensitive mutations have been identified and mutations in this enzyme have prominent cell cycle defects, as well as a number of other defects of protein metabolism.

E2 Carrier Proteins/Conjugating Enzymes

Ubiquitin is next transferred to one of a large family of ubiquitin carrier proteins (more often called ubiquitin conjugating enzymes) by a transthiolesterification with a cysteine on E2. The E2 family of proteins has at least 10 members in *S. cerevisiae* (Jentsch, 1992) and at least as many in higher organisms. These proteins are characterized by a central catalytic core of about 15 kDa and variable N- and/or C-terminal extensions. It is thought that these extensions are involved in substrate specificity or localization. For instance, reticulocyte E2-25K will catalyze the formation of a K48-linked polyubiquitin chain, presumably due to its C-terminal extension (Chen and Pickart, 1990). Similarly, RAD6 will polyubiquitinate histones in a reaction requiring the acidic C-terminal tail (Sung et al., 1988). Other E2s contain sequences which localize the enzyme to a membrane (Sommer and Jentsch, 1993) or to a DNA/protein complex (Bailly et al., 1994). Since some members of this family directly catalyze the ubiquitination of proteins, they have been referred to as ubiquitin conjugating enzymes.

E3 Ligases

Other members of the E2 family should probably be called ubiquitin carrier proteins to emphasize the fact that they require the participation of other proteins to ubiquitinate proteins. These other proteins are generically referred to as E3 ligases.

There appear to be at least two types: catalysts like E6-AP (Huibregtse et al., 1991) which accept the ubiquitin as a thiol ester and pass it on to the substrate amine, and adapter proteins like UBR1 which may bring substrate and an activated E2 in proximity for E2 catalyzed ubiquitination (Bartel et al., 1990). This is a subtle mechanistic point as both mechanisms require the participation of E1, E2, and E3.

Among the E3 ligases related to E6-AP (the E3 involved in p53 degradation) are at least three with WW protein-protein interaction domains (Sudol et al., 1995). Thus, *S. cerevisiae* Rsp5, mammalian Nedd-4, and nematode 56G7 sequences contain both the homology of E6-AP carboxyl terminus (HECT) domain and multiple WW domains. The latter sequences bind a consensus xPPxY sequence which resembles the SH3 domain (Chen and Sudol, 1995). It is tempting to suggest that these may be the E3 enzymes involved in ubiquitination of membrane associated proteins via interactions with SH3-like adapter proteins. Indeed, RSP5p is required for efficient degradation of the amino acid and uracil permeases in *S. cerevisiae* (Hein et al., 1995) and Nedd-4 binds to the epithelial sodium channel which is deleted in Liddle's syndrome (Staub et al., 1996).

B. Removal of Ubiquitin From Proteins

By now it should be apparent that ubiquitination of proteins is a versatile covalent modification, with multiple roles in cellular regulation. In many respects, ubiquitination resembles phosphorylation. It should not be surprising then, that deubiquitinating enzymes (DUBS) exist and are apparently important in regulation of ubiquitin-dependent processes.

The processing of primary gene products, the removal of ubiquitin or polyubiquitin from target proteins, and the disassembly of the polyubiquitin chain all require accurate proteolytic processing at the C-terminal glycine of ubiquitin. These enzymes are generically referred to as (DUBS) (Wilkinson, 1995; Wilkinson and Hochstrasser, 1997). Two gene families are known. The Family 1 enzymes are known as ubiquitin C-terminal hydrolases (UCH) and are a small group with significant sequence similarity to a neural-specific human protein called PGP9.5 (Wilkinson et al., 1989). These enzymes are all small thiol proteases with tissue specific expression in mammals. They are probably involved in processing the ubiquitin gene products and in salvaging ubiquitin which has been adventitiously trapped by attack of small cellular nucleophiles on the various thiol ester intermediates in the synthetic pathway (Wilkinson, 1995; Larsen et al., 1997). Family 2 enzymes form a much larger group of thiol proteases seemingly unrelated to the Family 1 enzymes. These are referred to as ubiquitin-specific proteases (UBPS) (Tobias and Varshavsky, 1991; Baker et al., 1992). The UBPs have specificity for cleavage of ubiquitin from larger proteins and vary in size from 50 to 250 kDa. Database searches of the *S. cerevisiae* genome reveal 16 such enzymes recognizable by the presence of at least six specific motifs (Wilkinson et al., 1995). A large number of putative UBPs are known in higher organisms, with more being reported daily. The substrate specificity and physiological roles are known for only a few.

Removal of Monomeric Ubiquitin

The first example of this class of deubiquitinating enzymes was characterized over 15 years ago (Andersen et al., 1981; Matsui et al., 1982) as an activity that deubiquitinated uH2A. This thiol protease is reported to have an apparent molecular weight of 38 kDa and was termed A24 lyase or isopeptidase, but as it still has not been purified, it is not clear to which family this enzyme belongs.

Using a screen for cleavage of a ubiquitin fusion peptide (Miller et al., 1989) or hydrolysis of ubiquitin ethyl ester (Mayer and Wilkinson, 1989; Wilkinson et al., 1989) the first members of the Family 1 UCHs were cloned and characterized. The crystal structure of a human UCH has been reported (Johnston et al., 1997). These proteins hydrolyze ubiquitin fusion proteins or adducts with small, unfolded leaving groups (Larsen et al., 1997). There are at least three mammalian isozymes and they are tissue specific, with particularly high levels of expression in nervous tissue (Wilkinson, 1995). There is only one such protein in *S. cerevisiae* and deletion of this gene is not lethal (Miller et al., 1989). Clearly, other enzymes must compensate, and these are the UBP enzymes first cloned by virtue of their ability to hydrolyze ubiquitin-protein conjugates (Tobias and Varshavsky, 1991; Baker et al., 1992).

Removal of Intact Chains

In principle, polyubiquitinated proteins could be deubiquitinated by hydrolysis of the isopeptide bond joining the chain to the target protein. There is no direct evidence for such activity, but it would be expected to be tightly regulated in order to avoid removal of the polyubiquitin signal before degradation of the target protein.

Trimming of Chains

There is some evidence for trimming of polyubiquitin chains from the distal end. Treatment of ubiquitin-lysozyme conjugates with reticulocyte lysate or partially purified UBP preparations leads to release of ubiquitin and a decrease in the molecular weight of the ubiquitinated protein (Hadari et al., 1992). Recent studies (Lam et al., 1997) suggest that the "trimming" activity copurifies with PA700, the regulator complex of the 26S proteasome (DeMartino et al., 1994). This may be an editing process whereby slowly degraded substrates are deubiquitinated before total proteolysis can occur (cf. NFκB), or it may simply be an intermediate step in the catabolism of the polyubiquitin chain (see below).

Disassembly of the Degradation Intermediates

The final step in the metabolism of polyubiquitinated proteins is the removal of the polyubiquitin chain from the degradation intermediates and the disassembly of the chain. The latter step is necessary since free polyubiquitin chains can bind to the

26S proteasome and interfere with proteolysis of other substrates. Two enzymes are thought to participate in the processing of these degradation intermediates. First, an *S. cerevisiae* protein called DOA4 appears to be responsible for releasing the polyubiquitin chain from residual peptides produced by proteasome-dependent proteolysis of ubiquitinated proteins (Papa and Hochstrasser, 1993). Following that, isopeptidase T (UBP14 in *S. cerevisiae*) degrades the free chain by hydrolysis of one subunit at a time from the proximal end of the chain (Stein et al., 1995; Wilkinson et al., 1995; Americk et al. 1997). Thus, the polyubiquitin chain is not destroyed or removed before the action of the proteasome is completed.

V. FATES OF UBIQUITINATED PROTEINS

Once proteins have been ubiquitinated, they are subject to a number of potential fates. The exact path taken must depend on the availability of specific binding proteins, the proximity of various organelles and subcellular particles that metabolize ubiquitin conjugates, the type of the polyubiquitin chain formed, and perhaps the presence of ubiquitin-like proteins.

A. Delivery to the Proteasome

The best understood fate of polyubiquitinated proteins is their degradation by the 26S proteasome. Thanks in part to the availability of moderately specific inhibitors (Palombella et al., 1994; Rock et al., 1994; Jensen et al., 1995), we now recognize several ubiquitin- and proteasome-dependent proteolytic processes. The proteasome will be covered in more detail elsewhere in this volume (chapters 4 and 5), so I will simply point out that the function of ubiquitination in this case seems to be to append a high-affinity binding site to the target protein so as to deliver it to the proteolytic machinery.

B. Delivery to the Lysosome

Ubiquitinated proteins are also found in the lysosome (chapter 7). Recently, it has become clear that ubiquitination, often following phosphorylation, is a signal for internalization of membrane-bound receptors and degradation by the endocytotic pathways. This ubiquitination seems to occur at the membrane and ubiquitin protein ligases such as RPS5 and NEDD4 have both been noted to contain the WW domain, a domain important for interacting with SH3 domains (Sudol et al., 1995). Also as noted above, UBC6 and PAS2 are membrane associated ubiquitin conjugating enzymes in *S. cerevisiae* (Sommer and Jentsch, 1993). The mechanism by which polyubiquitin induces internalization is unknown, although it seems possible that it may involve clustering of the ubiquitinated receptors via the action of specific polyubiquitin binding proteins.

C. Release From Sites of DNA Damage

A model for the role of ubiquitin-dependent proteolysis in DNA repair can be proposed. Briefly, it has been postulated that polymerases, both DNA and RNA, are stalled at sites of DNA damage. Ubiquitin system components may form complexes with specific DNA damage binding proteins, thereby directing the ubiquitination of polymerase complexes or other associated proteins. Disassembly of these stalled complexes would then allow access by the DNA repair machinery in order to repair the damage. Evidence for this model includes: observation of a RAD6/RAD18 complex and a RAD6/UBR1 complex (Bailly et al., 1994); the binding of a ubiquitin-like protein (RAD23) to the transcription factor TFIIH and to the DNA damage recognition factor RAD14 (Guzder et al., 1995); the binding of human Ubc9 protein to the Rad51 recombination protein and the localization of both to the synaptonemal complex (Kovalenko et al., 1996); and, finally, the observation of a rat DNA binding protein with significant sequence similarity with the E6-AP family of ubiquitin protein ligases (GenBank:U08214)

D. Vectoral Transport

While there is only scant evidence for ubiquitin-dependent vectoral transport, it remains an intriguing possibility. It has been reported that mitochondrial import of monoamine oxidase A and B requires ATP and is inhibited by anti-ubiquitin antibodies (Zhaung and McCauley, 1989; Zhuang et al., 1992). Also, the ubiquitin-ribosomal proteins must be transported into the nucleus for ribosome assembly. In *S. cerevisiae*, the UBI3 deletion mutant can be rescued by a plasmid encoding only the ribosomal protein, but only on a high copy number plasmid (Finley et al., 1989). This led to the suggestion that the ubiquitin leader sequence has a role in either transport or assembly. It may be that the nuclear localization signals on the ribosomal proteins are exposed in the fusion protein, but not in the processed protein.

In general, ubiquitination of proteins could serve to block or expose various translocation signals, as well as to furnish a posttranslational signal that could be utilized for translocation. Cleavage of ubiquitin or ubiquitin chains by deubiquitinating enzymes (see above) could remove the signal after transport. This appears to be exactly the mechanism utilized in regulating nuclear import (Matunis et al., 1996). The Ran GTPase has been shown to be involved in nuclear import and is thought to mediate the binding and release of the substrate/karyopherin complex to the docking sites in the central channel of the nuclear pore complex. Hydrolysis of GTP and cycling of the G-protein presumably requires association of the RanGAP1 activating proteins. This protein is normally cytoplasmic, but upon being modified by a ubiquitin-like protein (GMP1), it is bound to the cytoplasmic side of the nuclear pore complex (NPC). Deubiquitination is correlated with its release from the NPC and probably occurs on the luminal site of the nuclear membrane. The loss of the GTPase activity of Ran upon dissociation of the RanGAP1 could be a signal for its export back to the cytoplasmic side, perhaps due to the exposure of a nuclear export signal.

E. Modulation of Structure or Activity

In at least a few cases, ubiquitination of proteins is more limited and it may serve as a reversible modification modulating structure or function. There will undoubtedly be other examples of this type of regulation.

Histones

Modification of histones was the first example of protein ubiquitination. Approximately 10% of the core octomers have ubiquitinated H2A (uH2A) associated with them. Ubiquitin is removed at mitosis and reconjugated subsequently. The consequences of such ubiquitination are not known, but it is thought that ubiquitination may alter chromatin structure to facilitate chromatin condensation or DNA synthesis (Bradbury, 1992).

IκBα Kinase

Processing and regulation of NFκB activity are ubiquitin-dependent processes. In addition to processing the p105 precursor to the mature p50 subunit, ubiquitination of IκBα complexed to NFκB in the cytoplasm leads to degradation of the inhibitor and subsequent translocation into the nucleus, The ubiquitination event is triggered by phosphorylation of IκBα in response to extracellular signals catalyzed by a large multisubunit kinase. Surprisingly, this kinase is activated by a reversible ubiquitination that does not involve proteolysis (Chen et al., 1996).

IgE Receptors

Upon antigen-induced engagement of the high-affinity IgE receptor, active receptors become polyubiquitinated. This modification is not dependent on phosphorylation and is rapidly reversed upon removal of the stimulus. The specificity, rapidity, and reversibility of the ubiquitination suggests a role in modulation of activity or signaling, but not in degradation of the receptor (Paolini and Kinet, 1993).

VI. PERSPECTIVE

In summary, ubiquitination of proteins is an important and wide-spread posttranslational modification. Essentially every aspect of protein metabolism involves ubiquitin or ubiquitin-like proteins. In turn, a wide variety of biological phenomena are affected. Ubiquitination and deubiquitination reactions are similar in many respects to the action of kinases and phosphatases, and every bit as complex.

The ubiquitination signal is a versatile targeting mechanism which directs proteins to a variety of fates. Ubiquitinated proteins are directed to the proteasome and/or the lysosome for degradation, In other instances, we suspect that different fates await ubiquitinated proteins, although these are not at all clear. Since ubiquitin can be conjugated at a number of different lysines, many types of polyubiquitin signals can be generated, and ubiquitin-like sequences increase the diversity of the signaling possibilities.

Finally, it is useful to think of the ubiquitin domain as a protein interaction domain. Ubiquitinated proteins (and ubiquitin-like proteins) are selectively bound by the cellular organelles such as the proteasome and the nucleus, are often found associated with the cytoskeleton, and are found in virtually every cellular compartment. The enzymes of the pathway also depend heavily on protein-protein interactions to achieve specificity. Ubiquitinating enzymes are directed to sites of DNA damage, to active receptor complexes and to the cell-cycle machinery by use of binding domains such as the ubiquitin domain, the WW domain, the UBA domain, and probably other binding motifs. The model then involves assembling the appropriate ubiquitination or deubiquitination complexes at the proper location at the right time by using "adaptors" which are very reminiscent of mechanisms of transcriptional control and signal transduction. There is no doubt that ubiquitination will turn out to be just as complex and important as these two paradigms of biological regulation.

REFERENCES

Alagramam, K., Naider, F., & Becker, J.M. (1995). A recognition component of the ubiquitin system is required for peptide transport in Saccharomyces cerevisiae. Mol. Microbiol. 15, 225-234.

Andersen, M.W., Ballal, N.R., Goldknopf, I.L., & Busch, H. (1981). Protein A24 lyase activity in nucleoli of thioacetamide-treated rat liver releases histone 2A and ubiquitin from conjugated protein A24. Biochemistry 20, 1100-1104.

Arnason, T., & Ellison, M.J. (1994). Stress resistance in Saccharomyces cerevisiae is strongly correlated with assembly of a novel type of multiubiquitin chain. Mol. Cell Biol. 14, 7876-7883.

Attaix, D., Taillandier, D., Temparis, S., Larbaud, D., Aurousseau, E., Combaret, L., & Voisin, L. (1994). Regulation of ATP-ubiquitin-dependent proteolysis in muscle wasting. [Review]. Reprod. Nutr. Dev. 34, 583-597.

Baboshina, O.V., & Haas, A.L. (1996). Novel multiubiquitin chain linkages catalyzed by the conjugating enzymes E2EPF and RAD6 are recognized by 26 S proteasome subunit 5. J. Biol. Chem. 271, 2823-2831.

Bailly, V., Lamb, J., Sung, P., Prakash, S., & Prakash, L. (1994). Specific complex formation between yeast RAD6 and RAD18 proteins: a potential mechanism for targeting RAD6 ubiquitin-conjugating activity to DNA damage sites. Genes Dev. 8, 811-820.

Baker, R.T., Tobias, J.W., & Varshavsky, A. (1992). Ubiquitin-specific proteases of Saccharomyces cerevisiae. Cloning of UBP2 and UBP3, and functional analysis of the UBP gene family. J. Biol. Chem. 267, 23364-23375.

Baldi, L., Brown, K., Franzoso, G., & Siebenlist, U. (1996). Critical role for lysines 21 and 22 in signal-induced, ubiquitin-mediated proteolysis of I kappa B-alpha. J. Biol. Chem. 271, 376-379.

Banerji, J., Sands, J., Strominger, J.L., & Spies, T. (1990). A gene pair from the human major histocompatibility complex encodes large proline-rich proteins with multiple repeated motifs and a single ubiquitin-like domain. Proc. Natl. Acad. Sci. USA 87, 2374-2378.

Barral, Y., Jentsch, S., & Mann, C. (1995). G1 cyclin turnover and nutrient uptake are controlled by a common pathway in yeast. Genes Dev. 9, 399-409.

Bartel, B., Wunning, I., & Varshavsky, A. (1990). The recognition component of the N-end rule pathway. EMBO J. 9, 3179-3189.

Beal, R., Deveraux, Q., Xia, G., Rechsteiner, M., & Pickart, C. (1996). Surface hydrophobic residues of multiubiquitin chains essential for proteolytic targeting. Proc. Natl. Acad. Sci. USA 93, 861-865.

Biederer, T., Volkwein, C., & Sommer, T. (1996). Degradation of subunits of the Sec61p complex, an integral component of the ER membrane, by the ubiquitin-proteasome pathway. EMBO J. 15, 2069-2076.

Biggins, S., Ivanovska, I., & Rose, M.D. (1996). Yeast ubiquitin-like genes are involved in duplication of the microtubule organizing center. J. Cell. Biol. 133, 1331-1346.

Bradbury, E.M. (1992). Reversible histone modifications and the chromosome cell cycle. Bioessays 14, 9-16.

Bregman, D.B., Halaban, R., van Gool, A.J. Henning, K.A., Friedberg, E.C., & Warren, S.L. (1996). UV-induced ubiquitinaiton of RNA polymerase II: A novel modification deficient in Cockayne's syndrome cells. Proc. Natl. Acad. Sci. USA 93, 11586-11590.

Busch, H. (1984). Ubiquitination of proteins. Methods Enzymol. 106, 238-262.

Cenciarelli, C., Hou, D., Hsu, K.C., Rellahan, B.L., Wiest, D.L., Smith, H.T., Fried, V.A., & Weissman, A.M. (1992). Activation-induced ubiquitination of the T cell antigen receptor. Science 257, 795-797.

Chan, Y.L., Suzuki, K., & Wool, I.G. (1995). The carboxyl extensions of two rat ubiquitin fusion proteins are ribosomal proteins S27a and L40. Biochem. Biophys. Res. Commun. 215, 682-690.

Chen, H.I., & Sudol, M. (1995). The WW domain of Yes-associated protein binds a proline-rich ligand that differs from the consensus established for Src homology 3-binding modules. Proc. Natl. Acad. Sci. USA 92, 7819-7823.

Chen, P., Johnson, P., Sommer, T., Jentsch, S., & Hochstrasser, M. (1993). Multiple ubiquitin-conjugating enzymes participate in the in vivo degradation of the yeast MAT alpha 2 repressor. Cell 74, 357-369.

Chen, Z., & Pickart, C.M. (1990). A 25-kilodalton ubiquitin carrier protein (E2) catalyzes multi-ubiquitin chain synthesis via lysine 48 of ubiquitin. J. Biol. Chem. 265, 21835-21842.

Chen, Z., Hagler, J., Palombella, V.J., Melandri, F., Scherer, D., Ballard, D., & Maniatis, T. (1995). Signal-induced site-specific phosphorylation targets I kappa B alpha to the ubiquitin-proteasome pathway. Genes Dev. 9, 1586-1597.

Chen, Z.J., Parent, L., & Maniatis, T. (1996). Site-specific phosphorylation of IkappaBalpha by a novel ubiquitination-dependent protein kinase activity. Cell 84, 853-862.

Ciechanover, A., Finley, D., & Varshavsky, A. (1984). The ubiquitin-mediated proteolytic pathway and mechanisms of energy-dependent intracellular protein degradation. J. Cell Biochem. 24, 27-53.

Ciechanover, A., DiGiuseppe, J.A., Bercovich, B., Orian, A., Richter, J.D., Schwartz, A.L., & Brodeur, G.M. (1991a). Degradation of nuclear oncoproteins by the ubiquitin system in vitro. Proc. Natl. Acad. Sci. USA 88, 139-143.

Ciechanover, A., Gropper, R., & Schwartz, A.L. (1991b). The ubiquitin-activating enzyme is required for lysosomal degradation of cellular proteins under stress. Biomed. Biochim. Acta 50, 321-332.

Ciechanover, A. (1994). The ubiquitin-proteasome proteolytic pathway. [Review]. Cell 79, 13-21.

Ciechanover, A., & Schwartz, A.L. (1994). The ubiquitin-mediated proteolytic pathway: mechanisms of recognition of the proteolytic substrate and involvement in the degradation of native cellular proteins. [Review]. FASEB J. 8, 182-191.

Cook, J.C., & Chock, P.B. (1992). Isoforms of mammalian ubiquitin-activating enzyme. J. Biol. Chem. 267, 24315-24321.

Cook, W.J., Jeffrey, L.C., Kasperek, E., & Pickart, C.M. (1994). Structure of tetraubiquitin shows how multiubiquitin chains can be formed. J. Mol. Biol. 236, 601-609.

Crane, D.I., Kalish, J.E., & Gould, S.J. (1994). The Pichia pastoris PAS4 gene encodes a ubiquitin-conjugating enzyme required for peroxisome assembly. J. Biol. Chem. 269, 21835-21844.

DeMartino, G.N., Moomaw, C.R., Zagnitko, O.P., Proske, R.J., Chu-Ping, M., Afendis, S.J., Swaffield, J.C., & Slaughter, C.A. (1994). PA700, an ATP-dependent activator of the 20 S proteasome, is an ATPase containing multiple members of a nucleotide-binding protein family. J. Biol. Chem. 296, 20878-20884.

Deshaies, R.J., Chau, V., & Kirschner, M. (1995). Ubiquitination of the G1 cyclin Cln2p by a Cdc34p-dependent pathway. EMBO J. 14, 303-312.

Dice, J.F. (1992). Selective degradation of cytosolic proteins by lysosomes. Ann. N.Y. Acad. Sci. 674, 58-64.

Dohmen, R.J., Stappen, R., McGrath, J.P., Forrova, H., Kolarov, J., Goffeau, A., & Varshavsky, A. (1995). An essential yeast gene encoding a homolog of ubiquitin-activating enzyme. J. Biol. Chem. 270, 18099-18109.

Driscoll, J. (1994). The role of the proteasome in cellular protein degradation. Histol. Histopathol. 9, 197-202.

Egner, R., & Kuchler, K. (1996). The yeast multidrug transporter Pdr5 of the plasma membrane is ubiquitinated prior to endocytosis and degradation in the vacuole. FEBS Lett. 378, 177-181.

Finley, D., Bartel, B., & Varshavsky, A. (1989). The tails of ubiquitin precursors are ribosomal proteins whose fusion to ubiquitin facilitates ribosome biogenesis. Nature 338, 394-401.

Galan, J.M., Moreau, V., Andre, B., Volland, C., & Haguenauer-Tsapis, R. (1996). Ubiquitination mediated by the Npi1p/Rsp5p ubiquitin-protein ligase is required for endocytosis of the yeast uracil permease. J. Biol. Chem. 271, 10946-10952.

Galcheva-Gargova, Z., Theroux, S.J., & Davis, R.J. (1995). The epidermal growth factor receptor is covalently linked to ubiquitin. Oncogene 11, 2649-2655.

Garrett, K.P., Aso, T., Bradsher, J.N., Foundling, S.I., Lane, W.S., Conaway, R.C., & Conaway, J.W. (1995). Positive regulation of general transcription factor SIII by a tailed ubiquitin homolog. Proc. Natl. Acad. Sci. USA 92, 7172-7176.

Glotzer, M., Murray, A.W., & Kirschner, M.W. (1991). Cyclin is degraded by the ubiquitin pathway. Nature 349, 132-138.

Glotzer, M. (1995). Cell cycle. The only way out of mitosis. [Review]. Curr. Biol. 5, 970-972.

Goebl, M.G., Goetsch, L., & Byers, B. (1994). The Ubc3 (Cdc34) ubiquitin-conjugating enzyme is ubiquitinated and phosphorylated in vivo. Mol. Cell Biol. 14, 3022-3029.

Goldberg, A.L., & Dice, J.F. (1974). Intracellular protein degradation in mammalian and bacterial cells. Ann. Rev. Biochem. 43, 835-869.

Gray, D.A., Inazawa, J., Gupta, K., Wong, A., Ueda, R., & Takahashi, T. (1995). Elevated expression of Unph, a proto-oncogene at 3p21.3, in human lung tumors. Oncogene 10, 2179-2183.

Grenfell, S.J., Trausch-Azar, J.S., Handley-Gearhart, P.M., Ciechanover, A., & Schwartz, A.L. (1994). Nuclear localization of the ubiquitin-activating enzyme, E1, is cell-cycle-dependent. Biochem. J. 300, 701-708.

Groettrup, M., Soza, A., Eggers, M., Kuehn, L., Dick, T.P., Schild, H., Rammensee, H.G., Koszinowski, U.H., & Kloetzel, P.M. (1996). A role for the proteasome regulator PA28alpha in antigen presentation. Nature 381, 166-168.

Guarino, L.A. (1990). Identification of a viral gene encoding a ubiquitin-like protein. Proc. Natl. Acad. Sci. USA 87, 409-413.

Guzder, S.N., Bailly, V., Sung, P., Prakash, L., & Prakash, S. (1995). Yeast DNA repair protein RAD23 promotes complex formation between transcription factor TFIIH and DNA damage recognition factor RAD14. J. Biol. Chem. 270, 8385-8388.

Haas, A.L., Ahrens, P., Bright, P.M., & Ankel, H. (1987). Interferon induces a 15-kilodalton protein exhibiting marked homology to ubiquitin. J. Biol. Chem. 262, 11315-11323.

Haas, A.L., Katzung, D.J., Reback, P.M., & Guarino, L.A. (1996). Functional characterization of the ubiquitin variant encoded by the baculovirus Autographa californica. Biochemistry 35, 5385-5394.

Hadari, T., Warms, J.V., Rose, I.A., & Hershko, A. (1992). A ubiquitin C-terminal isopeptidase that acts on polyubiquitin chains. Role in protein degradation. J. Biol. Chem. 267, 719-727.

Hatfield, P.M., & Vierstra, R.D. (1992). Multiple forms of ubiquitin-activating enzyme E1 from wheat. Identification of an essential cysteine by in vitro mutagenesis. J. Biol. Chem. 267, 14799-14803.

Hegde, A.N., Goldberg, A.L., & Schwartz, J.H. (1993). Regulatory subunits of cAMP-dependent protein kinases are degraded after conjugation to ubiquitin: a molecular mechanism underlying long-term synaptic plasticity. Proc. Natl. Acad. Sci. USA 90, 7436-7440.

Hein, C., Springael, J.Y., Volland, C., Haguenauer-Tsapis, R., & Andre, B. (1995). NPI1, an essential yeast gene involved in induced degradation of Gap1 and Fur4 permeases, encodes the Rsp5 ubiquitin-protein ligase. Mol. Microbiol. 18, 77-87.

Hermida-Matsumoto, M.L., Chock, P.B., Curran, T., & Yang, D.C. (1996). Ubiquitinylation of transcription factors c-Jun and c-Fos using reconstituted ubiquitinylating enzymes. J. Biol. Chem. 271, 4930-4936.

Hershko, A., & Ciechanover, A. (1992). The ubiquitin system for protein degradation. Ann. Rev. Biochem. 61, 761-807.

Hicke, L., & Riezman, H. (1996). Ubiquitination of a yeast plasma membrane receptor signals its ligand-stimulated endocytosis. Cell 84, 277-287.

Hiller, M.M., Finger, A., Schweiger, M., & Wolf, D.H. (1996). ER degradation of a misfolded luminal protein by the cytosolic ubiquitin-proteasome pathway. Science 273, 1725-1728.

Hochstrasser, M., Ellison, M.J., Chau, V., & Varshavsky, A. (1991). The short-lived MAT alpha 2 transcriptional regulator is ubiquitinated in vivo. Proc. Natl. Acad. Sci. USA 88, 4606-4610.

Hochstrasser, M. (1995). Ubiquitin, proteasomes, and the regulation of intracellular protein degradation. Curr. Opin. Cell Biol. 7, 215-223.

Hofmann, K., & Bucher, P. (1996). The UBA domain: a sequence motif present in multiple enzyme classes of the ubiquitination pathway. Trends Biochem. Sci. 21, 172-173.

Holloway, S.L. (1995). Sister chromatid separation in vivo and in vitro. [Review]. Curr. Opin. Genet. Dev. 5, 243-248.

Huang, M.E., Chuat, J.C., & Galibert, F. (1995). Analysis of a 42.5 kb DNA sequence of chromosome X reveals three tRNA genes and 14 new open reading frames including a gene most probably belonging to the family of ubiquitin-protein ligases. Yeast 11, 775-781.

Huibregtse, J.M., Scheffner, M., & Howley, P.M. (1991). A cellular protein mediates association of p53 with the E6 oncoprotein of human papillomavirus types 16 or 18. EMBO J. 10, 4129-4135.

Jensen, T.J., Loo, M.A., Pind, S., Williams, D.B., Goldberg, A.L., & Riordan, J.R. (1995). Multiple proteolytic systems, including the proteasome, contribute to CFTR processing. Cell 83, 129-135.

Jentsch, S., McGrath, J.P., & Varshavsky, A. (1987). The yeast DNA repair gene RAD6 encodes a ubiquitin-conjugating enzyme. Nature 329, 131-134.

Jentsch, S. (1992). The ubiquitin-conjugation system. Annu. Rev. Genet. 26, 179-207.

Johnston, S.C., Larsen, C.N., Cook, W.J., Wilkinson, K.D., & Hill, C.P. (1997). Crystal structure of a deubiquitinating enzyme (human UCH-L3) at 1.8 Å resolution. EMBO J. 16, 3787-3796.

King, R.W., Peters, J.M., Tugendreich, S., Rolfe, M., Hieter, P., & Kirschner, M.W. (1995). A 20S complex containing CDC27 and CDC16 catalyzes the mitosis-specific conjugation of ubiquitin to cyclin B. Cell 81, 279-288.

Klotzbucher, A., Stewart, E., Harrison, D., & Hunt, T. (1996). The 'destruction box' of cyclin A allows B-type cyclins to be ubiquinated, but not efficiently destroyed. EMBO J. 15, 3053-3064.

Kolling, R., & Hollenberg, C.P. (1994). The ABC-transporter Ste6 accumulates in the plasma membrane in a ubiquitinated form in endocytosis mutants. EMBO J. 13, 3261-3271.

Kornitzer, D., Raboy, B., Kulka, R.G., & Fink, G.R. (1994). Regulated degradation of the transcription factor Gcn4. EMBO J. 13, 6021-6030.

Kovalenko, O., Plug, A.W., Haaf, T., Gonda, D.K., Ashley, T., Ward, D.C., Radding, C.M., & Golub, E.I. (1996). Mammalian ubiquitin-conjugating enzyme Ubc9 interacts with Rad51 recombination protein and localizes in synaptonemal complexes. Proc. Natl. Acad. Sci. USA 93, 2958-2963.

Kumar, S., Yoshida, Y., & Noda, M. (1993). Cloning of a cDNA which encodes a novel ubiquitin-like protein. Biochem. Biophys. Res. Commun. 195, 393-399.

Labrecque, S., & Matlashewski, G.J. (1995). Viability of wild type p53-containing and p53-deficient tumor cells following anticancer treatment: the use of human papillomavirus E6 to target p53. Oncogene 11, 387-392.

Laing, J.G., & Beyer, E.C. (1995). The gap junction protein connexin43 is degraded via the ubiquitin proteasome pathway. J. Biol. Chem. 270, 26399-26403.

Lam, Y.A., Xu, W., DeMartino, G.N., & Cohen, R.E. (1997). Editing of ubiquitin conjugates by an isopeptidase in the 26S proteasome. Nature 385, 737-740.

Larsen, C.N., Krantz, B.A., & Wilkinson, K.D. (1998). Substrate Specificity of Deubiquitinating Enzymes: Ubiquitin C-terminal Hydrolases. Biochemistry (In press).

Lechner, M.S., & Laimins, L.A. (1994). Inhibition of p53 DNA binding by human papillomavirus E6 proteins. J. Virol. 68, 4262-4273.

Lee, H.W., Smith, L., Pettit, G.R., & Smith, J.B. (1997). Bryostatin 1 and phorbol ester downmodulate protein kinase C-α and –ε via the ubiquitin/proteasome pathway in human fibroblasts. Mol. Pharm. 51, 439-447.

Lee, H.W., Smith L., Petit, G.R., Vinitsky, A., & Smith, J.B. (1996). Ubiquitination of protein kinase C-α and degradation by the proteasome. J. Biol. Chem. 271, 20973-20976.

Lehner, P.J., & Cresswell, P. (1996). Processing and delivery of peptides presented by MHC class I molecules. Curr. Opin. Immunol. 8, 59-67.

Linnen, J.M., Bailey, C.P., & Weeks, D.L. (1993). Two related localized mRNAs from *Xenopus laevis* encode ubiquitin-like fusion proteins. Gene 128, 181-188.

Loeb, K.R., & Haas, A.L. (1994). Conjugates of ubiquitin cross-reactive protein distribute in a cytoskeletal pattern. Mol. Cell Biol. 14, 8408-8419.

Loetscher, H., Schlaeger, E.J., Lahm, H.W., Pan, Y.C., Lesslauer, W., & Brockhaus, M. (1990). Purification and partial amino acid sequence analysis of two distinct tumor necrosis factor receptors from HL60 cells. J. Biol. Chem. 265, 20131-20138.

Low, P., Doherty, F.J., Fellinger, E., Sass, M., Mayer, R.J., & Laszlo, L. (1995). Related organelles of the endosome-lysosome system contain a different repertoire of ubiquitinated proteins in Sf9 insect cells. FEBS Lett. 368, 125-131.

Lowe, J., McDermott, H., Loeb, K., Landon, M., Haas, A.L., & Mayer, R.J. (1995). Immunohistochemical localization of ubiquitin cross-reactive protein in human tissues. J. Pathol. 177, 163-169.

Madura, K., & Varshavsky, A. (1994). Degradation of G alpha by the N-end rule pathway. Science 265, 1454-1458.

Magnani, M., Serafini, G., Antonelli, A., Malatesta, M., & Gazzanelli, G. (1991). Evidence for a particulate location of ubiquitin conjugates and ubiquitin-conjugating enzymes in rabbit brain. J. Biol. Chem. 266, 21018-21024.

Magnani, M., Crinelli, R., Antonelli, A., Casabianca, A., & Serafini, G. (1994). The soluble but not mitochondrially bound hexokinase is a substrate for the ATP- and ubiquitin-dependent proteolytic system. Biochim. Biophys. Acta 1206, 180-190.

Masutani, C., Sugasawa, K., Yanagisawa, J., Sonoyama, T., Ui, M., Enomoto, T., Takio, K., Tanaka, K., van der Spek, P.J., & Bootsma, D. (1994). Purification and cloning of a nucleotide excision repair complex involving the xeroderma pigmentosum group C protein and a human homologue of yeast RAD23. EMBO J. 13, 1831-1843.

Matsui, S., Sandberg, A.A., Negoro, S., Seon, B.K., & Goldstein, G. (1982). Isopeptidase: a novel eukaryotic enzyme that cleaves isopeptide bonds. Proc. Natl. Acad. Sci. USA 79, 1535-1539.

Matunis, M.J., Coutavas, E., & Blobel, G. (1996). A novel ubiquitin-like modification modulates the partitioning of the Ran-GTPase-activating protein RanGAP1 between the cytosol and the nuclear pore complex. . J. Cell Biol. 135, 1457-1470.

Mayer, A.N., & Wilkinson, K.D. (1989). Detection, resolution, and nomenclature of multiple ubiquitin carboxyl-terminal esterases from bovine calf thymus. Biochemistry 28, 166-172.

McGee, T.P., Cheng, H.H., Kumagai, H., Omura, S., & Simoni, R.D. (1996). Degradation of 3-hydroxy-3-methylglutaryl-CoA reductase in endoplasmic reticulum membranes is accelerated as a result of increased susceptibility to proteolysis. J. Biol. Chem. 271, 25630-25638.

Meisner, H., & Czech, M.P. (1995). Coupling of the proto-oncogene product c-Cbl to the epidermal growth factor receptor. J. Biol. Chem. 270, 25332-25335.

Meyer, E.M., West, C.M., & Chau, V. (1986). Antibodies directed against ubiquitin inhibit high affinity [3H]choline uptake in rat cerebral cortical synaptosomes. J. Biol. Chem. 261, 14365-14368.

Meyer, E.M., West, C.M., Stevens, B.R., Chau, V., Nguyen, M.T., & Judkins, J.H. (1987). Ubiquitin-directed antibodies inhibit neuronal transporters in rat brain synaptosomes. J. Neurochem. 49, 1815-1819.

Michalek, M.T., Grant, E.P., Gramm, C., Goldberg, A.L., & Rock, K.L. (1993). A role for the ubiquitin-dependent proteolytic pathway in MHC class I-restricted antigen presentation. Nature 363, 552-554.

Michiels, L., Van der Rauwelaert, E., Van Hasselt, F., Kas, K., & Merregaert, J. (1993). Ubi cDNA encodes a ubiquitin-like-S30 fusion protein and is expressed as an antisense sequence in the Finkel-Biskis-Reilly murine sarcoma virus. Oncogene 8, 2537-2546.

Miller, H.I., Henzel, W.J., Ridgeway, J.B., Kuang, W., Chisholm, V., & Liu, C. (1989). Cloning and Expression of a Yeast Ubiquitin-protein Cleaving Activity in *Escherichia coli*. Biotechnology 7, 698-704.

Miyazawa, K., Toyama, K., Gotoh, A., Hendrie, P.C., Mantel, C., & Broxmeyer, H.E. (1994). Ligand-dependent polyubiquitination of c-kit gene product: a possible mechanism of receptor down modulation in M07e cells. Blood 83, 137-145.

Molinari, M., & Milner, J. (1995). p53 in complex with DNA is resistant to ubiquitin-dependent proteolysis in the presence of HPV-16 E6. Oncogene 10, 1849-1854.

Mori, S., Claesson-Welsh, L., Okuyama, Y., & Saito, Y. (1995). Ligand-induced polyubiquitination of receptor tyrosine kinases. Biochem. Biophys. Res. Commun. 213, 32-39.

Murnane, J.P. (1995). Cell cycle regulation in response to DNA damage in mammalian cells: a historical perspective. [Review]. Cancer Metastasis Rev. 14, 17-29.

Murray, A. (1995). Cyclin ubiquitination: the destructive end of mitosis. [Review]. Cell 81, 149-152.

Nakamura, M., Xavier, R.M., Tsunematsu, T., & Tanigawa, Y. (1995). Molecular cloning and characterization of a cDNA encoding monoclonal nonspecific suppressor factor. Proc. Natl. Acad. Sci. USA 92, 3463-3567.

Nakamura, M., Xavier, R.M., & Tanigawa, Y. (1996). Ubiquitin-like moiety of the monoclonal nonspecific suppressor factor beta is responsible for its activity. J. Immunol. 156, 532-538.

Narasimhan, J., Potter, J.L., & Haas, A.L. (1996). Conjugation of the 15-kDa interferon-induced ubiquitin homolog is distinct from that of ubiquitin. J. Biol. Chem. 271, 324-330.

Nefsky, B., & Beach, D. (1996). Pub1 acts as an E6-AP-like protein ubiquitin ligase in the degradation of the cdc25. EMBO J. 15, 1301-1312.

Nirmala, P.B., & Thampan, R.V. (1995). Ubiquitination of the rat uterine estrogen receptor: dependence on estradiol. Biochem. Biophys. Res. Commun. 213, 24-31.

Olvera, J., & Wool, I.G. (1993). The carboxyl extension of a ubiquitin-like protein is rat ribosomal protein S30. J. Biol. Chem. 268, 17967-17974.

Ozkaynak, E., Finley, D., Solomon, M.J., & Varshavsky, A. (1987). The yeast ubiquitin genes: a family of natural gene fusions. EMBO J. 6, 1429-1439.

Pagano, M., Tam, S.W., Theodoras, A.M., Beer-Romero, P., Del Sal, G., Chau, V., Yew, P.R., Draetta, G.F., & Rolfe, M. (1995). Role of the ubiquitin-proteasome pathway in regulating abundance of the cyclin-dependent kinase inhibitor p27 [see comments]. Science 269, 682-685.

Palombella, V.J., Rando, O.J., Goldberg, A.L., & Maniatis, T. (1994). The ubiquitin-proteasome pathway is required for processing the NF-kappa B1 precursor protein and the activation of NF-kappa B. Cell 78, 773-785.

Panchamoorthy, G., Fukazawa, T., Miyake, S., Soltoff, S., Reedquist, K., Druker, B., Shoelson, S., Cantley, L., & Band, H. (1996). p120cbl is a major substrate of tyrosine phosphorylation upon B cell antigen receptor stimulation and interacts in vivo with Fyn and Syk tyrosine kinases, Grb2 and Shc adaptors, and the p85 subunit of phosphatidylinositol 3-kinase. J. Biol. Chem. 271, 3187-3194.

Paolini, R., & Kinet, J.P. (1993). Cell surface control of the multiubiquitination and deubiquitination of high-affinity immunoglobulin E receptors. EMBO J. 12, 779-786.

Papa, F.R. & Hochstrasser, M. (1993). The yeast *DOA4* gene encodes a deubiquitinating enzyme related to a product of the human tre-2 oncogene. Nature 366, 313-319.

Parakh, K.A., & Kannan, K. (1993). Demonstration of a ubiquitin binding site on murine haemopoietic progenitor cells: implication of ubiquitin in homing and adhesion. Br. J. Haematol. 84, 212-218.

Qu, D., Teckman, J.H., Omura, S., & Perlmutter, D.H. (1996). Degradation of a mutant secretory protein, alpha1-antitrypsin Z, in the endoplasmic reticulum requires proteasome activity. J. Biol. Chem. 271, 22791-22795.

Reilly, L.M., & Guarino, L.A. (1996). The viral ubiquitin gene of Autographa californica nuclear polyhedrosis virus is not essential for viral replication. Virology 218, 243-247.

Rivett, A.J., Palmer, A., & Knecht, E. (1992). Electron microscopic localization of the multicatalytic proteinase complex in rat liver and in cultured cells. J. Histochem. Cytochem. 40, 1165-1172.

Rivett, A.J. (1993). Proteasomes: Multicatalytic proteinase complexes. Biochem. J. 291, 1-10.

Rock, K.L., Gramm, C., Rothstein, L., Clark, K., Stein, R., Dick, L., Hwang, D., & Goldberg, A.L. (1994). Inhibitors of the proteasome block degradation of most cell proteins and the generation of peptides presented on MHC Class I molecules. Cell 78, 761-771.

Rubin, D.M., & Finley, D. (1995). Proteolysis. The proteasome: a protein-degrading organelle? [Review]. Curr. Biol. 5, 854-858.

Scheffner, M., Werness, B.A., Huibregtse, J.M., & Levine, A.J. (1990). The E6 oncoprotein encoded by human papillomavirus types 16 and 18 promotes the degradation of p53. Cell 63, 1129-1136.

Scheffner, M., Huibregtse, J.M., Vierstra, R.D., & Howley, P.M. (1993). The HPV-16 E6 and E6-AP complex functions as a ubiquitin-protein ligase in the ubiquitination of p53. Cell 75, 495-505.

Scheffner, M., Huibregtse, J.M., & Howley, P.M. (1994). Identification of a human ubiquitin-conjugating enzyme that mediates the E6-AP-dependent ubiquitination of p53. Proc. Natl. Acad. Sci. USA 91, 8797-8801.

Scherer, D.C., Brockman, J.A., Chen, Z., Maniatis, T., & Ballard, D.W. (1995). Signal-induced degradation of I kappa B alpha requires site-specific ubiquitination. Proc. Natl. Acad. Sci. USA 92, 11259-11263.

Schwartz, A.L., Brandt, R.A., Geuze, H., & Ciechanover, A. (1992). Stress-induced alterations in autophagic pathway: relationship to ubiquitin system. Am. J. Physiol. 262, C1031-C1038.

Sepp-Lorenzino, L., Ma, Z., Lebwohl, D.E., Vinitsky, A., & Rosen, N. (1995). Herbimycin A induces the 20 S proteasome- and ubiquitin-dependent degradation of receptor tyrosine kinases. J. Biol. Chem. 270, 16580-16587.

Serdaroglu, P., Askanas, V., & Engel, W.K. (1992). Immunocytochemical localization of ubiquitin at human neuromuscular junctions. Neuropathol. Appl. Neurobiol. 18, 232-236.

Shen, Z., Pardington-Purtymun, P.E., Comeaux, J.C., Moyzis, R.K., & Chen, D.J. (1996). UBL1, a human ubiquitinlike protein associating with human RAD51/RAD52 proteins. Genomics 36, 271-279.

Siegelman, M., Bond, M.W., Gallatin, W.M., St.John, T., Smith, H.T., Fried, V.A., & Weissman, I.L. (1986). Cell surface molecule associated with lymphocyte homing is a ubiquitinated branched-chain glycoprotein. Science 231, 823-829.

Smith, J.B., Lee, H.-W., & Smith, L. (1996). Production of nonphosphorylated, ubiquinated protein kinase C (PKC) from autophosphorylated enzyme in human fibroblasts. FASEB J. 10, A1397.

Soltoff, S.P., & Cantley, L.C. (1996). p120cbl is a cytosolic adapter protein that associates with phosphoinositide 3-kinase in response to epidermal growth factor in PC12 and other cells. J. Biol. Chem. 271, 563-567.

Sommer, T., & Jentsch, S. (1993). A protein translocation defect linked to ubiquitin conjugation at the endoplasmic reticulum. Nature 365, 176-179.

Sorimachi, H., Saido, T.C., & Suzuki, K. (1994). New era of calpain research. Discovery of tissue-specific calpains. [Review]. FEBS Lett. 343, 1-5.

Spence, J., Sadis, S., Haas, A.L., & Finley, D. (1995). A ubiquitin mutant with specific defects in DNA repair and multiubiquitination. Mol. Cell Biol. 15, 1265-1273.

Spencer, S.A., Hammonds, R.G., Henzel, W.J., Rodriguez, H., Waters, M.J., & Wood, W.I. (1988). Rabbit liver growth hormone receptor and serum binding protein. Purification, characterization, and sequence. J. Biol. Chem. 263, 7862-7867.

Srivenugopal, K.S., Yuan, X.H., Friedman, H.S., & Ali-Osman, F. (1996). Ubiquitination-dependent proteolysis of O6-methylguanine-DNA methyltransferase in human and murine tumor cells following inactivation with O6-benzylguanine or 1,3-bis(2-chloroethyl)-1-nitrosourea. Biochemistry 35, 1328-1334.

Stancovski, I., Gonen, H., Orian, A., Schwartz, A.L., & Ciechanover, A. (1995). Degradation of the proto-oncogene product c-Fos by the ubiquitin proteolytic system in vivo and in vitro: identification and characterization of the conjugating enzymes. Mol. Cell Biol. 15, 7106-7116.

Staub, O., Dho, S., Correa, J., Ishikawa, T., McGlade, J., & Rotin, D. (1996). WW domains of Nedd4 bind to the proline-rich PY motifs in the epithelial Na+ channel deleted in Liddle's syndrome. EMBO J. 15, 2371-2381.

Stein, R.L., Chen, Z., & Melandri, F. (1995). Kinetic studies of isopeptidase T: modulation of peptidase activity by ubiquitin. Biochemistry 34, 12616-12623.

Sudakin, V., Ganoth, D., Dahan, A., Heller, H., Hershko, J., Luca, F.C., Ruderman, J.V., & Hershko, A. (1995). The cyclosome, a large complex containing cyclin-selective ubiquitin ligase activity, targets cyclins for destruction at the end of mitosis. Mol. Cell Biol. 6, 185-197.

Sudol, M., Bork, P., Einbond, A., Kastury, K., Druck, T., Negrini, M., Huebner, K., & Lehman, D. (1995). Characterization of the mammalian YAP (Yes-associated protein) gene and its role in defining a novel protein module, the WW domain. J. Biol. Chem. 270, 14733-14741.

Sung, P., Prakash, S., & Prakash, L. (1988). The RAD6 protein of *Saccharomyces cerevisiae* polyubiquitinates histones, and its acidic domain mediates this activity. Genes Dev. 2, 1476-1485.

Suzuki, K., Saido, T.C., & Hirai, S. (1992). Modulation of cellular signals by calpain. [Review]. Ann. N.Y. Acad. Sci. 674, 218-227.

Tanaka, K., Tamura, T., Yoshimura, T., & Ichihara, A. (1992). Proteasomes: protein and gene structures. New. Biol. 4, 173-187.

Tanaka, K. (1994). Role of proteasomes modified by interferon-gamma in antigen processing. [Review]. J. Leukoc. Biol. 56, 571-575.

Tautz, N., Meyers, G., & Thiel, H.J. (1993). Processing of poly-ubiquitin in the polyprotein of an RNA virus. Virology. 197, 74-85.

Tobias, J.W., & Varshavsky, A. (1991). Cloning and functional analysis of the ubiquitin-specific protease gene UBP1 of Saccharomyces cerevisiae. J. Biol. Chem. 266, 12021-12028.

Toniolo, D., Persico, M., & Alcalay, M. (1988). A "housekeeping: gene on the X chromosome encodes a protein similar to ubiquitin. Proc. Natl. Acad. Sci. USA 85, 851-855.

Tsurumi, C., Ishida, N., Tamura, T., Kakizuka, A., Nishida, E., Okumura, E., Kishimoto, T., Inagaki, M., Okazaki, K., Sagata, N., & et al, (1995). Degradation of c-Fos by the 26S proteasome is accelerated by c-Jun and multiple protein kinases. Mol. Cell Biol. 15, 5682-5687.

Vassilev, A.P., Rasmussen, H.H., Christensen, E.I., Nielsen, S., & Celis, J.E. (1995). The levels of ubiquitinated histone H2A are highly upregulated in transformed human cells: partial colocalization of uH2A clusters and PCNA/cyclin foci in a fraction of cells in S-phase. J. Cell Sci. 108, 1205-1215.

Vijay-Kumar, S., Bugg, C.E., Wilkinson, K.D., Vierstra, R.D., Hatfield, P.M., & Cook, W.J. (1987). Comparison of the three-dimensional structures of human, yeast, and oat ubiquitin. J. Biol. Chem. 262, 6396-6399.

Wang, Y., Yeung, Y.G., Langdon, W.Y., & Stanley, E.R. (1996). c-Cbl is transiently tyrosine-phosphorylated, ubiquitinated, and membrane-targeted following CSF-1 stimulation of macrophages. J. Biol. Chem. 271, 17-20.

Ward, C.L., Omura, S., & Kopito, R.R. (1995). Degradation of CFTR by the ubiquitin-proteasome pathway. Cell 83, 121-127.

Watkins, J.F., Sung, P., Prakash, L., & Prakash, S. (1993a). The *Saccharomyces cerevisiae* DNA repair gene RAD23 encodes a nuclear protein containing a ubiquitin-like domain required for biological function. Mol. Cell Biol. 13, 7757-7765.

Watkins, J.F., Sung, P., Prakash, S., & Prakash, L. (1993b). The extremely conserved amino terminus of RAD6 ubiquitin-conjugating enzyme is essential for amino-end rule-dependent protein degradation. Genes Dev. 7, 250-261.

Wiebel, F.F., & Kunau, W.H. (1992). The Pas2 protein essential for peroxisome biogenesis is related to ubiquitin-conjugating enzymes. Nature 359, 73-76.

Wilkinson, K.D. (1988). Purification and structural Properties of ubiquitin. In Ubiquitin. (Rechsteiner, M., ed.) pp. 5-38. Plenum Press, New York.

Wilkinson, K.D., Lee, K.M., Deshpande, S., Duerksen-Hughes, P.J., Boss, J.M., & Pohl, J. (1989). The neuron-specific protein PGP 9.5 is a ubiquitin carboxyl-terminal hydrolase. Science 246, 670-673.

Wilkinson, K.D., Deshpande, S., & Larsen, C.N. (1992). Comparisons of neuronal (PGP 9.5) and non-neuronal ubiquitin C-terminal hydrolases. Biochem. Soc. Trans. 20, 631-637.

Wilkinson, K.D. (1995). Roles of ubiquitinylation in proteolysis and cellular regulation. [Review]. Annu. Rev. Nutr. 15, 161-189.

Wilkinson, K.D., Tashayev, V.L., O'Connor, L.B., Larsen, C.N., Kasperek, E., & Pickart, C.M. (1995). Metabolism of the polyubiquitin degradation signal: structure, mechanism, and role of isopeptidase T. Biochemistry 34, 14535-14546.

Wilkinson, K.D., & Hochstrasser, M. (1997). The deubiquitinating enzymes. In Ubiquitin. (Finley, D.J., & Peters, J.-M., eds.). Plenum, New York. In press.

Yaglom, J., Linskens, M.H., Sadis, S., Rubin, D.M., Futcher, B., & Finley, D. (1995). p34Cdc28-mediated control of Cln3 cyclin degradation. Mol. Cell Biol. 15, 731-741.

Yarden, Y., Escobedo, J.A., Kuang, W.J., Yang-Feng, T.L., Daniel, T.O., Tremble, P.M., Chen, E.Y., Ando, M.E., Harkins, R.N., Francke, U., Fried, V.A., Ullrich, A., & Williams, L.T. (1986). Structure of the receptor for platelet-derived growth factor helps define a family of closely related growth factor receptors. Nature 323, 226-232.

York, I.A., & Rock, K.L. (1996). Antigen processing and presentation by the class I major histocompatibility complex. Annu. Rev. Immunol. 14, 369-396.

Zhaung, Z.P., & McCauley, R. (1989). Ubiquitin is involved in the in vitro insertion of monoamine oxidase B into mitochondrial outer membranes. J. Biol. Chem. 264, 14594-14596.

Zhuang, Z.P., Marks, B., & McCauley, R.B. (1992). The insertion of monoamine oxidase A into the outer membrane of rat liver mitochondria. J. Biol. Chem. 267, 591-596.

THE 20S PROTEASOME:
SUBUNITS AND FUNCTIONS

Keiji Tanaka and Nobuyuki Tanahashi

I. INTRODUCTION

Proteasomes are present in all eukaryotic cells examined to date. Moreover, proteasome-like complexes with similar structural properties to those of eukaryotes were recently found in some archaebacterial and eubacterial cells, such as *Ther-*

Advances in Molecular and Cell Biology
Volume 27, pages 105-128.

ISBN: 0-7623-0387-5

moplasma acidophilum and *Rodococcus*, respectively, although their exact subunit compositions differ considerably from each other (Lupas et al., 1994; Baumeister et al., 1998). This suggests that proteasomes are present in all living cells. Proteasomes have been implicated as major cytosolic proteinase complexes responsible for the nonlysosomal proteolytic pathway. They are present as both 20S and 26S forms in cells, although the latter form has as yet been identified only in eukaryotic cells (Coux et al., 1996). The 20S proteasome is a multicatalytic proteinase, which is the catalytic unit of the 26S proteasome, the eukaryotic ATP-dependent protease complex responsible for degrading a variety of cellular proteins that have acquired a specific degradation signal(s) such as a multiubiquitin chain (Coux et al., 1996; Hochstrasser, 1997). The 20S proteasome with a molecular mass of 700–750 kDa is composed of a set of small subunits with molecular masses of 21–32 kDa (Orlowski, 1990; Rivett, 1993; Tanaka, 1995). By electron microscopy, the 20S proteasome appears to be a cylindrical particle with a dimeric structure of two distinct rings. On the other hand, the 26S proteasome consists of a central 20S proteasome and two large V-like terminal modules, having possible regulatory roles, composed of multiple subunits of 25–112 kDa attached to the central part in opposite orientations. It appears to be a dumbbell-shaped particle with a molecular mass of approximately 2,000 kDa (for reviews, see Lupas et al., 1993; Peters, 1994; Dubiel et al., 1995; Coux et al., 1996; Baumeister et al., 1998). Recently, the involvements of proteasomes in various biological processes in cells have been determined by genetic and biochemical analyses (Jentsch and Schienker, 1995; Rubin and Finley, 1995). We have been examining the primary structures of eukaryotic 20S proteasomes and their pathophysiological functions. Here we focus attention on the structural and functional aspects of eukaryotic 20S proteasomes (hereafter referred to as the 20S proteasome or simply as the proteasome, unless otherwise mentioned).

II. THE PROTEASOME GENE FAMILY

The primary structures of many subunits of proteasomes from various eukaryotes have been deduced from the nucleotide sequences of their cDNAs or genes isolated by recombinant DNA techniques. These subunits from human and budding yeast are listed in Table 1. All the proteasomal genes examined so far encode previously unidentified proteins. The primary structures of these subunits show considerably high inter-subunit homology within species, and high evolutional conservation in various eukaryotes, suggesting that they may have originated from a common ancestral gene (Tanaka, 1995).

The archaebacterial 20S proteasome is a cylindrical particle consisting of four rings, each of which is comprised of seven identical α- or β -subunits: the rings are assembled in the order $\alpha_7\beta_7\beta_7\alpha_7$ (Lupas et al., 1994). The 20S proteasome of the eubacterion *Rodococcus* consists of four subunits, two α-type subunits, α1 and α2, subunits and two β-type, β1 and β2, judging from sequence similarities to the ar-

Table 1. Subunits of the 20S Proteasome

Human Gene	MW	pI	Chromosoml Locus	GDB Symbol	Yeast Gene	Essential
α-Type Subunits						
HC2*	29,555	6.16		PSMA1	PRE5	+
Pros30*	30,239	6.56	11q15.1	PSMA1		
HC3	25,898	7.29	6q27	PSMA2	Y7	+
HC8	28,433	5.06	14q23	PSMA3	PRS1	+
HC9	29,483	7.69		PSMA4	Y13	–
Zeta	26,425	4.59		PSMA5	PUP2	+
Pros27 (Iota)	27,374	5.97	14q13	PSMA6	PRS2	+
XAPC7-S*	27,900	8.69		PSMA7	PRE6	+
XAPC7-L*				PSMA7		
β-Type Subunits						
HC5	26,489	8.20	7p12-13	PSMB1	PRS3	+
HC7	22,836	6.61		PSMB2	PRE1	+
HC10	22,931	6.15		PSMB3	PUP3	+
HN3	29,192	5.63		PSMB4	PRE4	+
X (MB1)	22,897	8.67	14q11.2	PSMB5	PRE2	+
Y (Delta)	25,315	4.65	17p13	PSMB6	PRE3	+
Z	29,965	7.61	9q34.11-q34.12	PSMB7	PUP1	+
LMP7-E1*	29,769	5.46	6p21	PSMB8		–
LMP7-E2*	30,354	7.18	6p21	PSMB8		–
LMP2	23,245	4.75	6p21	PSMB9		–
MECL1	28,936	7.73	16q22.1	PMSB10		–

Notes: * These three pairs of subunits (HC2 and Pros30, XAPC7-S and XAPC7-L, LMP7-E1 and LMP7-E2) are almost identical. Therefore, their mRNAs may have arisen by alternative splicing or use of different transcription initiation sites. GDB, human Genome Database.

chaebacterial 20S proteasome (Tamura et al., 1995). However, its gross four-ringed, cylindrical structure is indistinguishable from that of the *Thermoplasma* proteasome, implying that the *Rodococcus* proteasome consists of a $(\alpha1)_7(\beta1)_7(\beta1)_7(\alpha1)_7$ and $(\alpha2)_7(\beta2)_7(\beta2)_7(\alpha2)_7$, or random mixtures of complexes of these four subunits (Tamura et al., 1995). Eukaryotic proteasome genes can also be divided into two distinct subgroups, α and β, with high similarities to archaebacterial α- and β-subunits, respectively (Tanaka, 1995; Coux et al., 1996). Seven subunits arranged in a ring are observed in negatively-stained electron micrographs of the 20S proteasome viewed from the top (Yoshimura et al., 1993). Immunoelectron microscopic studies showed that the 20S eukaryotic proteasome is a complex dimer and that the α-type subunits are located at the outer ends of a cylindrical particle consisting of four discs (Kopp et al., 1995). Phylogenetic analysis of the various eukaryotic proteasomal genes studied have shown shows that the α-type and β-type subunits can each be classified into seven subgroups independently (Scherrer and

A

α-Subunit Subfamily of the Human Proteasome

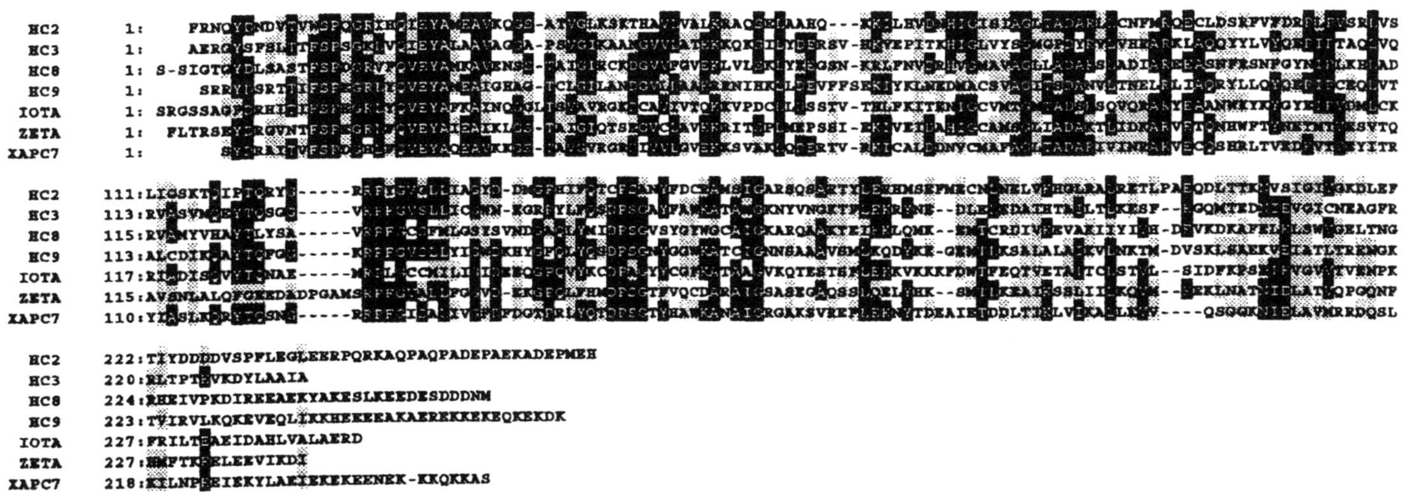

B

β-Subunit Subfamily of the Human Proteasome

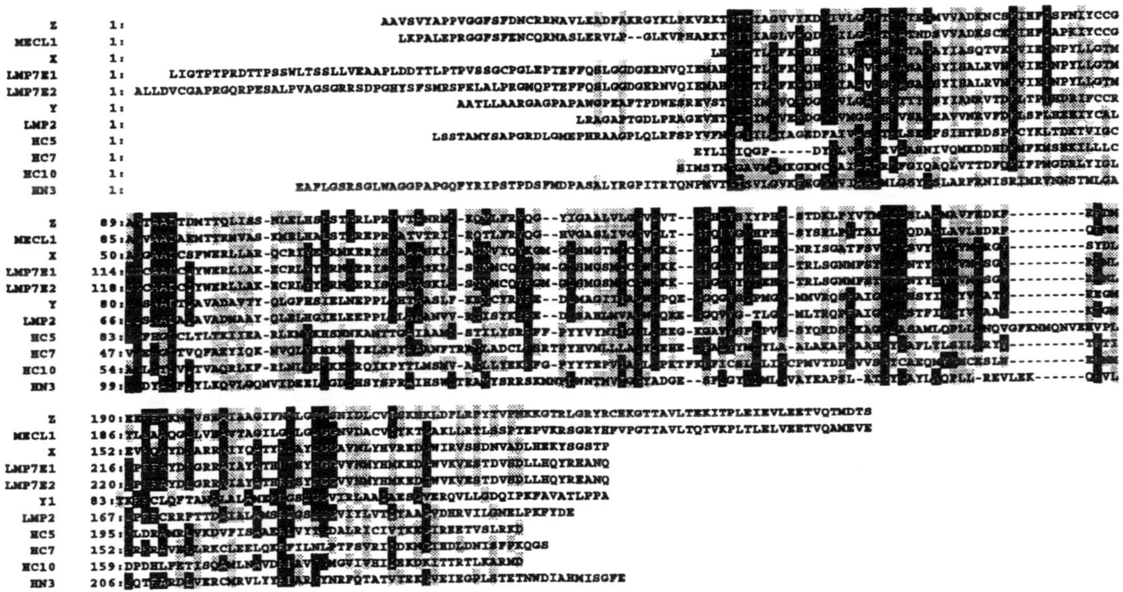

Figure. 1. Two types of subunits of the human 20S proteasome classified by their homologies with the α- and β-subunits of the 20S proteasome from *Thermoplasma acidophilum*. (**A**) Sequence alignment of the α-type subfamily. (**B**) Sequence alignment of the β-type subfamily. Gaps shown by dashes are inserted to achieve maximum sequence homology. Identical and conserved amino acid residues are shown by black and gray shading, respectively. Residue numbers are indicated.

Bey, 1994; Hilt and Wolf, 1995). Based on these findings, the basal molecular organization of eukaryotic proteasomes is concluded to be $\alpha_{n(1-7)}\beta_{n(1-7)}\beta_{n(1-7)}\alpha_{n(1-7)}$, where "n" indicates the number of the seven heterogeneous subunits (Tanaka, 1995). Recently, high-ordered structural analysis of yeast proteasomes by X-ray crystallography showed that 14 (7α- and 7β-) independent subunits were found to be assembled in a single complex (Groll et al., 1997).

Most, but not all, α-subunits contain a cluster of basic amino acid residues that presumably act as a nuclear localization signal (NLS), and a complementary NLS (cNLS) sequence consisting of an acidic amino acid cluster, which may antagonize the NLS-dependent nuclear import of proteasomes into the nucleus (Tanaka et al., 1990). Based on these findings we proposed a model for nuclear translocation of proteasomes assuming the existence of proteasomes in two conformations: one in which the NLS is exposed, and the other in which it is masked due to electrostatic interactions with the cNLS (Tanaka et al., 1990). We also proposed that phosphorylation may act as a switch between the two conformations, because several proteasomal α-subunits contain conserved potential phosphorylation sites. Consistent with this proposal, the nuclear, but not the cytosolic, proteasome subunit C3 was recently found to be phosphorylated specifically (Benedict et al., 1995), and the NLSs of human and *Thermoplasma* proteasomal α-subunits were found to be functional *in vitro* (Nederlof et al., 1995), suggesting that the α-subunits contribute to determination of the intracellular location of proteasomes. Indeed, immunocytochemical studies have revealed that proteasomes are colocalized in the cytoplasm and nucleus in a wide variety of cells. The nuclear location of proteasomes is greatly increased in rapidly-proliferating tumor cells and mature T lymphocytes activated by various mitogens, and this location in the nucleus changes drastically in a cell cycle-dependent fashion in mammalian and ascidian cells, implying a specific role of proteasomes in cell cycle progression (Ichihara and Tanaka, 1995; Coux et al., 1996).

On the other hand, some β-subunits are thought to be involved in catalytic functions, based on findings by yeast genetic and protein-chemical analyses of proteasomes (see below). Alignment of the primary structures of the seven α-subunits of human proteasomes showed that they have relatively high homologies except in their C-terminal parts, which show approximately 30% identities (Figure 1a). The identities of 10 human β-subunits appears to be approximately 15% (Figure 1b), with the amino acid sequences of their N- and C-terminal regions differing considerably. Interestingly, most β-type proteasomal subunits, but not α-type subunits, are synthesized as precursor proteins and processed to the mature forms during their assembly to form the β-ring complex by removal of their N-terminal prosequences perhaps autocatalytically (Chen and Hochstrasser, 1996; Seemüller et al., 1996), suggesting that proteasomes are zymogen-type proteases, like other intracellular proteases such as calpains and lysosomal catheptic proteases. No similarity has been found in the presumptive precursor regions of β-subunits, suggesting that evolutional conservation of mature regions is required for construction of their tertiary complex and/or functions.

The chromosomal locations of the multiple human proteasome genes examined so far have different loci (Table 1). Thus, as they are not present as a cluster of multiple genes, there must be some mechanism(s) controlling their coexpression. However, the structural basis of their coordinated expressions has not yet been determined by sequence analyses of the 5'-flanking regions and functional analyses of three human proteasome subunit genes examined so far (Tanaka, 1995; Coux et al., 1996).

It is of particular interest that the structural features of proteasomes resemble those of chaperonins, Hsp60 family chaperones, which are cylindrical particles with a central cavity (Weissman et al., 1995). Moreover, eubacterial and archaebacterial chaperonins, such as GroEL and thermophilic factor 55 (TF55), respectively, are composed of two subunits forming homo-polymer complexes of approximately 700 kDa, whereas the eukaryotic cytoplasmic chaperonin termed CCT (chaperonin containing TCP-1, t-complex polypeptide 1) is hetero-oligomeric, consisting of at least eight homologous subunit species sharing approximately 30% identity, which are highly conserved in eukaryotes from yeast to mammals (Kubota et al., 1995). Furthermore, the recently discovered chaperonin CCT is abundant in the cytosol of eukaryotes like the proteasome with which it constitutes: 1–2% of the total cell proteins. The size, shape, subunit organization, intracellular distribution, and molecular evolution of CCT is similar to those of the proteasome, implying that both play critical roles in the life cycles of all cells. However, the functions of these two particles are completely opposite: chaperonins function as molecular chaperones assisting in protein folding and promoting protein assembly, whereas proteasomes constitute a multisubunit machinery contributing to destruction of target proteins with aberrant structures.

III. IMMUNOPROTEASOMES INDUCED BY γ-INTERFERON

As listed in Table 1, seven human cDNAs for α-type subunits have been cloned, each showing a high homology with its counterpart of budding yeast, consistent with the assumed subunit organization of the α-ring. Similarly, seven genes for β-subunits have been cloned, in accord with the molecular structure of the β-ring, which is also formed from seven subunits. On the other hand, 10 human cDNAs of β-subunits have been found so far (see Table 1). The existence of 10 β-subunits seems curious considering that one β-type ring is composed of seven subunits in vertebrates.

Recently, three additional human proteasomal β-genes have been implicated to be involved in immunoregulation, based on the findings that two new proteasomal subunits designated as low molecular mass polypeptide (LMP2 and LMP7) are encoded in the major histocompatibilty complex (MHC) class II region linked to two TAP1/TAP2 (transporter associated with antigen processing). LMP2 and LMP7 were found to be two polymorphic proteins of a large complex which was later demonstrated to be identical to the proteasomal complex (for details, see reviews by Heemels and Ploegh, 1995; Monaco and Nandi, 1995; York and Rock, 1996). These two LMP proteins were found to belong to the β-type subunit subfamily (see, Figure 1b), and their expressions were greatly induced by a major immunomodulatory cytokine, γ-interferon (γ-IFN), like the expressions of many MHC and TAP genes, implying their involvement in the immune response. Subsequently, many investigators including us noticed that γ-IFN changes the subunit compositions of human and mouse proteasomes (Yang et al., 1992; Aki et al., 1993; Brown et al., 1993; Boes et al., 1994; Früh et al., 1994). Figure 2 shows the subunit patterns of human proteasomes with or without γ-IFN treatment, analyzed by two dimensional polyacrylamide gel electrophoresis (2D-PAGE) (Hisamatsu et al., 1996). γ-IFN

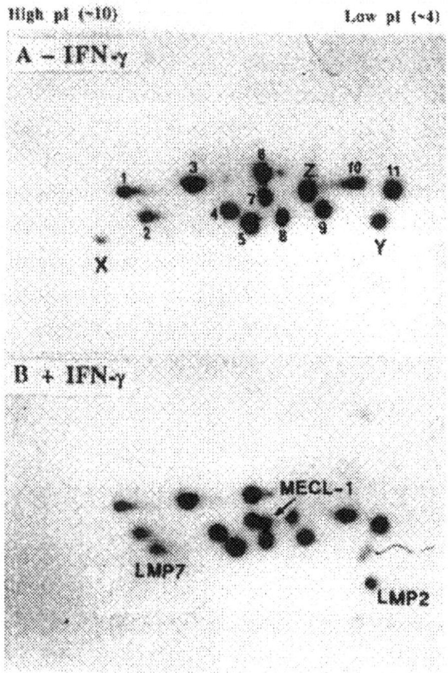

Figure 2. Effect of γ-IFN on the subunit pattern of newly synthesized proteasomes in HeLa cells. Cells were labeled for three days with ^{3}H-leucine in the absence (panel-**A**) or presence (panel-**B**) of γ-IFN. Then cell extracts were treated with antiproteasomal antibodies, and the resulting immunoprecipitates were examined by 2D-PAGE and fluorography (Hisamatsu et al., 1996). Note that the spots of proteasome subunits LMP2, LMP7, and MECL-1 and the three spots of X, Y, and Z were up- and downregulated, respectively, by γ-IFN treatment. Other proteasomal subunits were identified by protein sequencing and immunoblot analysis, and are numbered as follows: 1, XAPC7; 2, C5; 3, C9; 4, C3; 5, C7; 6, C2; 7, Iota; 8, C10; 9, N3; 10, C8, and 11, Zeta.

strongly induced not only the expressions of LMP2 and LMP7, but also that of MECL-1 (equivalent to LMP10; Nandi et al., 1996), which was cloned as a unique gene within a tight cluster of five unrelated human genes on chromosome 16q22.1 (Larsen et al., 1993). However, the degrees of inducibility and/or the basal expression levels differed, depending on the type of cells and tissues examined (Akiyama et al., 1994; Groettrup et al., 1996; Hisamatsu et al., 1996).

It is curious that only three subunits were greatly increased in a γ-IFN-dependent fashion. We found that three subunits designated as X, Y, and Z were downregulated by γ-IFN (Figure 2), although the expressions of almost all other subunits remained unchanged (Akiyama et al., 1994; Hisamatsu et al., 1996). Similar results were obtained by Nandi et al. (1996) and Groettrup et al. (1996). Surprisingly, molecular cloning of the cDNAs encoding X (identical to MB1, Belich et al., 1994, or LMP17), Y (equivalent to delta, LMP9), and Z (equivalent to LMP19, Nandi et al., 1996, or MC14, Groettrup et al., 1996) showed that X, Y, and Z are novel proteasomal subunits with high homologies to LMP7, LMP2, and MECL-1, respectively (Figure 3). Indeed, the three γ-IFN dependent, downregulated subunits, X, Y, and Z, are remarkably similar to their respective γ-IFN-responsive subunits LMP7 (68% identity), LMP2 (57% identity), and MECL1 (63% identity, excluding the N-terminal region). The high identities of these three pairs of subunits differ markedly from those of all other β-type subfamily subunits, which show approximately 15% identity (Figure 1b). The γ-IFN-dependent reciprocal expressions of these three highly homologous pairs of subunits suggest that γ-IFN may induce subunit replacements of X, Y, and Z by LMP7, LMP2, and MECL-1, respectively, as shown in the model in Figure 4. Moreover, γ-IFN-inducible proteasomal genes may have appeared by gene duplication from their constitutive counterparts in association with the *MHC* and *TAP* genes during acquirement of the MHC class I-restricted immune system. Recently, Kandil et al. (1996) reported that cartilaginous fish, the most primitive class of vertebrates in which the *MHC* has been identified, have both *LMP7* and *X* genes. In contrast, only cDNA clones with residues characteristics of *X* were identified in hagfishes and lampreys, two jawless fish in which no *MHC* has been identified. Thus, the *LMP7* gene may have emerged after the appearance of jawless fish. Accordingly, some β-type subunits appear to have been replaced by very homologous, but different, gene products, possibly generated by gene duplication, in response to extracellular signals, such as γ-IFN, during vertebrate evolution.

The LMP2 and LMP7 genes are located at p21 on human chromosome 6 (Monaco and Nandi, 1995), differing in location from the MECL-1 gene at 16q22.1 (Larsen et al., 1993). Moreover, X (MB1), Y (delta), and Z have distinct loci at 14q11.2, 17p13, and 9q34.11-q34.12, respectively (Belich et al., 1994; Hisamatsu et al., 1996). The location of subunit Z gene on chromosome 9 is noteworthy, because inspection of human loci adjacement to the Z locus provides evidence that the paracentromeric region of chromosome 9 and the MHC region on chromosome 6 probably arose by duplication at an early stage of vertebrate evolution (Kasahara et al., 1996). Traces of this duplication are also evident in homologous mouse chro-

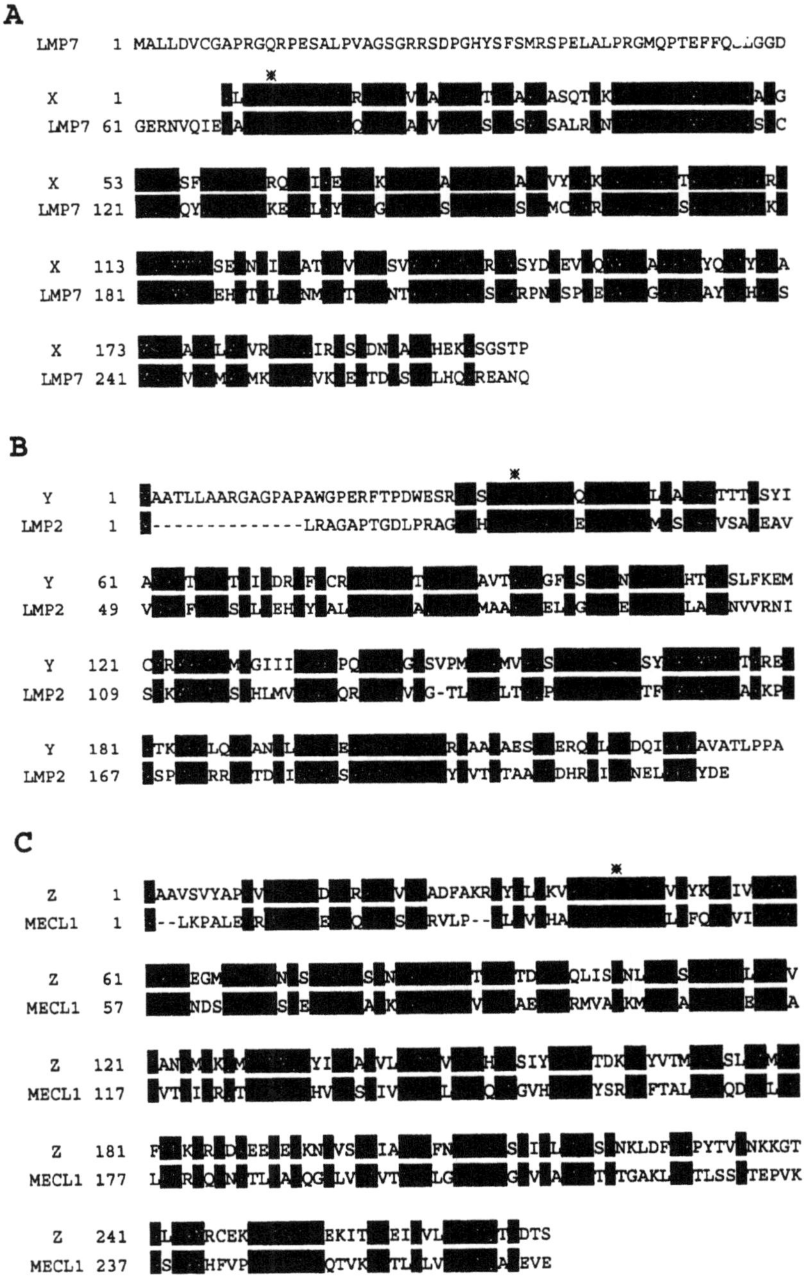

Figure 3. (**A**, **B**, and **C**) Comparison of the amino acid sequences of X , Y, and Z with those of LMP7, LMP2, and MECL-1, respectively, from human proteasomes. Identical amino acids are boxed in black. The asterisk indicates a putative processing site (see text).

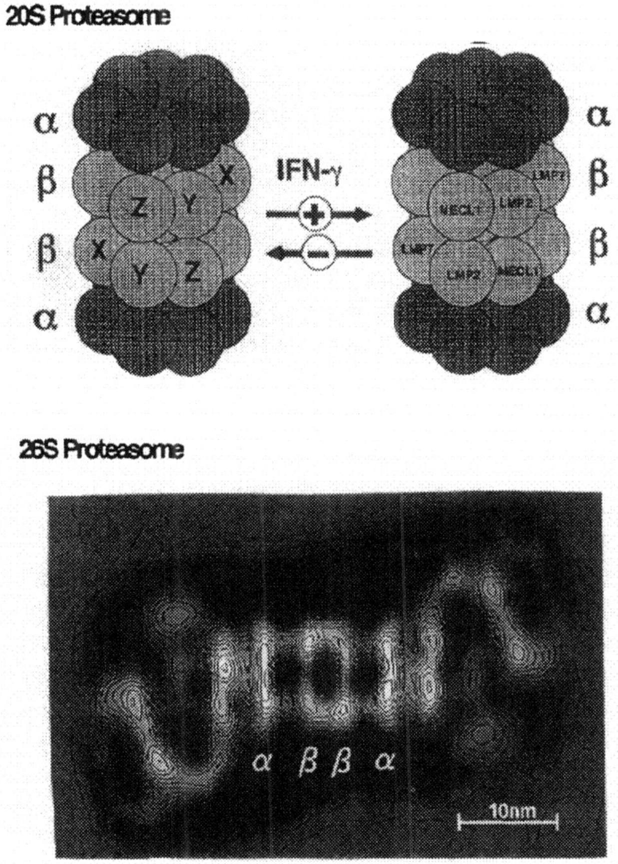

Figure 4. Hypothetical models of the housekeeping-type 20S X/Y/Z-type proteasome and the γ-IFN-induced LMP7/LMP2/MECL-1-type immunoproteasome (**Top** panel). α and β, α- and β-subunit rings, respectively, of the 20S proteasome. The locations of the γ-IFN regulated subunits are only tentative, because their exact topology is still unknown. **Bottom** panel: Structure of the 26S proteasome analyzed by electron microscopy (Yoshimura et al., 1993).

mosomal regions (chromosome 2 and 17). These findings have implications in understanding the genomic organization of present-day MHC and provide insights into the origin of MHC.

The protein levels of LMP2, LMP7, and MECL-1 increase in response to γ-IFN, which are only roughly correlated with the amounts of their mRNAs (Hisamatsu et al., 1996). However, the mRNA levels of X, Y, and Z are not greatly affected by γ-IFN, suggesting that complete loss, or marked decrease in the expressions of subunits X, Y, and Z induced by γ-IFN is due to their posttranslational degradations. The amounts of these three proteins were found to decrease gradually 2–3 days after stimulation with γ-IFN (Hisamatsu et al., 1996). Moreover, Aki et al. (1993) showed that γ-IFN has no effect on the subunit pattern of pre-existing proteasomes. Although the induction of proteasomes containing LMP7, LMP2, and MECL-1 by γ-IFN is rapid, the overall change from pre-existing X/Y/Z-type proteasomes is probably slow, because of the large size and slow turnover of the pool of constitutive proteasomes (Rivett, 1993). Therefore, the most likely explanations for the mechanism of subunit substitution are preferential incorporation of γ-IFN-inducible LMP2, 7 and MECL-1 subunits and rapid degradation of unassembled Y, X, and Z subunits (Tanaka et al., 1997). Recent studies have provided insights into the molecular mechanisms underlying the assembly of the homogenous immunoproteasomes containing all three IFN-γ-inducible subunits (Griffin et al., 1998).

In an attempt to clarify the role of LMPs in antigen presentation, mice with disrupted *LMP2* or *LMP7* genes were generated by homologous recombination. Mice lacking *LMP7* showed reduced surface levels of MHC-class I molecules and impaired presentation of viral antigens, establishing the role of LMP7 for supplying peptides necessary for MHC class I assembly (Fehling et al., 1994). Deletion of the *LMP2* gene did not result in any profound phenotypic change, such as that in cell surface class I expression, unlike disruption of the *LMP7* gene, but macrophages and spleen cells of *LMP2*-deficient mice exhibited a clear defect in presentation of viral antigens, such as influenza nucleoprotein (Van Kaer et al., 1994). These findings clearly indicate that γ-IFN-induced modification of the proteasome composition increases the efficiency of immune surveillance. Functional analysis of mice carrying a deletion of the third γ-IFN-inducible subunit, MECL-1, is currently underway.

Studies on the effect of γ-IFN are of particular interest for revealing the exact role of the proteasome in the immune system, because γ-IFN is a major immunomodulatory cytokine. Results have shown that γ-IFN alters the proteolytic specificity of proteasomes, increasing their activities for cleavage of peptide bonds on the carboxyl side of basic and hydrophobic amino acid residues of fluorogenic substrates, but decreasing their activities for peptides containing acidic amino acid residues, without affecting their activity for ATP-, ubiquitin-dependent proteolysis. However, conflicting results showing a reduction in chymotryptic activity by γ-IFN have been reported (Goldberg et al., 1996). These modifications of peptidase activity suggest that, during protein breakdown, the proteasome from γ-IFN-treated cells should generate more peptides with hydrophobic or basic carboxyl termini and fewer peptides with acidic carboxyl termini. As a hydrophobic or basic carboxy-terminal residue is required for strong binding of a peptide to MHC-molecules (Rammensee et al., 1995), this change in cleavage specificity would favor the peptides with carboxyl termini most commonly found bound to class I molecules.

An N-terminal threonine residue of the β-subunit is essential for the proteolytic function of the proteasome, so the proteasome has been defined as a threonine protease with a different catalytic mechanism from those known for other proteases

(Fenteany et al., 1995; Löwe et al., 1995; Seemüller et al., 1995). Thus of particular interest is the finding that the γ-IFN-induced subunits X, Y, and Z and the counterparts they replace, LMP7, LMP2, and MECL-1, respectively, also have two N-terminal threonine residues (see Figure 3, asterisks), suggesting that all three pairs of homologous subunits are catalytically active or form a part of the catalytic center. All the active subunits with two N-terminal threonine residues, which have been conserved during evolution, seem to be regulated by γ-IFN, perhaps accounting for changes in the proteolytic functions of the proteasome. We have proposed that proteasomes containing these γ-IFN-inducible subunits should be named immunoproteasomes to emphasize their specialized functions in antigen presentation (Tanaka et al., 1997). Immunoproteasomes may be better suited for immunological processing of endogenous antigens. The altered molecular organization of the proteasome induced by γ-IFN may be responsible for acquisition of its functional changes in a variety of heterogeneous complexes.

IV. BIOLOGICAL FUNCTIONS OF PROTEASOMES

The proteasome was initially identified as a multicatalytic proteinase with diversed peptidase activities and a weakly alkaline optimum pH (for reviews, see Orlowski, 1990; Rivett, 1993). Later, it was found to play a major role in the non-lysosomal pathway of ATP-dependent breakdown not only of naturally occurring unstable regulatory proteins but also of proteins with aberrant structures that may be generated in cells by mutations or various environmental stresses, such as fever, oxidation, and exposure to heavy metals (Coux et al., 1996). In addition to these selective proteolyses, the proteasome appears to be involved in the unregulated destruction of constitutively long-lived proteins (Rock et al., 1994). The selective degradations of most of these proteins are mediated by the ubiquitin pathway, which is a major system for intracellular protein degradation (Coux et al., 1996; Hochstrasser, 1997 see Wilkinson, chapter 3). Covalent attachment of ubiquitin to target proteins is catalyzed by a multi-enzymatic system, consisting of E1 (activating), E2 (conjugating), and E3 (ligating) enzymes (Herschko and Ciechanover, 1992). Actually, the proteasome has been demonstrated to be responsible for the degradations of a variety of functional proteins with a multiubiquitin chain as a degradation signal (Ciechanover, 1994; Jentsch and Schienker, 1995; Rubin and Finley, 1995; Hochstrasser, 1997). Moreover, the proteasome has been shown also to degrade non-ubiquitinated, ornithine decarboxylase (ODC) *in vitro* (Murakami et al., 1992) and *in vivo* (Murakami et al., 1996), when associated with its specific inhibitor protein, named antizyme, which may expose the degradation signal of the ODC moiety. Therefore, the proteasome may be able to attack diverse target proteins by recognizing various potential degradation signals.

In general, it seems difficult to determine the biological roles of intracellular proteases. Recently, yeast proteasomal mutants and membrane permeable inhibitors have been used to determine the *in vivo* functions of proteasomes. Wolf and his colleagues have separated budding yeast mutants lacking the chymotrypsin-like and peptidyl-glutamyl peptide degrading activities of proteasomes (for review, see Hilt and Wolf, 1995 and Thumm and Wolf, chapter 2). These mutants have contributed greatly to our understanding of the involvement of proteasomes in the degradations of many unstable key enzymes and N-end rule substrates. Genetic analyses thus seem useful for determining proteasomal functions *in vivo*. However their application to multicellular organisms appear difficult, although the knock-out system in mice is available.

To ascertain the role of proteasomes in living cells, various peptide aldehyde inhibitors have recently been devised, because they are likely to across plasma membranes. N-Acetyl-Leu-Leu-Norleucinal (Ac-LLnL-H, calpain inhibitor I) was initially proposed as a specific inhibitor of calpains, but later found to inhibit the proteasome (Rock et al., 1994). The following peptidyl aldehydes that are capable of inhibiting proteasomal activities more strongly have also been synthesized: N-carbobenzoxyl-Gly-Pro-Phe-leucinal (Z-GPFL-H; Vinitsky et al., 1994), Z-Ile-Glu(O-t-Bu)-Ala-Leu-H (PSI; Traenckner et al., 1994), Z-LLnV-H (MG115; Rock et al., 1994), and Z-LLL-H (MG132; Coux et al., 1996). Indeed, treatments with these inhibitors resulted in abnormal accumulation of ubiquitinated cellular proteins of very large sizes, implying that they induce large defects in the ubiquitin-proteasome pathway. However, caution must be exercised in their use for inferring proteasome functions, because they inhibit not only proteasomes but also cysteine proteases such as calpains and lysosomal cathepsins. Therefore, a negative control using Ac-LLM-H (calpain inhibitor II, Figueiredo-Pereira et al., 1994), Z-LnL-H (calpeptin, Tsujinaka et al., 1989), Z-LL-H (Tsubuki et al., 1996), or E-64 (cysteine protease inhibitor), which do not affect proteasomal activity, is required in determining the involvement of the proteasome. In contrast to these peptidyl aldehyde compounds, a new microbial metabolite, lactacystin, was found to be a selective proteasome inhibitor that does not affect other proteases examined so far (see next section). However, it should be noted that all these reagents strongly inhibit chymotrypsin-like and trypsin-like activities of 20S and 26S proteasomes, while not affecting the peptidyl glutamyl peptide degrading activity appreciably. Moreover, they also only partially inhibit ATP-dependent degradation of ubiquitinated lysozyme (Yokosawa, personal communication) and ODC in the presence of antizyme (Murakami et al., 1996) by the purified 26S proteasome. However, they are very effective in living cells, inducing almost complete loss of proteasome-mediated proteolysis. The reason why their effects *in vitro* and *in vivo* apparently differ is unknown.

There is accumulating evidence that the proteasome is involved in the destruction of bulk proteins with rapid and slow turnover rates in a wide variety of biological processes such as cell cycle progression, apoptosis, metabolic regulation, immune responses, protein quality control, signal transduction, and stress responses. Recent results indicating proteolytic processes mediated by the proteasome are discussed below.

A. Cell Cycle Control

The first indication of the involvement of the proteasome in cell proliferation was obtained by gene disruption analysis of the budding yeast. Disruption of a chromosomal gene encoding one proteasomal subunit by homologous recombination had a lethal effect on haploid yeast, indicating that the proteasome is essential for proliferation of yeast (Fujiwara et al., 1990). Subsequently, deletion of each subunit individually—except subunit Y13—was found to have a similar lethal effect, as listed in Table 1 (Hilt and Wolf, 1995; Tanaka, 1995). These findings suggest that each subunit may play a distinct and indispensable role in cell growth, or more probably, that all these subunits are required for the proper functional assembly of the proteasome complex. Other classes of yeast mutants have been separated to examine mutants for abnormalities in cell cycle progression. These mutants have mainly been mutants of regulatory subunits of the 26S proteasome (see chapter 5), such as *mts2*$^{+}$ (Gordon et al., 1993) and *CIM5/CIM3* (Ghislain et al., 1993) encoding ATPase subunits. These mutations resulted in marked stabilization of cyclin B, with arrest in the M phase of the cell cycle. These mutant cells have contributed greatly to our knowledge of the *in vivo* functions of the 26S proteasome (for review, see Hilt and Wolf, 1995).

Consistent with these genetic findings evidence that the proteasome is a main regulator of the cell cycle in eukaryotic cells has been obtained by biochemical analyses. Ac-LLnL-H inhibited cyclin B degradation and induced cell cycle arrest (Sherwood et al., 1993). Lactacystin also arrested cell cycle progression in the G1 and G2 phases in osteosarcoma cells (Fenteany et al., 1994) and Neuro 2a cells (Katagiri et al., 1995). The latter findings are apparently consistent with our report that Nin1p, a non-ATPase regulatory subunit of the 26S proteasome of yeast, is required for activation of Cdc28p kinase at the G1/S and G2/M boundaries (Kominami et al., 1995). Presumably, the ubiquitin-proteasome pathway plays a pivotal role in cell cycle regulation by causing destruction of cyclin dependent protein kinase (CDK) inhibitors such as p40^{sic1} (Schwob et al., 1994) and p27 (Pagano et al., 1995) at the time of G1/S transition, and by degrading cyclin B (Murray, 1995) and cut2 (Funabiki et al., 1996) required for M-phase progression (for reviews, see King et al., 1996; Hershko, 1997). Thus there is accumulating evidence that the ubiquitin-proteasome system has key roles in progression at various stages of the cell cycle. The main reason for this may be that proteolysis can catalyze rapid and irreversible biological reactions, and that such definite regulations appear to be essential for cell cycle traverse.

B. Neuritogenesis and Apoptosis Induced by Z-LLL-H and Lactacystin

Kawashima and his collaborators first found that like nerve growth factor (NGF), some peptidyl aldehyde derivatives have strong activity for initiating neurite outgrowth from PC12 cells (Saito et al., 1990). Of these derivatives, Ac-

LLnL-H and Z-LLL-H (which was later found to be identical to MG132) were the most effective. Since Ac-LLnL-H was found to be calpain inhibitor I, calpain was initially thought to be responsible for neuronal differentiation. To identify the protease sensitive to the stronger inducer Z-LLL-H, the enzymatic degradation of Z-LLL-MCA was examined. Surprisingly the enzyme purified was found to be the proteasome, not calpain (Tsubuki et al., 1993). Very recently, treatment with Z-LLL-H was also found not only to block cell cycle progression but also to induce apoptotic cell death (Shinohara et al., 1996). The appreciable accumulation of p53 caused by Z-LLL-H is speculated to induce apoptosis, since p53 is known to have a potential role in triggering programmed cell death.

On the other hand, Õmura and his group found a new microbial metabolite, named lactacystin, which induced neuritogenesis of neuroblastoma Neuro 2a cells (Õmura et al., 1991). Lactacystin also resulted in cell cycle progression (Feneteany et al., 1994). However, the mechanism by which lactacystin exerts these biological effects was determined only recently. Schreiber and his colleagues tried to find cellular molecules capable of binding with ^{3}H-lactacystin (Fenteany et al., 1995). They found that ^{3}H-lactacystin bound covalently to two proteasomal β-subunits, mainly to subunit X and to a lesser extent to another subunit that was later identified as subunit Z (Hisamatsu et al., 1996). Thus lactacystin modifies the N-terminal threonine residue of subunit X, and possibly also that of Z, specifically and covalently, resulting in almost complete inhibition of the proteolytic activity (Fenteany et al., 1995). These findings are consistent with the proposal that the proteasome is a novel type of threonine protease (Seemüller et al., 1995). Interestingly, lactacystin, like tumor necrosis factor-α (TNF-α), was found to induce apoptosis of malignant hematopoietic cells, (Imajoh-Ohmi et al., 1995). It is noteworthy that these two potential proteasomal inhibitors, Z-LLL-H and lactacystin, have been found in similar systems, and exert various similar cellular effects including inductions of neural cell differentiation, cell cycle arrest, and apoptosis. On the other hand, recently the proteasome has been shown to play an essential role in thymocyte apoptosis (Grimm et al., 1996) and in the programmed cell death of NGF-derived sympathetic neurons (Sadoul et al., 1996), because proteasome inhibitors resulted in almost complete block of these processes (see Mason and Beyotte chapter 9). It is unknown why the same inhibitors caused such apparent discrepancy, but these effects may depend on states of the cells used. Further study is required to clarify this issue.

C. Ubiquitination of Integral Membrane Proteins

The ubiquitin-proteasome pathway for proteolysis (see chapter 3) is believed mainly to play physiological roles in the cytoplasm and nucleus. But recently this proteolysis machinery has also been shown to operate on various membrane proteins. The first report of ubiquitination of a membrane protein was that of the lymphocyte homing receptor, and thereafter various other receptors, such as those of platelet-derived growth factor (PDGF), epidermal growth factor (EGF), and growth hormone, were found to be ubiquitinated in a ligand-dependent fashion, although the role of their ubiquitinations is unknown. Recently, Mori et al. (1995) reported that degradation of the PDGFβ-receptor was considerably suppressed by addition of proteasome inhibitors, such as lactacystin and MG132, implying that ubiquitination of the PDGFβ-receptor acts as a targeting signal for its proteolysis mediated by the proteasome. In addition, treatment with herbimycin A, an inhibitor of tyrosine kinase, induced ubiquitin-dependent proteolysis of EGF- and insulin-receptors, and these proteolyses were also inhibited by peptidyl aldehyde proteasome inhibitors, suggesting involvement of the proteasome in ligand-dependent EGF-receptor proteolysis (Sepp-Lorenzino et al., 1995). Thus, ligand-dependent ubiquitination appears to play critical role in a signal termination by proteolysis.

Moreover, the Cl^- channel protein, cystic fibrosis transmembrane conductance regulator (CFTR) (Jensen et al., 1995; Ward et al., 1995), an unstable yeast protein translocator, Sec61p, and its associated protein Sss1p (Biederer et al., 1996), which are located in the endoplasmic reticulum (ER), are degraded by the proteasome-ubiquitin pathway, because their degradations are sensitive to proteasomal inhibitors, but not to lysosomotropic agents (see chapter 6). These observations suggest that the cytoplasmic domain of membrane-spanning proteins in both plasma and ER membranes is a site of ubiquitination, resulting in the formation of targeting signals for the proteasome. Surprisingly, newly synthesized both membrane and soluble proteins that fail to fold or assemble properly in the ER were shown to be exported into the cytosol through the Sec61 translocon machinery and then degraded by the proteasome, suggesting involvement of the proteasome in the "quality control" of the ER (for reviews, see Lord, 1996; Bonifacino, 1996, and chapter 6).

Until recently, the lysosomal pathway has been thought to be the major route responsible for degradation of membrane proteins, but now it appears that some membranous proteins are degraded by the proteasome, as described above. However, it is unknown whether all ubiquitinated membrane proteins are degraded by the proteasome, and recently, the accumulation of considerable amounts of ubiquitin in lysosomes has been observed in various patho-physiological conditions (Ciechanover, 1994). Consistent with these findings, results showed that the unstable ABC transporter proteins Ste6 (Kolling and Hollenberg, 1994) and Pdr5 (Egner and Kuchler, 1996) in the budding yeast were ubiquitinated, and Ste6 was greatly stabilized in mutant *end4* cells carrying an endocytosis defect and a *pep4* mutant lacking lysosomal protease A. However, the instabilities of these proteins and also of uracil permease (Galan et al., 1996) and Ste2, a G-protein linked receptor protein (Hicke and Riezman, 1996) are unaffected in the proteasomal mutant cells, indicating that all these membrane proteins are primarily degraded by lysosomes. These findings suggest that ubiquitination of some membrane proteins serves as a sorting signal for their entrapping in endosomes and/or lysosomes in a ligand-stimulated manner.

There are many questions in relation with these apparently conflicting reports. Why are some ubiquitinated proteins degraded by the proteasome and others in the vacuolar system? The type of ubiquitination may play a role in this distinction. To ascertain this difference, the conjugating E2 and ligating E3 enzymes responsible for ubiquitination of the respective proteins should be identified. Recently, one proteasomal regulatory subunit (see chapter 5), MBP1, was identified as a multiubiquitin receptor (van Nocker et al., 1996), providing a molecular basis for elucidating how the proteasome can recognize ubiquitinated target proteins for their selective breakdown. Actually, addition of excess *Arabidopsis* MBP1 protein resulted in competitive suppression of cyclin B degradation in a *Xenopus* egg extract (Deveraux et al., 1995), indicating that it acts as a functional multiubiquitin receptor. However, it is not known whether the same MBP1 is responsible for recognition of ubiquitinated membrane proteins. Therefore, in order to understand the significance of ubiquitination in lysosomal proteolysis (see chapter 7) it is essential to examine whether the putative ubiquitin receptor is actually located in endosomal and/or lysosomal membranes. So far, lysosomes have been thought to mediate only nonselective proteolysis through fusion with autophagic vacuoles. But if there are ubiquitin receptors in their membranes, they may also be involved in selective proteolysis entrapping ubiquitinated proteins from not only membrane proteins but also possibly the cytosol.

D. Limited Proteolysis by the Proteasome

The ubiquitin-proteasome system has been shown to be responsible for not only catalysis of exhaustive degradation of target proteins but also generation of antigenic peptides for presentation of the MHC class I-restricted antigen, implying a distinct role of the proteasome for processing as well as for protein-degradation. But the former function might be an obligatory role for use of the products by the immune system (Tanaka et al., 1997). Moreover, the proteasome was found to have a new function as a real processing enzyme, namely activation of transcriptional factor NF-κB through conversion of the p105 precursor to the mature p50 mature protein (Palombella et al., 1994). Transcriptional factor NF-κB is synthesized as a p105 precursor, which is processed by degrading its C-terminal domain containing ankyrin repeats, and the resulting p50 N-terminal domain is associated with a partner p65 protein to function as the active transcriptional factor NF-κB. Interestingly, the p105 precursor is multiubiquitinated before the processing reaction, which was inhibited by proteasomal inhibitors. It is still unknown how the proteasome cleaves the C-terminal half specifically (Pahl and Baeuerle, 1996).

NF-κB is one of the most important transcriptional factors, having various physiological roles in inflammation and immune responses. It is unique, in that almost all other transcriptional factors are nuclear proteins that are transported into the nucleus immediately after their synthesis, whereas, under unstimulated conditions, p50/p65 is present in the cytoplasm as an inactive complex in association with another specific inhibitor protein, I-κB, which masks the nuclear targeting signal of p50/p65 molecules. Recent work demonstrated that a variety of signals such as

TNF-α induced phosphorylation, ubiquitination, and subsequent degradation of I-κB, which could promote the transfer of NF-κB into the nucleus, suggesting that proteolysis is a key regulatory process for signal transduction (Chen et al., 1996). Selective proteasome inhibitors such as PSI (Traenckner et al., 1994), MG132, and lactacystin almost completely inhibited rapid signal-dependent I-κB degradation (Palombella et al., 1994), and induced accumulation of multi-ubiquitinated I-κB, implying that the proteasome-ubiquitin pathway is involved in various processes mediated by the NF-κB system (Finco and Baldwin, 1995).

Notch is also an integral membrane protein receptor, which on association with another membrane-spanning ligand protein, delta, causes activation of various genes responsible for cell-fate determination during development. Interestingly, the cytoplasmic portion of the Notch receptor containing ankylin repeats was found to be detached from plasma membranes in a signal-dependent fashion, and the mediator generated was transported into the nucleus to activate transcription of target genes, suggesting that proteolytic cleavage plays a critical role in Notch signaling (Goodbourn, 1995; Pahl and Baeuerle, 1996). Generation of the cytoplasmic domain of Notch was strongly inhibited by MG132 (Z-LLL-H), suggesting that the proteasome is involved in the Notch receptor signaling pathway through limited proteolysis (Kopan et al., 1996).

Sterol regulated element binding protein (SREBP-1) is a transcriptional ER membrane receptor. Triggering of signal transduction induced by sterol depletion was found to be initiated by cleavage of the cytoplasmic domain of the SREBP-1 receptor, and the active p68 transcriptional factor generated was translocated into the nucleus to activate various target genes responsible for sterol uptake and sterol synthesis, such as low density lipoprotein (LDL) receptor and 3-hydroxy-3-methylglutaryl-CoA (HMG-CoA) synthase (Pahl and Baeuerle, 1996). The mechanism of this cleavage of the SREBP-1 receptor is unknown, but the active p68 transcriptional factor was shown to be degraded rapidly in the nucleus, and this degradation to be inhibited by Ac-LLnL-H (Wang et al., 1994), suggesting involvement of the proteasome in the SREBP-1 signal termination pathway.

V. PERSPECTIVES

In this review, we summarized the structural and functional features of proteasomes, but at present there is only limited information on this enzyme complex and there are many unresolved questions. For instance, what is the molecular mechanism of its subunit assembly? To understand the roles and precise functions of individual subunits, it will be necessary to construct an *in vitro* resolution-association system. So far the α- and β-subunits of the *Thermoplasma* proteasome have been coexpressed in *Escherichia. coli* with production of an active complex like the native enzyme in considerably large amounts. A similar approach should be tried for eukaryotic proteasomes. Moreover, in considering catalytic functions, there are several critical questions. For example, although the eukaryotic proteasomes have three threonine catalytic subunits, what are the functions of the other four β-subunits? To answer this question, the tertiary structure of mammalian proteasomes including immunoproteasomes should be examined by X-ray crystallography, as has been done for the *Thermoplasma* and yeast proteasomes. The high-ordered structure of eukaryotic proteasomes must be known to understand why they are composed of heterogeneous populations. Finally, a genetic approach must be introduced to clarify the biological functions of proteasomes in living cells of higher organisms. Knock out mice have been useful for clarification of the roles of mammalian proteasomes, and may be of increasing value for understanding their biological functions in multicellular organisms.

REFERENCES

Aki, M., Shimbara, N., Takashina, M., Akiyama, K., Kagawa, S., Tamura, T., Tanahashi, N., Yoshimura, T., Tanaka, K., & Ichihara, A. (1993). Interferon-γ induces different subunit organizations and functional diversity of proteasomes. J. Biochem. 115, 257-269.

Akiyama, K., Yokota, K., Kagawa, S., Shimbara, N., Tamura, T., Akioka, A., Nothwang, H. G., Noda, C., Tanaka, K., & Ichihara, A. (1994). cDNA cloning and interferon-γ down-regulation of proteasomal subunits X and Y. Science. 265, 1231-1234.

Beumeister, W., Walz, J., Zühl, F., & Seemüller, E. (1998). The proteasome: Paradigm of a self-compartmentalizing protease. Cell 92, 367-380.

Belich, M.P., Glynne, R.J., Senger, G., Sheer, D., & Trowsdale, J. (1994). Proteasome components with reciprocal expression to that of the MHC-encoded LMP proteins. Curr. Biol. 4, 769-776.

Benedict, C. M., Ren, L., & Clawson, G. A. (1995). Nuclear multicatalytic proteinase α subunit RC3: Different size, tyrosine phosphorylation, and susceptibility to antisense oligonucleotide treatment. Biochemistry 34, 9587-9598.

Biederer, T., Volkswein, C., & Sommer, T. (1996). Degradation of subunits of the Sec61p complex, an integral component of the ER membrane, by the ubiquitin-proteasome pathway. EMBO J. 15, 2069-2076.

Boes, B., Hengel, H., Ruppert, T., Multhaup, G., Koszinowski, U.H., & Kloetzel, P.M. (1994). Interferon gamma stimulation modulates the proteolytic activity and cleavage site preference of 20S proteasomes. J. Exp. Med. 179, 901-909.

Bonifacino, J.S. (1996). Reversal of fortune for nascent proteins. Nature 384, 405-406.

Brown, M.G., Driscoll, J., & Monaco, J.J. (1993). MHC-linked low-molecular mass polypeptide subunits define distinct subsets of proteasomes. Implications for divergent function among distinct proteasome subsets. J. Immunol. 151, 1193-1204.

Chen, Z.J., Parent, L., & Maniatis, T. (1996). Site-specific phosphorylation of IκBα by a novel ubiquitination-dependent protein kinase activity. Cell 84, 853-862.

Chen, P., & Hochstrasser, M. (1996). Autocatalytic subunit processing couples active site formation in the 20S proteasome to completion of assembly. Cell 86, 961-972.

Ciechanover, A. (1994). The ubiquitin-proteasome proteolytic pathway. Cell 79, 13-21.

Coux, O., Tanaka, K., & Goldberg, A.L. (1996). Structure and functions of the 20S and 26S proteasomes. Annu. Rev. Biochem. 65, 801-847.

Deveraux, Q., van Novker, S., Mahaffey, D., Vierstra, R., & Rechsteiner, M. (1995). Inhibition of ubiquitin-mediated proteolysis by the Arabidopsis 26S protease subunit S5a. J. Biol. Chem. 270, 29660-29663

Dubiel, W., Ferrell, K., & Rechsteiner, M. (1995). Subunits of the regulatory complex of the 26S proteasome. Mol. Biol. Rep., 21, 27-34.

Egner, R., & Kuchler, K. (1996). The yeast multidrug transporter Pdr5 of the plasma membrane is ubiquitinated prior to endocytosis and degradation in the vacuole. FEBS Lett. 378, 177-181.

Fehling, H.J., Swat, W., Laplace, C., Kuhn, R., Rajewsky, K., Muller, U., & von Boehmer, H. (1994). MHC class I expression in mice lacking the proteasome subunit LMP-7. Science 265, 1234-1237.

Fenteany, G., Standaert, R.F., Reichard, G. A., & Schreiber, S.L. (1994). A β-lactone related to lactacystin induces outgrowth in a neuroblastoma cell line and inhibits cell cycle progression in an osteosarcoma cell line. Proc. Natl. Acad. Sci. USA 91, 3358-3362.

Fenteany, G., Standaert, R.F., Lane, W.S., Choi, S., Corey, E.J., & Schreiber, S.L. (1995). Inhibition of proteasome activities and subunit-specific amino-acid terminal threonine modification by lactacystin. Science 268, 726-731.

Figueiredo-Pereira, M.E., Banik, N., & Wilk, S. (1994). Comparison of the effect of calpain inhibitors on two extralysosomal proteinases: The multicatalytic proteinase complex and m-calpain. J. Neurochem. 62, 1989-1994.

Finco, T.S., & Baldwin, A.S. (1995). Mechanistic aspects of NF-κB regulation: The emerging role of phosphorylation and proteolysis. Immunity 3, 263-272.

Früh, K., Gossen, M., Wang, K., Bujard, H., Peterson, P.A., & Yang, Y. (1994). Displacement of housekeeping proteasome subunits by MHC-encoded LMPs: a newly discovered mechanism for modulating the multicatalytic proteinase complex. EMBO J. 13, 3236-3244.

Funabiki, H., Yamano, H., Kumada, K., Nagano, K., Hunt, T., & Yanagida, M. (1996). Cut2 proteolysis required for sister-chromatid separation in fission yeast. Nature 381, 438-441.

Fujiwara, T., Tanaka, K., Orino, E., Yoshimura, T., Kumatori, A., Tamura, T., Chung, C. H., Nakai, T., Yamaguchi, K., Shin, S., Kakizuka, A., Nakanishi, S., & Ichihara, A. (1990). Proteasomes are essential for yeast proliferation : cDNA cloning and gene disruption of two major subunits. J. Biol. Chem. 265, 16604-16613.

Galan, J.M., Moreau, V., &re, B., Volland, C., & Hagueauer-Tsapis, R. (1996). Ubiquitination mediated by the Npi1p/Rsp5p ubiquitin-protein ligase is required for endocytosis of the yeast uracil permease. J. Biol. Chem. 271, 10946-10952.

Ghislain, M., Udvardy, A., & Mann, C. (1993). *S. cerevisiae* 26S protease mutants arrest cell division in G2/metaphase. Nature 366 358-362.

Goldberg, A.L., Gaczynska, M., Grant, E., Michalek, M., & Rock, K.L. (1995). Functions of the proteasome in antigen presentation. Cold Spring Harb. Symp. Quant. Biol. 60, 479-490.

Goodbourn, S. (1995). Notch takes a short cut. Nature 377, 288-289.

Gordon, C., McGurk, G., Dillon, P., Rosen, C., & Hastie, N.D. (1993). Defective mitosis due to a mutation in the gene for a fission yeast 26S protease subunit. Nature 366 3655-3657.

Grimm, L.M., Goldberg, A.L., Poirier, G.G., Schwartz, L.M., & Osborne, B.A. (1996). Proteasomes play an essential role in thymocyte apoptosis. EMBO J. 15, 3835-3844.

Groettrup, M., Kraft, R., Kostka, S., Standera, S., Stohwasser, R., & Kloetzel, P.-M. (1996). A third interferon-γ-induced subunit exchange in the 20S proteasome. Eur. J. Immunol. 26, 863-869.

Griffin, T.A., Nandi, D., Cruz, M., Fehling, H.J., Van Kaer, L., Monaco, J.J., & Colbert, R.A. (1998). Immunoproteasomes assembly: Cooperative incorporation of interferon-γ– (IFN-γ)–inducible subunits. J. Exp. Med. 187, 97-104.

Groll, M., Ditzel, L., Lowe, J., Stock, D., Bochtler, M., Bartunik, H.D., & Huber, R. (1997). Structure of 20S proteasomes from yeast at 2.4 Å resolution. Nature 386, 463-471.

Heemels, M.T., & Ploegh, H. (1995). Generation, translocation, and presentation of MHC class I-restricted peptides. Annu. Rev. Biochem. 64, 463-491.

Hershko, A. (1997). Roles of ubiquitin-dependent proteolysis in cell cycle control. Curr. Opin. Cell. Biol. 9, 788-799.

Hershko, A., & Ciechanover, A. (1992). The ubiquitin system for protein degradation. Annu. Rev. Biochem. 61, 761-807.

Hicke, L., & Riezman, H. (1996). Ubiquitination of a yeast plasma membrane receptor signals its ligand-stimulated endocytosis. Cell 84, 277-287.

Hilt, W., & Wolf, D.H. (1995). Proteasomes: destruction as a programme. Trends Biochem. Sci., 21, 96-102.

Hisamatsu, H., Shimbara, N., Saito, Y., Kristensen, P., Hendil, K. B., Fujiwara, T., Takahashi, E., Tanahashi, N., Tamura, T., Ichihara, A., & Tanaka, K. (1996). Newly identified pair of proteasomal subunits regulated reciprocally by interferon-γ. J. Exp. Med. 183, 1807-1816.

Hochstrasser, M. (1997). Ubiquitin-dependent protein degradation: Annu. Rev.Genet. 30, 405-439.

Ichihara, A., & Tanaka, K. (1995). Roles of proteasomes in cell growth. Mol. Biol. Rep. 21, 49-52.

Imajoh-Ohmi, S., Kawaguchi, T., Sugiyama, S., Tanaka, K., Omura, S., & Kikuchi, H. (1995). Lactacystin, a specific inhibitor of the proteasome, induces apoptosis in human monocyte U937 cells. Biochem. Biophys. Res. Commun. 217, 1070-1077.

Jensen, T.J. Loo, M.A., Pind, S., Williams, D.B., Goldberg, A.L., & Riordan, J. R. (1995). Multiple proteolytic systems, including the proteasome contribute to CFTR processing. Cell 83, 129-135.

Jentsch, S., & Schienker, S. (1995). Selective protein degradation: A journey's end within the proteasome. Cell 82, 881-884.

Kandil, E., Namikawa, C., Nonaka, M., Greenberg, A. S., Flajnik, M. F., Ishibashi, T., & Kasahara, M. (1996). Isolation of low molecular mass polypeptide complementary DNA clones from primitive vertebrates. Implication for the origin of MHC class I-restricted antigen presentation. J. Immunol. 156, 4246-4253.

Kasahara, M., Hayashi, M., Tanaka, K., Inoko, H., Sugaya, K., Ikemura, T., & Ishibashi, T. (1996). Chromosomal localization of the proteasome Z gene reveals an ancient chromosomal duplication involving the major histocompatibility complex. Proc. Natl. Acad. Sci. USA 93, 9096-9101.

Katagiri, M., Hayashi, M., Matsuzaki, K., Tanaka, H., & Omura, S. (1995). The neuritogenesis inducer lactacystin arrests the cell cycle at both G0/G1 and G2 phases in neuro 2a cells. J. Antibiot. 48, 344-346.

King, R.W., Deshaies, R.J., Peters, J.-M., & Kirschner, M.W. (1996). How proteolysis drives the cell cycle. Science 274, 1652-1659.

Kolling, R., & Hollenberg, C.P. (1994). The ABC-transporter Ste6 accumulates in the plasma membrane in a ubiquitinated form in endocytosis mutants. EMBO J. 13, 3261-3271.

Kominami, K., DeMartino, G.N., Moomaw, C.R., Slaughter, C.A. Shimbara, N., Fujimuro, M., Yokosawa, H., Hisamatsu, H., Tanahashi, N., Shimizu, Y., Tanaka, K., & Toh-e, A. (1995). Nin1p, a regulatory subunit of the 26S proteasome, is necessary for activation of Cdc28p kinase of *Saccharomyces cerevisiae*. EMBO J. 14, 3105-3115.

Kopan, R., Schroeter, E.H., Weintraub, H., & Nye, J.S. (1996). Signal transduction by activated mNotch: Importance of proteolytic processing and its regulation by the extracellular domain. Proc. Natl. Acad. Sci. USA 93, 1683-1688.

Kopp, F., Kristensen, P., Hendil, K. B., Johnsen, A., Sobek, A., & Dahlmann, B. (1995). The human proteasome subunit HsN3 is located in the inner rings of the complex dimer. J. Mol. Biol. 248, 264-272.

Kubota, H., Hynes, G., & Willison, K. (1995). The chaperonin containing *t*-complex polypeptide (TCP-1). Multisubunit machinery assisting in protein folding an assembly in the eukaryotic cytosol. Eur. J. Biochem. 230, 3-16.

Larsen, F., Solheim, J., Kristensen, T., Kolsto, A.B., & Prydz, H. (1993). A tight cluster of five unrelated human genes on chromosome 16q22.1. Hum. Mol. Gen. 2, 1589-1595.

Lord, J.M. (1996). Protein degradation: go outside and see the proteasome. Curr. Biol. 6, 1067-1069.

Löwe, J., Stock, D., Jap, B., Zwickl, P., Baumeister, W., & Huber, R. (1995). Crystal structure of the 20S proteasome from the archaeon T. acidophilum at 3.4 Å resolution. Science 268, 533-539.

Lupas, A., Koster, A.J., & Baumeister, W. (1993). Structural features of 26S and 20S proteasomes. Enzyme Protein 47, 252-273.

Lupas, A., Zwickl, P., & Baumeister, W. (1994). Proteasome sequences in eubacteria. Trends Biochem. Sci. 19, 544-545.

Monaco, J.J., & Nandi, D. (1995). The genetics of proteasomes and antigen processing. Annu. Rev. Genet. 29, 729-754.

Mori, S., Tanaka, K., Ōmura, S., & Saito, Y. (1995). Degradation process of ligand-stimulated platelet-derived growth factor β-receptor involves the ubiquitin-proteasome proteolytic pathway. J. Biol. Chem 270, 29447-29452.

Murakami, Y., Matsufuji, S., Kameji, T., Hayashi, S., Igarashi, K., Tamura, T., Tanaka, K., & Ichihara, A. (1992). Ornithine decarboxylase is degraded by the 26S proteasome without ubiquitination. Nature 360, 597-599.

Murakami, Y., Tanahashi, N., Tanaka, K., Ōmura, S., & Hayashi, S. (1996). Proteasome pathway is operating for degradation of ornithine decarboxylase in intact cells. Biochem. J. 317, 77-80.

Murray, A.W. (1995). Cyclin ubiquitination: The destructive end of mitosis. Cell 81, 149-152.

Nandi, D., Jiang, H., & Monoco, J.J. (1996). Identification of MECL-1 (LMP-10) as the third IFN-γ-inducible proteasome subunit. J. Immunol. 156, 2361-2364.

Nederlof, P.M., Wang, H-R., & Baumeister, W. (1995). Nuclear localization signals of human and Thermoplasma proteasomal α subunits are functional *in vitro*. Proc. Natl. Acad. Sci. USA 92, 12060-12064.

Ōmura, S., Fujimoto, T., Otoguro, K., Matsuzaki, K., Moriguchi, R., Tanaka, H., & Sasaki, Y. (1991). Lactacystin, a novel microbial metabolite, induces neuritogenesis of neuroblastoma cells. J. Antibiot. 44, 113-116.

Orlowski M. (1990). The multicatalytic proteinase complex, a major extralysosomal proteolytic system. Biochemistry 29, 10289-97.

Pagano, M., Tam, S.W., Theodoras, A.M., Beer-Romero, P., Del Sal, G., Chan, V., Yew, P.R., Draetta, G.F., & Rolfe, M. (1995). Role of the ubiquitin-proteasome pathway in regulating abundance of the cyclin dependent kinase inhibitor p27. Science 269, 682-685.

Pahl, H.L., & Baeuerle, P.A. (1996). Control of gene expression by proteolysis. Curr. Opin. Cell Biol. 8, 340-347.

Palombella, V. J., Rando, O. J., Goldberg, A. L., & Maniatis, T. (1994). The ubiquitin-proteasome pathway is required for processing the NF-kappa B1 precursor protein and the activation of NF-kappa B. Cell 78, 773-785.

Peters, J.-M. (1994). Proteasomes: protein degradation machines of the cells. Trends Biochem. Sci. 19, 377-382.

Rammensee, H-G., Friede, T., & Stefanovic S. (1995). MHC ligands and peptide motifs: first listing. Immunogenetics 41, 178-228.

Rivett, A.J. (1993). Proteasomes: multicatalytic proteinase complexes. Biochem. J. 291, 1-10.

Rock, K.L., Gramm, C., Rothstein, L., Clark, K., Stein, R., Dick, L., Hwang, D., & Goldberg, A.L. (1994). Inhibitors of the proteasome block the degradation of most cell proteins and the generation of peptides presented on MHC class I molecules. Cell 78, 761-771.

Rubin, D.M., & Finley, D. (1995). The proteasome: a protein-degrading organelle? Curr. Biol. 8, 854-858.

Saito, Y., Tsubuki, S., Ito, H., & Kawashima, S. (1990). The structure-function relationship between peptide aldehyde derivatives in initiation of neurite outgrowth of PC12h cells. Neurosci. Lett. 120, 1-4.

Sadoul, R., Fernandez, P.-A., Quiquerez, A.-L., Schröter, M., Becherer, J.D., Irmler, M., Tschopp, J., & Martinou, J.-C. (1996). Involvement of the proteasome in the programmed cell deth of NGF-deprived sympathetic neurons. EMBO J. 15, 3845-3852.

Scherrer, K., & Bey, F. (1994). The prosomes (multicatalytic proteinases, proteasomes) and their relationship to the untranslated messenger ribonucleoproteins, the cytoskeleton, and cell differentiation. Prog. Acid Res. Mol. Biol. 49, 1-64.

Schwob, E., Boehm, T., Mendenhall, M.D., & Nasmyth, K. (1994). The B-type cyclin kinase inhibitor $p40^{SIC1}$ controls the G1 to S transition in *S. cerevisiae*. Cell 79, 233-244.

Seemüller, E., Lupas, A., Stock, D., Löwe, J., Huber, R., & Baumeister, W. (1995). Proteasome from *Thermoplasma acidophilum*: a threonine protease. Science 268, 579-582.

Seemüller, E., Lupas, A., & Baumeister, W. (1996). Autocatalytic processing pf the 20S proteasome. Nature 382, 468-470.

Sepp-Lorenzino, L., Ma, Z., Lebwohl, D.E., Vinitsky, A., & Rosen, N. (1995). Herbimycin A induces the 20S proteasome- and ubiquitin-dependent degradation of receptor tyrosine kinases. J. Biol. Chem. 270, 16580-16587.

Sherwood, S.W., Kung, A.L., Roitelman, J., Simoni, R.D., & Schimke, R.T. (1993). In vivo inhibition of cyclin B degradation and induction of cell cycle arrest in mammalian cells by the neutral cysteine protease inhibitor N-acetylleucylleucylnorleucinal. Proc. Natl. Acad. Sci. USA 90, 3353-3357.

Shinohara, K., Tomioka, M., Nakano, H., Tohe, S., Ito, H., & Kawashima, S. (1996). Apoptosis induction resulting from proteasome inhibition. Biochem. J. 317, 385-388.

Tamura T., Nagy, I., Lupas, A., Lottspeich, F., Cejka, Z., Schoofs, G., Tanaka, K., Mot, R.D., & Baumeister, W. (1995). The first characterization of a eubacterial proteasome: the 20S complex of Rhodococcus sp. strain NI86/21. Curr. Biol. 5, 766-774.

Tanaka, K., Yoshimura, T., Tamura, T., Fujiwara, T., Kumatori, A., & Ichihara, A. (1990). Possible mechanism of nuclear translocation of proteasomes. FEBS Lett. 271, 41-46.

Tanaka K., Tanahashi, N., Tsurumi, C., & Shimbara, N. (1997). Proteasomes and antigen processing. Adv. Immunol. 64, 1-38.

Tanaka, K. (1995). Molecular biology of proteasomes. Mol. Biol. Rep. 21, 21-26.

Traenckner, E. B.-M., Wilk, S., & Baeuerle, P. A. (1994). A proteasome inhibitor prevents activation of NF-κB and stabilizes a newly phosphorylated form of IκB-α that is still bound to NF-κB. EMBO J. 13, 5433-5441.

Tsubuki, S., Kawasaki, H., Saito, Y., Miyashita, N., Inimata, M., & Kawashima, S. (1993). Purification and characterization of Z-Leu-Leu-Leu-MCA degrading protease expected to regulate neurite formation: Novel catalytic activity in the proteasome. Biochem. Biophys. Res. Commun. 15, 1195-1201.

Tsubuki, S., Saito, Y., Tomioka, M., Ito, H., & Kawashima, S. (1996). Differential inhibition of calpain and proteasome activities by peptidyl aldehydes of di-leucine and tri-leucine. J. Biochem.119, 572-576.

Tsujinaka, T., Kajiwara, Y., Kambayashi, Y., Sakon, M., Higuchi N, Tanaka, T., & Mori, T. (1989). Synthesis of a new cell penetrating calpain inhibitor (calpeptin). Biochem. Biophys. Res. Commun. 153, 1201-1208.

Van Kaer, L., Ashton-Richardt, P.G., Eichelberger, M., Gaczynska, M., Nagashima, K., Rock, K.L., Goldberg, A.L., Doherty, P.C., & Tonegawa, S. (1994). Altered peptidase and viral-specific T cell response in LMP2 mutant mice. Immunity 1, 533-541.

van Nocker, S., Deveraux, Q., Rechsteiner, M., & Vierstra, R. (1996). *Arabidopsis* MBP1 gene encodes a conserved ubiquitin recognition component of the 26S proteasome. Proc. Natl. Acad. Sci. USA 93: 856-860.

Vinitsky, A., Cardozo, C., Sepp-Lorenzino, L., Michaud, C., & Orlowski, M. (1994). Inhibition of the proteolytic activity of the multicatalytic proteinase complex (proteasome) by substrate-related peptidyl aldehydes. J. Biol. Chem. 47, 29860-29866.

Wang, X., Sato, R. Brown, M.S., Hua, X., & Goldstein, J.L. (1994). SREBP-1, a membrane-bound transcription factor released by sterol-regulated proteolysis. Cell 77, 53-62.

Ward, C., Omura, S., & Kopito, R.R. (1995). Degradation of CFTR by the ubiquitin-proteasome pathway. Cell 83, 121-127.

Weissman, J. S., Sigler, P. B., & Horwich A. L. (1995). From the cradle to the grave: Ring complexes in the life of a protein. Science 268, 523-524.

Yang, Y., Waters, J.B., Früh, K., & Peterson, P.A. (1992). Proteasomes are regulated by interferon-γ: Implications for antigen processing. Proc. Natl. Acad. Sci. USA 89, 4928-4932.

York, I.A., & Rock, K.L. (1996). Antigen processing and presentation by the class I major histocompatibility complex. Annu. Rev. Immunol. 14,369-397.

Yoshimura, T., Kameyama, K., Takagi, T., Ikai, A., Tokunaga, F., Koide, T., Tanahashi, N., Tamura, T., Cejka, Z., Baumeister, W., Tanaka, K., & Ichihara, A. (1993). Molecular characterization of the "26S" proteasome complex from rat liver. J. Struct. Biol. 111, 200-211.

THE 19S REGULATORY COMPLEX OF THE 26S PROTEASOME

Wolfgang Dubiel and Martin Rechsteiner

Advances in Molecular and Cell Biology
Volume 27, pages 129-163.

ISBN: 0-7623-0387-5

I. INTRODUCTION

It is becoming increasingly evident that proteolysis serves to regulate a variety of biochemical processes within eukaryotic cells. Rapid destruction of cyclins is required for cell cycle progression through both G1 and mitosis (Deshaies, 1995). Downregulation of key metabolic enzymes, such as ornithine decarboxylase (Hayashi and Murakami, 1995) or fructose- 1,6-biphosphatase (Schork et al., 1995), is mediated by destruction of the protein catalyst. Likewise, a host of transcription factors are rapidly degraded proteins (Scheffner et al., 1993; Richter-Ruoff et al., 1994; Tsurumi et al., 1995). Despite the widespread occurrence of proteolytic events in the cytosol and nucleus, few proteases have been localized to these compartments. At present, only the calpains, the ICE family of cysteine proteases and the proteasomal system have been implicated in the degradation of cytosolic and nuclear proteins. Cytoskeletal proteins are thought to be the principal substrates of the calpains (Suzuki et al., 1995), and members of the ICE family appear to be involved either in the processing of certain cytokines, for example interleukin-1β, or in a generalized proteolysis that results in cell death (Henkart, 1996). The 26S proteasome probably degrades most cytosolic and nuclear proteins; at present it has been implicated in the degradation of almost 20 specific proteins (Hilt and Wolf, 1996). The list of natural substrates for this large proteolytic particle will surely become much longer.

The 26S proteasome (or 26S protease) was discovered by its ability to degrade ubiquitylated lysozyme (Hough et al., 1986). During purification, the 26S proteasome was accompanied by a smaller 20S protease (the multicatalytic protease or 20S proteasome) that did not hydrolyze ubiquitin (Ub) conjugates or require nucleotides for cleavage of fluorogenic peptides (Hough et al., 1987). The 20S proteasome is a ~700 kDa complex possessing at least five endopeptidase activities that cleave bonds at the carboxyl sides of basic, hydrophobic, acidic, small neutral, and branched chain amino acid residues (Cardozo et al., 1994). The crystal structure of the 20S proteasome from the archacbacterium *Thermoplasma acidophilum* reveals a cylindrical particle assembled from four stacked rings. Each ring contains seven subunits which can be classified as α or β based upon its primary amino acid sequence (Löwe et al., 1995; see chapter 4) The end rings are composed of α subunits and the inner rings of β subunits. Mutagenic data indicate that the catalytic activity resides in the β subunits (Seemüller et al., 1995) and difference Fourier analysis of inhibitor complexes localizes the protease active sites to the inner compartment of the cylindrical structure (Löwe et al., 1995). Archaebacterial and eukaryotic proteasomes appear to share the same α7-β7-β7-α7 ring structure. They are very similar by electron microscopic analysis (Dahlmann et al., 1989), although not all previous studies concluded that the eukaryotic proteasome possesses sevenfold symmetry (Kleinschmidt et al., 1983; Arrigo et al., 1988; Baumeister et al., 1988) Unlike the archaebacteria, which have only one kind of α and one kind of β subunit, eukaryotcs encode at least 14 different subunits that can be grouped as α-like or as β-like based upon sequence analysis (Tanaka et al., 1992; Heinemeyer et al., 1994). Substrates can access the proteasome active sites through the central channel that passes along the sevenfold axis of the cylindrical complex (Wenzel and Baumeister, 1995). This restricted access to the active sites can explain how indiscriminate degradation of cellular proteins is avoided and why the proteasome is inactive against folded proteins.

We now know that the 26S proteasome contains both the "cylindrical" proteasome and a "spherical" 19S regulatory complex (19S RC). The proteasome provides the active sites for the peptide bond hydrolysis; the 19S RC confers substrate recognition and the energy dependent reactions needed for the degradation of intact proteins. In this chapter, we will review properties of the 26S protease with emphasis on components in the regulatory complex. We also cover the recently discovered 11S activators of the proteasome (Dubiel et al., 1992a; Ma et al., 1992). More detailed information about the 20S proteasome is presented in chapter 4 of this book.

II. ELECTRON MICROSCOPY OF THE 26S PROTEASOME

The 26S proteasome was visualized in the electron microscope more than 25 years ago. In their studies on aminoacyltransferases, Shelton et al. (1970) examined particles obtained by sucrose density gradient centrifugation of HeLa cell extracts. They described electron microscopic images of a mushroom-shaped particle that sedimented at 26S. Based on those micrographs and the subunit composition of the rabbit reticulocyte 26S protease, Hough et al. (1988) proposed a model for the 26S proteasome in which a large "spherical" particle was attached to two rings of subunits provided by the multicatalytic protease or 20S proteasome (see Figure 1A).

It now seems that this initial model is incorrect since subsequent electron microscopic analyses indicate that all four rings of the proteasome are present in the 26S protease. For example, Ikai et al. (1991) described the rat liver 26S protease as a barbell-shaped structure measuring 40 nm in length. The central "bar" portion of the enzyme contains 4 rings that in their model are provided by the 20S proteasome. Attached to each end of the proteasome is a larger structure 12 nm in length and 20 nm in diameter (see Figure 1B). Peters et al. (1991) came to a similar model for the 26S protease from *Xenopus* oocytes. That is, the enzyme is 40 nm long consisting of a central core formed by the 20S proteasome to which large particles were attached at each end. Follow-up studies by Peters in collaboration with Baumeister have provided a higher resolution image of the *Xenopus* 26S protease (Peters et al., 1993). Using digital image analysis these authors describe a barbell-shaped structure in which the two particles attached at each end of the protease are highly asymmetric having the appearance of a

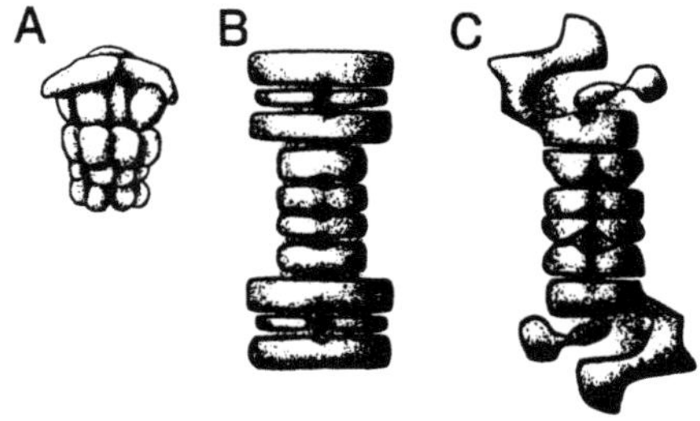

Figure 1. The evolution of models for the 26S protease. The original 1988 model for the 26S protease is shown at the left (**A**). The 1991 model shown in panel **B** is redrawn from Ikai et al. (1991). The 1993 and current model for the 26S proteasome is shown at the right (**C**).

Chinese dragon head motif (see Figure 1C). Interestingly, in those particles with two attached dragon heads, the dragons face in opposite directions implying that the regulatory complexes bind the proteasome ends in a highly specific manner. Equivalent images were obtained with rat liver 26S proteasomes (Yoshimura et al., 1993). Thus, four recent electron microscopic studies on the 26S protease reveal a particle in which the ends of an intact 20S proteasome associate with one or two equivalent regulatory complexes. It seems clear that the central proteasome provides peptide cleavage functions. The analysis of the crystal structure of the 20S proteasome from *Thermaplasma acidophilum* (Löwe et al., 1995) and results from site-directed mutagenesis (Seemüller et al., 1995) have suggested that it is a novel threonine protease rather than a serine protease. There is a prevailing belief that the terminal 19S RCs confer substrate recognition, substrate unfolding and the ability to transfer polypetide chains to the protease active sites located within the central axis of the proteasome. If this is a correct view of the functions performed by the regulatory complex, then it is a sophisticated biochemical assembly. The presence of more than 15 distinct regulatory complex subunits certainly supports such an idea, and in the following section we examine the properties and possible functions for these subunits.

III. COMPOSITION AND MOLECULAR BIOLOGY OF THE 19S REGULATORY COMPLEX

A. The Non-ATPases

The subunit composition of the 19S RC has been reviewed (Dubiel et al., 1995a; Tanaka,1995). However, new subunits have been sequenced and characterized in the past few months. Today we know that the complex is composed of at least 10 non-ATPases and six ATPases. The components of the 19S RC from human erythrocytes and their homologues in other species are summarized in Table 1.

Table 1. Subunits of the Human 19S Regulatory Complex and Their Homologue in Other Species

Subunit		*Origin*	*Acc. Number*	*MM*
S1		Human		110
	p112	Human		
	p112	Bovine		
	SEN3	*S. cerevisiae*	LO6321	
S2		Human		100
	p97	Bovine	D78151	
	TRAP2	Human	U12596	
S3		Human		61
	p58	Bovine		
	p91A	Mouse	M25149	
	SUN2	*S. Cerevisiae*		
S4*		Human	L02426	52
	mts2	*S. pombe*		
	YTA5	*S. cerevisiae*	X81070	
	Yhs4p	*S. cerevisiae*	L17040	
TBP1*		Human		50
	YTA1	*S. cerevisiae*		
S5a		Human		50
	M:BP1	*A. thaliana*		
	p54	*Drosophila*		
	SUN1	*S. cerevisiae*		
S5b		Human		50
S6*		Human		48
	TBP7	Human		
	p48	Bovine		
	TBP2	*D. discooideum*	L16578	
	MS73	*M. sexta*	Z38135	
	YTA2	*S. cerevisiae*	X73570	
	Ynt1p	*S. cerevisiae*	U06229	
S7*				47
	MSS1	Human	D11094	
	MSS1	*M. laevis*	X80157	
	p48	*Drosophila*		
	CIM5/YTA3	*S. cerevisiae*	X73571	

Conintued

Table 1. (Continued)

Subunit		*Origin*	*Acc. Number*	*MM*
S8*		Human		45
	TRip1	Human	L38820	
	P45	Human		
	P45	Bovine		
	P43	*Drosophila*		
	TBP10	*D. discoideum*	L69123	
	CIM3/SUG1/TBPT	*S. cerevisiae*	X66400	
	LET1	*S. pombe*	U02280	
	PfTBP	*P. falciparum*	X77914	
S10a/HUMORF07		Human	D14663	44
S10b*		Human		42
	SUG2	*S. cerevisiae*		
S11		Human		42
S12		Human		36
	Mov34	Mouse	M64641	
	Mov34	*Drosophila*	M64643	
S14		Human		30
	p31	Human		
	p31	Bovine		
	NIN1	*S. cerevisiae*	S27434	
	mts3	*S. pombe*		
S15		Human		
	p28	Bovine		

Note: * ATPase.

Most of the subunits of the 19S RC studied to date are essential. The *Saccharomyces cerevisiae* gene *SEN3* which encodes for a homologue of human subunit 1 (S1) has been cloned and characterized (DeMarini et al., 1995). The gene has been localized to chromosome IX. Mutant sen3-1 cells accumulate Ub conjugates and increase the levels of Sen1, a nuclear protein involved in tRNA splicing. The authors identified Sen3 as a component of the yeast 26S protease indicating that the Ub system might have a role in RNA processing. Recently a cDNA encoding human Sen3 has been characterized (Yokota et al., 1996) and our peptide sequences of subunit 1 (S1) match 100% with that sequence.

The Sen3 protein has similarities with a protein called p55.11 (Boldin et al., 1995) or tumor necrosis factor receptor (TNFR)-associated protein 2, TRAP2

(Song et al., 1995). It has been shown that TRAP2 is identical to p97 of the PA700 complex (Tsurumi et al., 1996). Our peptide sequence analysis of subunit 2 (S2) supports the fact that TRAP2 is a component of the 26S protease (see Figure 2). All our peptides match 100% with the published sequence of subunit 2.

Why does a subunit of the 26S protease interact with the TNFR? The cytokine TNF is produced by macrophages during infections and malignancies (Beutler and Cerami, 1989). By interacting with the two TNFRs (TNFR-1, 55 kDa and TNFR-2, 75 kDa) it promotes immunity, antiviral responses, T cell growth, NF-κB activation, and apoptosis. TNFR-1 has a conserved region of about 80 amino acids at its C-terminus, which signals for cell death and is, therefore, called the death domain (Tartaglia et al., 1993). Self-associations of TNFR-1 death domains or their binding to a death domain motif present in Fas/APO1/CD95 cell surface receptor induces apoptosis. In two-hybrid assays TRAP2=p55.11=S2 binds to a region of TNFR-1 outside the death domain (Boldin et al., 1995). The consequence of the complex formation is not yet clear. It is possible that subunit 2 of the 19S RC selects the cytosolic domain of the TNFR-1 for degradation by the 26S protease which would prevent apoptosis. This might be a first direct indication that the 26S enzyme interferes with a signal transduction pathway. Alternatively TRAP2 could act as a component of the 26S protease as well as a ligand of TNFR-1. Subunit 3 (S3) is identical to the mouse tumor antigen P91A (DeMartino et al., 1994). Recently a homologue in *S. cerevisiae* has been identified called *SUN2* (Tanaka, 1995). *SUN1*, which is a homologue of human S5a, and *SUN2* are multicopy suppressors of *nin1-1*, a thermosensitive (ts) mutant of *NIN1* in yeast, suggesting physical association (Tanaka, 1995) in the 19S RC. Nin1 protein is a component of *S. cerevisiae* 26S protease needed for both G1/S and G2/M transitions in cell cycle, perhaps by activating the cdc28p kinase via degradation of the inhibitor $p40^{SIC1}$ (Kominami et al., 1995). *NIN1* is the budding yeast homologue of the fission yeast *mts34*$^{+}$ gene (see Table 1) and a conditional lethal mutant, *mts3-1*, has been isolated in *S. pombe* which is defective in metaphase to anaphase transition at the restrictive temperature (Gordon et al., 1996). The 26S protease isolated from *mts3-1* strains after growing cells at the restrictive temperature does not contain mts3 protein and cannot degrade Ub conjugates in an ATP-dependent manner (Seeger et al., 1996). Mts3^{+} protein function is obviously essential for Ub conjugate degradation. Characterization of the 50 kDa non-ATPase (S5a) which binds Ub-lysozyme conjugates and free multiUb chains (Deveraux et al., 1994, 1995a,b) has made the most progress since the last reviews. Human S5a has been cloned and sequenced (Ferrell et al., 1996) and homologue have been identified in *A. thaliana*, MBP1 (van Nocker et al., 1996a), in *Drosophila*, p54 (Haracska and Udvardy, 1995), and in *S. cerevisiae* (van Nocker et al., 1996b). The function of S5a in selecting proteins conjugated to polyUb chains for proteolysis will be reviewed below. The fact that S5a is almost identical to a human protein called antisecretory factor (AF) that is secreted by the pituitary and inhibits cholera toxin induced intestinal fluid secretion (Johansson et al., 1995) might indicate that S5a is also a protein with dual functions. Alternatively, free S5a has been

```
                                                o †                 ▼                   o                             o     ▼
Trap2                                           S G G T D E K P S G K G R R D A G D K D K E L E L S E E D K Q L Q D E L V M L V   40
26Sp97  M E E G G R D K A P V Q P Q Q S P A A A P G G T D E K P S G K E R R D A G D K D K E Q E L S E E D K Q L Q D E L E M L A   60
S2pep1    E E G G R D K A P V Q P Q Q S P A A A P X G T D E K P

Trap2   E R L G E K D T S L Y R P A L E E L R R Q I R S S T T S M T S V P K P L K F L R P H Y G K L K E I Y E N M A P G E N K R  100
26Sp97  E R L G E K D T S L Y R P A L E E L R R Q I R S S T T S M T S V P K P L K F L R P H Y G K L K E I Y E N M A P G E N K R  120

Trap2   F A A D I I S V L A M T M S G E R E C L K Y R L V G S Q E E L A S W G H E Y V R H L A G E V A K E W Q E L D D A E K V Q  160
26Sp97  F A A D I I S V L A M T M S G E R E C L K Y R L V G S Q E E L A S W G H E Y V R H L A G E V A K E W Q E L D D A E K V Q  180

                                                                                          o
Trap2   R E P L L T L V K E I V P Y N M A H N A E H E A C D L L M E I E Q V D M L E K D I D E N A S A K V C L Y L T S C V N Y V  220
26Sp97  R E P L L T L V K E I V P Y N M A H N A E H E A C D L L M E I E Q V D M L E K D I D E N A Y A K V C L Y L T S C V N Y V  240
S2pep2                                                            E I E Q V D

                                                                                o o ▼
Trap2   P E P E N S A L L R C A L G V F R K F S R F P E A L R L A L M L N D M E L V E D S - S S C K D V V V Q K Q M A F M L G R  279
26Sp97  P E P E N S A L L R C A L G V F R K F S R F P E A L R L A L M L N D M E L V E D I F T S C K D V V V Q K Q M A F M L G R  300

                                                                                                            †
Trap2   H G V F L E L S E D V E E Y E D L T E I M S N V Q L N S N F L A L A R E L D I M E P K V P D D I Y K T H L E N N R F G G  339
26Sp97  H G V F L E L S E D V E E Y E D L T E I M S N V Q L N S N F L A L A R E L D I M E P K V P D D I Y K T H L E N N R F G G  360
S2pep3                                                                                  E P K V P D D I Y K T H L E X N R F

                                                                                                            ▼
Trap2   S G S Q V D S A R M N L A S S F V N G F V N A A F G Q D K L L T D D G N K W L Y K N K D H G M L S A A A S L G M I L L W  399
26Sp97  S G S Q V D S A R M N L A S S F V N G F V N A A F G Q D K L L T D D G N K W L Y K N K D H G M L S A A A S L A M I L L W  420

Trap2   D V D G G L T Q I D K Y L Y S S E D Y I K S G A L L A C G I V N S G V R N E C D P A L A L L S D Y V L H N S N T M R L G  459
26Sp97  D V D G G L T Q I D K Y L Y S S E D Y I K S G A L L A C G I V N S G V R N E C D P A L A L L S D Y V L H N S N T M R L G  480
```

Figure 2. Alignment of TRAP2 protein, the 26S proteasome subunit p97, and peptide sequences obtained from subunit two of the 19S regulatory complex. TRAP2 sequence (acc. number: U12596) has been published by Song et al. (1995) and the 26S proteasome subunit p97 (26Sp97) has been submitted to Genbank by Kato et al. (1995; acc. number: D78151). The five CNBr-peptides obtained from human subunit two match to both sequences. The crosses mark positions of ambiguity in our peptides. Homologous (closed triangles) and nonhomologous (open circles) amino acids are indicated.

shown to be a potent inhibitor of Ub-dependent proteolysis (Deveraux et al., 1995b). It occurs in excess in *Drosophila* oocytes (Haracska and Udvardy, 1995) and in *S. cerevisiae* (van Nocker et al., 1996b), and it might be a natural regulator of Ub-dependent proteolysis. Perhaps S5a=AF interferes with intestinal fluid secretion by inhibiting the Ub pathway. Although S5a binds Ub conjugates, it is not the only subunit of the 19S RC which is essential for conjugate processing. As mentioned above, a mutation in mts3=S14 leads to the inability of the *S. pombe* 26S protease to degrade Ub conjugates. In addition, mutations in ATPases of the 19S RC lead to accumulation of Ub conjugates (see below).

The non-ATPases S10 (now known as 10a) and S12 have been identified and characterized (Dubiel et al., 1995a,b) and we are currently analyzing a full-length clone encoding human subunit 11 (S11). Subunit 11 possesses no significant homology to proteins in current databases. An antibody made against a recombinant fragment of S11 specifically recognizes S11 in the 26S proteasome. The antibody also reacts with glycerol gradient fractions that do not contain the 26S enzyme or 19S RC. The possibility of the existence of 19S RC precursor complexes is currently being examined.

Although mutations in the non-ATPase subunits demonstrate their physiological importance, the lack of homologies to known proteins does not indicate the biochemical functions of these 26S protease components. Possibly the non-ATPases are unique proteins responsible for specific interactions within the 19S RC and/or with certain substrates or E3s. There is a region common to subunits 1, 2, 5a, 10, and 12 which is rich in glutamate and lysine residues called the KEKE motif (Realini et al., 1994a). It has been suggested that KEKE motifs are responsible for protein-protein interactions.

B. The ATPases

Six ATPases, S4 (Dubiel et al., 1992b), MSS1 (Dubiel et al., 1993), TBP7 (Dubiel et al., 1994), SUG1 (Akiyama, et al., 1995; Rubin et al., 1996), TBP1, and SUG2 (DeMartino et al., 1996) have been identified in the 19S RC. These polypeptides belong to a novel ATPase family called the ATPase-associated activity (AAA)-family (Confalonieri and Duguet, 1995). Members of the AAA-family have been found in all eukaryotic cells examined and are involved in a wide variety of important cellular functions. SUG1 has been identified as both a component of the transcriptional complex (Swaffield et al., 1995) and as a ligand, Trip1, of the thyroid-hormone receptor (Lee et al., 1995). Again, the polypeptide might have dual functions: as a subunit of the 26S proteasome and as a component of other complexes. Alternatively, it is known that proteolysis is involved in regulation of transcription and in signal transduction pathways. In this regard, it has recently been reported that SUG1 is indeed a member of the 26S proteasome; furthermore, in direct conflict with the results of Swaffield et al. (1995) SUG1 was not detected in transcriptional complexes (Rubin et al., 1996). Temperature sensitive mutants with point mutations in 19S RC ATPases have been isolated in *S. cerevisiae* genes, *CIM3* and *CIM5* (Ghislain et al., 1993), and in *S. pombe, mts2* (Gordon et al., 1993). These yeast cells are defective in chromosome segregation and arrest in mitosis. We have recently examined the effect of the mts2 mutation on the structure and function of the 26S proteasome from yeast. The enzyme isolated from mts2-1 cells grown at the restrictive temperature was not able to degrade Ub lysozyme conjugates in an ATP-dependent manner indicating that the ATPases of the 19S RC are not redundant (Seeger et al., 1996). There is evidence for physical interactions between different ATPases (Gorbea et al., manuscript submitted); possibly they form a ring-shaped structure as described for another AAA-polypeptide, p97 (Peters et al., 1994). Mts2^{+}=S4 interacts with MSS1=S7 (Gordon et al., 1993) and TBP1 with TBP7=S6 and the formation of TBP1/TBP7 heterodimers requires the N-terminal coiled-coil domains (Ohana et al., 1993). Data are not yet available on interactions between the ATPases and the non-ATPases within the 19S RC. Nor is it clear which subunits in the regulatory complex make direct contact with the 20S proteasomes.

C. ATPase and Kinase Activities

ATP has at least two distinct roles in the function of the 26S protease: first, the assembly of the 26S complex and, second, the degradation of substrates by the assembled complex. Although the exact mode of action of ATP remains unknown, on the basis of our knowledge about subunit composition and sequence homologies, ATP binding and hydrolysis is expected to be connected with the 19S RC rather than the 20S proteasome. It was shown in 1988 certain subunits of the 19S RC can be labeled either by crosslinking with α^{32}P ATP or by phosphorylation with γ^{32}P ATP (Hough et al., unpublished observations), indicating the existence of ATPase and kinase activities. ATP can be replaced only by CTP in the assembly process, whereas all four nucleoside triphosphates stimulate Ub conjugate degradation (Hershko and Ciechanover, 1992). These data indicate that there may be distinct ATP-binding sites involved in the two processes. Mutations in 19S RC ATPases led to the inability to degrade, but not to assemble, Ub conjugates (Seeger et al., 1996). A *S. cerevisiae* S4 homologue, Yhs4p, has been expressed in *Escherichia coli* (Lucero et al., 1995). Interestingly, the isolated Yhs4p ATPase possesses a nucleotide specificity similar to that of the assembled 26S proteasome and is inhibited by N-ethylmaleimide (NEM). Since polyanions activated the ATPase activity of Yhs4p, a modulation of the ATPase through interactions with negatively charged regions of neighboring proteins in the complex has been suggested (Lucero et al., 1995). Phosphorylated subunits have been identified both in 20S and 26S proteasomes. Since the protein kinases inhibitor, staurosporine, inhibited the assembly of the 26S protease from 20S proteasome and 19S RC, the involvement of phosphorylation in assembly has been suggested. The phosphorylated subunits in the 19S RC were S6=TBP7 and perhaps S10b=SUG2 (Yang et al., 1995). In fact, a kinase activity which coprecipitates with the 26S complex and phosphorylates SUG1 has been identified (Satoh et al., 1995). However, sequence homologies of the known 19S RC subunits with kinases have not yet been found.

D. Putative Mechanism of 19S Regulatory Complex Activities

It is believed that the l9S RC is responsible for binding, unfolding and perhaps transporting substrates. These assumptions are based on the facts that a non-ATPase, S5a, is sufficient for Ub conjugate binding and that polypeptide chains can only make their way into the proteolytic compartment of the 20S proteasome if they are completely unfolded (Wenzel and Baumeister, 1995). The variety of non-ATPases in the complex leads to speculations that they provide additional binding sites for other substrates of the 26S protease. However, so far there is no experimental data to support such an attractive idea. Unfolding of substrate proteins is a prerequisite for most protease catalyzed reactions including the 26S protease. The access to the proteolytic active sites inside the 20S proteasome is presumably controlled by a channel which functions by size exclusion. Whereas small peptides without secondary structures can pass into the proteolytic chamber formed by the two β rings, proteins are resistant to degradation by free 20S proteasome and are presumably unfolded by the 19S RC. Analogous to ClpA (Kessel et al., 1995), it has been suggested that the six ATPases of the regulatory complex form a hexameric ring which binds to the proteolytic core. A reverse chaperonin cycle as described for GroEL (Hartl, 1996) with emphasis on unfolding is a conceivable mechanism. How the unfolded proteins then pass to the proteolytic compartment is unknown. We assume that energy is necessary for this process and the ATPases might "pump" the unfolded proteins in this direction. Polyubiquitin chains, with high probability, do not go through the 20S proteasome. Therefore, they have to be cleaved off before the protein substrates reach their final destination. Isopeptidase T (Wilkinson et al., 1995) and an ATP-dependent isopeptidase activity found in isolated 26S protease complexes (Eytan et al., 1993) might be involved in the Ub recycling. In summary, the 19S complex modulates the 20S proteasome activities by selecting substrates and presumably transferring the polypeptide chains to the β subunits in the proteasome core. These functions require substrate recognition domains and metabolic energy.

IV. REGULATION OF 26S PROTEASOME ACTIVITY BY THE ASSEMBLY/DISASSEMBLY PROCESS

Regulation of the Ub pathway (see chapter 3) occurs on the level of the Ub conjugating enzymes (E2s) and the Ub ligases (E3s). Members of the E2 family specifically interact with E3s which target certain protein substrates and control the Ub conjugate formation. There is increasing evidence that the pathway is also regulated by the 26S protease assembly/disassembly process. The interconversion between 20S and 26S proteasomes switches Ub conjugate degradation on and off.

Early work on the identification and characterization of the 26S protease indicated that ATP is important for the stability of the enzyme and that the 20S proteasome shares subunits with the 26S complex (Hough et al., 1987). Later it was shown by many groups that the 20S proteasome is the proteolytic core of the 26S protease (for review see Hershko and Ciechanover, 1992). Hershko and co-workers separated three factors—CF-1, CF-2, and CF-3 (for conjugate-degrading factors 1–3)—with apparent molecular masses of 600, 250, and 650, respectively, that are all required for the degradation of proteins conjugated to Ub (Ganoth et al., 1988). CF-1 and CF-2 contain ATP-binding sites and CF-3 has been identified as the 20S proteasome (Eytan et al., 1989). The three factors assemble to the 26S protease in a reaction which requires ATP hydrolysis. On the other hand, the 19S regulatory complex from rabbit reticulocytes. (Hoffman et al., 1992) or the *Drosophila* μ particle (Udvardy, 1993) and bovine PA700 (Ma et al., 1994) associate with the 20S proteasome to form a 26S protease which degrades suc-Leu-Leu-Val-Tyr-MCA in an ATP-dependent manner. However, Ub conjugate degrading activity could not be reconstituted from the purified μ particle and the 20S proteasome. Consequently, it has been suggested that an additional protein factor, perhaps CF-2, is necessary reassembly of an Ub conjugate degrading complex (Udvardy, 1993). Considering their characteristics, the 19S RC and CF-1 seem to be very similar, if not identical. Just recently a protein complex has been identified and characterized which activates the peptidase activity of the 20S proteasome/PA700 complex up to eightfold (DeMartino et al., 1996) and which resembles CF-2 by its molecular mass and by its content of ATPases (TBP1 and SUG2).

Unfortunately, Ub conjugate degradation has not yet been measured in the presence or absence of this modulator of 26S proteasome activity. In any event, there seem to be at least two forms of 26S enzyme, one with and one without the modulator. It is possible that these two forms are identical with the two isoforms of the 26S protease identified in preparations from reticulocytes (Hoffman et al., 1992) and rat liver (Ugai et al., 1993) and that they have different functions in cells.

Intracellular calcium mobilization has been suggested to regulate the activity of the 26S protease during the metaphase-anaphase transition in the ascidian meiotic cell cycle by stimulating the association between the 20S proteasome and the 19S RC (Kawahara and Yokosawa, 1994). It seems to be a calcium/calmodulin-dependent event indicating the involvement of a calcium/calmodulin-dependent kinase in the regulation of 26S protease activity/assembly (Sawada et al., 1996). Data with *Xenopus* egg activation show that there is a calcium-induced assembly of the 26S proteasome which might be regulated during the cell cycle (Aizawa et al., 1996).

The assembly/disassembly cycle of the 26S proteasome could be imagined as follows (see Figure 3) a protein kinase, perhaps a calcium/calmodulin-dependent component of the 19S RC, phosphorylates SUG1 and/or other ATPases of the com-

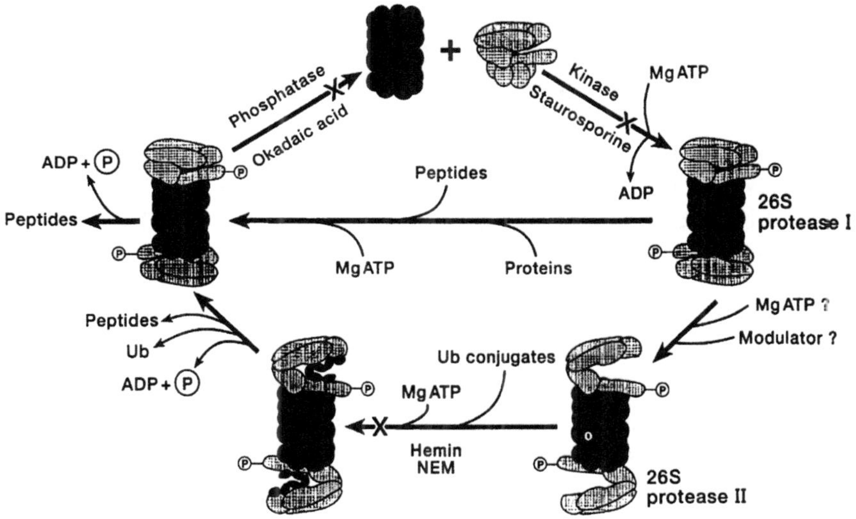

Figure 3. Putative mechanism of 26S proteasome assembly/disassembly. The 20S proteasome associates with the 19S RC in an ATP-dependent reaction. A kinase which is sensitive to staurosporine (Yang et al., 1995) might be involved in the formation of 26S protease I. This form of the 26S complex degrades peptides and perhaps yet unknown proteins in an ATP-dependent manner, but it is unable to cleave Ub conjugates. It interacts with the modulator (which is perhaps identical to CF2) to form the 26S protease II which is able to degrade Ub conjugates. Both the ATPase and the protease activities of the two 26S protease forms can be inhibited by hemin and NEM. The modulator might dissociate from 26S protease II as observed during 26S protease preparations. A phosphatase, sensitive to ocadaic acid, could lead to the disassembly of the 26S protease by dephosphorylation of a 19S RC component. The different shapes of the 19S RC indicate different putative conformational stages. See text for more detail.

plex, causing a conformational change of the ATPase ring which subsequently directly or indirectly associates with the α rings of the 20S proteasome. The interaction of the two subcomplexes leads to activation of the 20S proteasome peptide bond hydrolyzing activity and the 19S RC ATPase activity. These two activities are closely linked and are both inhibited by NEM, hemin, and vanadate (Kanayama et al., 1992; DeMartino et al., 1994). The assembly-induced ATPase activity can be due to the six ATPases identified in the 19S RC and is essential for Ub conjugate degradation. To switch to a protease form which degrades Ub conjugates (26S protease II) the modulator might be necessary. The 26S protease I degrades peptides and perhaps yet unknown proteins in an ATP-dependent manner. Dephosphorylation of the 19S RC leads to disassembly and to the inactivation of the complex.

The assembly of the 20S proteasome and the 19S RC are not fully understood. Coexpressed α and β subunits of the *Thermoplasma* 20S proteasome assemble to a proteolytically active enzyme in *E. coli*. The propeptide of the β subunit is not essential for this process. α subunits, which do not have propeptides, assemble into heptameric rings in the absence of β subunits. β subunits alone are monomeric and posttranslational processing does not occur in the absence of α subunits (Zwickl et al., 1994). In eukaryotes β5-type subunits are also synthesized as proproteins which are processed in 13S to 16S precursor complexes indicating the existence of a controlled assembly pathway for the 20S proteasome (Frenzel et al., 1994; Yang et al., 1995). It has been proposed that seven α-type subunits first associate with each other to a heptameric ring to which seven β-type subunit precursors subsequently assemble. The order of the ring structure is controlled by the α subunits. At this time exchangeable subunit homologue such as LMP2 and LMP7 can enter the complex (Früh et al., 1994). It is during the assembly of the intermediate complexes that the prosequences of the β subunits are cleaved off. The function of the prosequences and how they are cleaved is not yet fully understood. They may play a role for correct protein folding, for determining the position in the ring structure, and for protecting cellular proteins against proteolysis by inhibition of the proteolytic active sites localized on the β subunits. In the last step the precursor complex consisting of seven α and seven β subunits forms a dimer in the order αββα, the functional 20S proteasome (Zwickl et al., 1994).

The assembly of the 19S RC has not yet been studied systematically. Preliminary experiments with human subunit 11 indicate that there might be other complexes in addition to the 19S RC containing the polypeptide (Seeger and Dubiel, unpublished observations). Whether these are precursor complexes or degradation products of the 19S RC remains to be seen.

V. SUBSTRATE RECOGNITION BY THE 26S PROTEASOME

A. Substrates of the 26S Enzyme

As noted in the introduction, an increasing number of intracellular proteins are being identified as 26S proteasome substrates. In 1991, there were less than a handful of natural substrates of the ubiquitin-dependent pathway (Rechsteiner, 1991); a recent review lists almost 20 (Hilt and Wolf, 1996). We assume that ubiquitin-dependent substrates are necessarily degraded by the 26S proteasome since it is the only enyzme identified so far that is capable of hydrolyzing ubiquitylated proteins. We recognize, however, that this assumption is difficult to prove. The substrates of the 26S proteasome are found in the nucleus, the cytosol, and surprisingly in cellular membranes. Most known substrates are short-lived proteins with important regulatory functions, e.g., the cycling, IκBα, an inhibitor of the transcription factor NFκB, and transcription factors such as MATα2 or GCN4, etc. Whether longer lived cellular proteins, like lactate dehydrogenase, are degraded by the 26S proteasome is a matter for debate. Nonetheless, eukaryotic cells express so many rapidly degraded proteins that we suspect it will soon be impossible to present an updated list of 26S proteasome substrates in a review of reasonable length. So rather than do this, we will instead discuss issues relating to substrate recognition and the mechanism of polypeptide degradation by the 26S enzyme.

B. The 26S Proteasome Is not a Ubiquitin-Dependent Enzyme

Although we assume that ubiquitylated proteins are degraded exclusively by the 26S proteasome, the converse is clearly not the case. That is, some proteins can be degraded by the 26S proteasome without their prior marking by ubiquitin. The first piece of evidence that the rapidly degraded enzyme, ornithine decarboxylase (ODC), is degraded in a ubiquitin independent reaction was obtained in 1989 by Ciechanover and his colleagues (Rosenberg-Hasson et al., 1989). Using antibodies to remove the ubiquitin activating enzyme, E1, from reticulocyte lysate, they observed that p53 was stable in depleted extract, but ODC was still rapidly degraded. This result was significantly extended by Murakami et al. (1992) who demonstrated that the purified 26S proteasome degrades ornithine decarboxylase in a reaction requiring ATP and the polyamine induced protein antizyme, but not ubiquitin or ubiquitin activating enzymes. In further support of the idea that ornithine decarboxylase degradation does not require ubiquitylation, Mahaffey et al. (manuscript in preparation) have found that proteolysis of full-length ornithine decarboxylase in reticulocyte lysate is not inhibited by addition of subunit 5a. This 26S proteasome component has been shown to bind Ub-lysozyme conjugates and free multiubiquitin chains and it is a powerful inhibitor of cyclin degradation in *Xenopus* egg extracts and the degradation of preformed ubiquitin conjugates by the human 26S proteasome (Deveraux et al., 1995b). Thus, three independent studies indicate that ODC is degraded in a ubiquitin-independent fashion, and one report clearly implicates the 26S proteasome.

The transcription factor, c-jun, provides a second example of a ubiquitin independent substrate for the 26S proteasome. Although ubiquitin can be conjugated to c-jun (Stancovski et al., 1995; Hermida-Matsumoto et al., 1996), Piechaczyk and his colleagues have presented evidence that this modification is not required for degradation of c-jun by the 26S proteasome (Jariel-Encontre et al., 1995). They found that recombinant c-jun was degraded in an ATP-dependent reaction that required the 26S proteasome; additional cofactors such as ubiquitin or antizyme, were not needed.

C. The Role of Ubiquitin in Substrate Recognition

The 26S proteasome was identified by its ability to degrade ubiquitin-lysozyme conjugates (Hough et al., 1986). This substrate consists of lysozyme molecules attached to multiubiquitin chains, and Chau et al. (1989) have shown that long ubiquitin polymers markedly increase the rate of β-galactosidase degradation. Some cellular proteins, for example arthrin, a ubiquitin conjugate of insect actin (Ball et al., 1987), are found conjugated to one or a lew ubiquitin moieties. In many cases, these proteins are not rapidly degraded. Their metabolic stability could reflect the fact that they are sequestered from the 26S proteasome; e.g., ubiquitin-histone H2A by binding to DNA or arthrin by its assembly in flight muscle sarcomeres. Alternatively, monoubiquitylated proteins may simply not be recognized by the 26S proteasome. A recent paper suggests, however, that this second possibility is not correct. Shaeffer and Kania (1995) isolated ^{125}I-α-globin molecules conjugated to one, two, or a mixture of three and four ubiquitins. Each of these conjugates was then incubated with 26S or 20S proteasomes. Whereas α-globin itself was not degraded by the 26S proteasome, the monoubiquitin adduct was proteolyzed at a substantial rate (24% per hour), and α-globin molecules attached to two or more ubiquitins were degraded threefold faster. Degradation of the globin conjugates by the 26S proteasome was ATP-dependent. The study is clouded to some extent by a surprisingly high rate of proteolysis of the globin conjugates by the 20S proteasome which normally does not degrade intact proteins. Nonetheless, the presented evidence strongly suggests that Ub_2-α-globin is a 26S proteasome substrate; Ub_1-α-globin may be one as well.

Several studies have shov n that multiubiquitin chains per se are not mandatory for substrate recognition by the 26S proteasome. Hershko and Heller (1985) blocked the amino groups on ubiquitin by reductive methylation and showed that the modified ubiquitin produced much lower molecular weight conjugates of lysozyme. They concluded that high molecular weight conjugates are produced by the formation of multinbiquitin chains in which one molecule of ubiquitin is linked to an amino group on another ubiquitin. (As mentioned above, this idea was confirmed by Chau et al. (1989) several years later). Despite the fact that multiubiquitin chains were not formed, the MeUb-lysozyme conjugates were degraded in the reconstituted extract at half the rate of lysozyme conjugates formed from unmodified ubiquitin. Studies using histone H3 as substrate confirmed that conjugates formed from methylated ubiquitin are degraded by the 26S proteasome (Haas et al., 1990). A subsequent paper from Hershko's group in which five proteins were used as substrates demonstrated that the requirement for multiubiquitin chains differs significantly among proteins (Hershko et al., 1991). For example, MeUb conjugates of lysozyme, S-protein and β-lactoglobulin were degraded at 30–40% the rate seen with native Ub conjugates. By contrast the degradation of α-lactalbumin, oxidized ribonuclease, and clam cyclin was significantly inhibited when methylated ubiquitin was added to the reaction mixes. The requirement for multiubiquitin chains thus appears to be substrate specific.

D. Recognition of Multiubiquitin Chains by the 26S Proteasome

It has been known for more than a decade that high molecular weight conjugates are preferred substrates for 26S proteasome. The molecular basis for this preference is beginning to emerge from the recent discovery of a 26S proteasome subunit

(S5a) that can bind multinbiquitin chains. Using far western blots, Deveraux et al. (1994) demonstrated that radiolabeled ubiquitin lysozyme conjugates bind the regulatory complex and the 26S protease but not the 20S proteasome. In a surprising result it was found that the ^{125}I-conjugates bound a subunit of the regulatory complex even after the latter had been resolved by SDS-PAGE. This subunit, now known as S5a, also binds free multiubiquitin chains. Moreover, when incubated with ubiquitin polymers of various lengths, S5a selected species containing four or more ubiquitin monomers and exhibited increased affinity for longer chains (Deveraux et al., 1994). These two properties of S5a, length dependent binding and insensitivity to denaturation, coupled with the x-ray structure of the ubiquitin tetramer (Cook et al., 1994), suggested a model in which tetramers and longer chains of ubiquitin bound loops on S5a (Deveraux et al., 1995a). Increased affinity for longer chains could be explained by the presence of multiple loops on the proteasome subunit and resistance to denaturation could be explained by assuming the loops were short enough to readily renature after SDS-PAGE and transfer of S5a to membranes. The model envisions that repeating grooves in the multiubiquitin chain would interact with repeated loops on S5a.

Several recent publications (Haracska and Udvardy, 1995; van Nocker et al., 1996a) and unpublished observations from one of our laboratories support the repeated loop-groove model. The sequence of subunit S5a obtained from three higher eukaryotes reveal repeated motifs in the C-terminal half of S5a (Ferrell et al., 1996). These motifs are characterized by a stretch of five hydrophobic residues followed by a conserved serine and flanked by several aspartates and/or glutamates. The presence of repeated hydrophobic stretches in S5a is highly relevant to substrate recognition in view of the observation of Beal et al. (1996), who have shown that the hydrophobic residues L8, I44, and V70 in ubiquitin are critical for targeting. Mutations of pairs of these residues to alanine had little effect on attachment of ubiquitin to substrates but severely inhibited degradation of the resulting conjugates. The same mutations blocked the binding of the ubiquitin chains to S5a. The side-chains implicated in this binding—L8, I44, and V70—form repeating patches on the chain surface. Thus, hydrophobic interactions between these patches and S5a apparently contribute to enhanced proteolytic targeting by multiubiquitin chains. Recent deletion analyses of recombinant S5a have identified two independent regions of the protein that bind multiubiquitin chains (Deveraux et al., in preparation). These regions correspond to the hydrophobic clusters in S5a, strongly supporting the idea that hydrophobic loops on S5a bind to hydrophobic patches in grooves on multinbiquitin chains.

To some extent, we are beginning to understand molecular details of the recognition of ubiquitylated substrates by the 26S proteasome. However, S5a cannot be the whole story since this subunit does not exhibit substantial affinity for ubiquitin monomers, dimers, or trimers, yet Ub_2-globulin is a good substrate. Of more importance, deletion of the gene for S5a in yeast is not lethal (Toh-e, personal communication; van Nocker et al., 1996b) indicating that other 26S proteasome components recognize ubiquitylated proteins. Also, we do not know how antizyme-ODC complexes or unmodified c- jun molecules are recognized by the 26S enzyme. Thus, there remains a great deal of work to be done.

VI. RFGULATORS OF THE 20S PROTEASOME: THE PROTEASOMAL SYSTEM AND ANTIGEN PRESENTATTON

A. 20S Proteasome Activators

As described above, the peptidase activity and substrate specificity of the 20S proteasome can be influenced by association with the regulatory complex. We and others discovered a second, more potent activator that differentially increases peptide hydrolysis by proteasome. In the presence of the activator? hydrolysis of some peptides increased 200-fold, others only threefold (Yukawa et al., l991; Di-Cola 1992; Dubiel et al., 1992a; Ma et al., 1992). We call this proteasome activator REG, whereas DeMartino and his colleagues call it PA28. REG sediments at 11S and binds the proteasome reversibly in the absence of nucleotides (Dubiel et al., 1992a). The activator purified from human red cells is composed of two subunits (REGα and REGβ) of molecular weight 28.6 kDa and 27.3 kDa, respectively. REGα and REGβ are also known as PA28α and PA28β (Ahn et al., 1995), and have been referred to as REG29k and REG31k because of their mobility on SDS-PAGE. REGα and REGβ have been expressed in *E. coli* and each is active independently of the other (Realini et al., 1994b; Realini et al., in preparation). Human REGα and REGβ share 47% sequence identity. They also share approximately 40% sequence identity with another human protein called Ki, which is an autoantigen prevalent in patients with systemic lupus erythematosus (Nikaido et al., 1990). Ki has a predicted molecular weight of 29.5 kDa. A cDNA encoding Ki has been cloned and the recombinant protein activates the proteasome (Realini et al., in preparation); accordingly we refer to Ki as REGγ to emphasize its relationship to REGα and REGβ. As shown in Figure 4, sequences from the three REG subunits are very similar throughout, except for a unique insert of l6–33 residues. It is tempting to speculate that each of these inserts imparts a distinct biological function such as substrate binding, subcellular localization, or differential activation of proteasome β subunits.

The insert unique to REGα is a lysine-glutamate rich region extending from lysine 70 to lysine 97. This segment of REGα binds Ca^{2+} ions and Ca^{2+} reversibly inhibits the activation of proteasome by REGα (Realini and Rechsteiner, 1995). This KEKE motif is particularly interesting because, as mentioned above, similar tracts of alternating glutamates and lysines are present in proteins known to associate with each other and with REG, namely two proteasome α subunits as well as 26S protease subunits l, 2, 5a, l0, and 12. KEKE motifs are also found in various chaperoning, including hsp 90 which associates with the proteasome (see below). It has

REGα	MAMLRVQPEAQAK	VDVFREDLCTKTENLLGS	YFPKKISELDAFLKEPAL	NEANLSNLKAPLDIPVPDP
REGβ	MAKPCGVRLSGEARKQ	VEVFRQNLFQEAEEFLYR	FLPQKIIYLNQLLQEDSL	NVADLTSLRAPLDIPIPDP
REGγ	MASLLKVDQEVKLK	VDSFRERITSEAEDLVAN	FFPKKLLELDSFLKEPIL	NIHDLTQIHSDMNLPVPDP

VKEKEKEERKKQQEKEDKDEKKKGEDEDK	GPPCGPVNCNEKIVVLLQRL
PPKDDEMETDKQEKKE	VPKCGFLPGNEKVLSLLALV
ILLTNSHDGLDGPTYKKRRLDECEEAFQGTKVF	VMPNGMLKSNQQLVDIIEKV

KPEIKDVIEQLNLVTTWL	QLQIPRIEDGNNFGVAVQ	EKVFELMTSLHTKLEGFH	TQISKYFSERGDAVTKAA
KPEVWTLKEKCILVITWI	QHLIPKIEDGNDFGVAIQ	EKVLERVNAVKTKVEAFO	TTISKYFSERGDAVAKAS
KPEIRLLIEKCNTVKMWV	QLLIPRIEDGNNFGVSIQ	EETVAELRTVESEAASYL	DQISRYYITRAKLVSKIA

KQPHVGDYRQLVHELDEA	EYRDIRLMVMEIRNAYAV	LYDIILKNFEKLKKPRGE	TKGMIY
KETHVMDYRALVHERDEA	AYGELRAMVLDLRAFYAE	LYHIISSNLEKIVTPKGE	EKPSMY
KYPHVEDYRRTVTEIDEK	EYISLRLIISELRNQYVT	LHDMILKNIEKIKRPRSS	NAETLY

Figure 4. Alignment of REG sequences. The sequences show the greatest variation in the region corresponding to the KEKE motif of REGα (bold).

been hypothesized that KEKE sequences mediate protein-protein interactions (Realini et al., 1994a) and this idea is supported by the observation that KEKE fusions cause heterologous proteins to cosediment with REGα (Realini and Rechsteiner, unpublished observations). Likewise, Nobel et al., (1994) have identified a KEKE region in MAP1B as being responsible for its binding to microtubules. Also, Lu and others (1994) have shown that the interaction of dihydropyridine receptors with the calcium release channel involves KEKE regions. The prevalence of KEKE sequences in precursors to peptides presented on major histocompatibility complex (MHC) class I molecules further suggests that KEKE interactions may mediate widespread associations including the assembly of proteolytic complexes and the targeting of peptides for presentation on MHC class I molecules (Realini et al., 1994a).

Structural Studies on REG

Electron microscopy studies by DeMartino and his colleagues show that bovine REG forms ring structures that can bind one or both ends of the proteasome (Grey et al., 1994). These studies indicate the formation of two types of REG ring structures, the smaller of which (~105 Å diameter x 75 Å) is consistent with images of REG attached to the end of proteasomes (see Figure 5). It has previously been argued that REG rings are hexameric, based upon gel filtration and centrifugation through glycerol gradients. The electron microscopy study also suggested that REG was hexameric from the apparent size of the REG cap on the end of the proteasome and from an assumption of the protein density within this volume (Grey et al., 1994). However, preliminary x-ray crystallographic results, equilibrium sedimentation, and cross-linking, demonstrate that, in fact, REGα has a heptameric ring structure (Johnston et al., in preparation). This result has significance for REG-proteasome interactions and supports the view that the eukaryotic proteasome will display

Figure 5. Structures of REG/proteasome. The drawing is based on images in Gray et al. (1994). (**A**) Averaged electron microscopic images of REG. (**B**) Averaged electron microscopic images of proteasome capped with REG at one or (**C**) both ends.

pseudo sevenfold symmetry, like the archaebacterial proteasome, rather than six- or eightfold symmetry as has been reported (Arrigo et al., 1988). Light scattering experiments suggest that REGγ is also a heptamer. Surprisingly, however, it appears that recombinant REGβ forms a dimer (Johnston et al., in preparation).

DeMartino and colleagues observed that REG is rapidly inactivated by carboxypeptidase(s) present in extracts from tissues other than red cells (Ma et al., 1993). Apparently, removal of two amino acids from the C-terminus of REG abolishes its binding to proteasome, but does not cause it to dissociate to monomers. Support for the importance of the C-terminus of REG in binding to the proteasome is provided by our preliminary data which show that mutation of the C-terminal Tyr to Cys results in a REGα variant that forms normal heptamers but has greatly reduced activity (Zhang et al., in preparation). We have also isolated a heptameric, but inactive, variant of REGα that is extended by an additional 20 residues at the C-terminus. It is likely that the C-terminal segment of REG binds to the proteasome. It is possible that REG is a two-domain protein with a C-terminal domain that binds the proteasome separated from the more distant N-terminal domain by the variable inserts shown in bold in Figure 4. To speculate further, the variable insert could line the central opening of the REG ring, and in that position play a direct role in the selection and delivery of substrates to the proteasome active sites. The central hole visible in electron microscopic images appears to pass completely through the middle of the REG oligomer (Grey et al., 1994). It is attractive to speculate that this is the funnel through which substrates pass. DeMartino and colleagues, however, caution that this apparent hole does not necessarily traverse the entire ring, and a central domain may be masked by deposition of stain in a central depression (Grey et al., 1994).

Possible Mechanisms for Proteasome Activation by REG Homologues

There are a number of possible mechanisms by which REG may alter proteasome activity. REG could act as a filter by allowing passage of only a subset of potential substrates, although this model cannot explain the increase in activity seen against most substrates. Moreover, small substrates of similar size and chemical character are hydrolyzed at very different rates, despite the expectation that they would have similar success in passing a REG gate. It seems unlikely that REGs function in an active delivery capacity because their effects on proteasome activity do not require energy. It is possible, however, that REG increases the local concentration of selected substrates at the mouth of the proteasome by presenting a large surface area that binds suitable substrates weakly. It is most attractive to believe that REG binding produces a conformational change at the active sites of the proteasome β subunits, some 50 Å distant. DeMartino's group has shown that REG converts substrate versus velocity plots from sigmoidal (cooperative) to hyperbolic (noncooperative) for certain oligopeptide substrates (Ma et al., 1992). Therefore, REG may function by loosening the contact between proteasome β subunits. Low concentrations of detergent (SDS) have a similar effect of increasing proteasome activity while relieving cooperativity (Arribas and Castano, 1990; Orlowski et al., 1991). Change at the proteasome β subunits induced by REG cannot be very large, however, because there is no apparent change in proteasome structure upon REG binding as seen by electron microscopy (Grey et al., 1994).

Proteasome Activators and Antigen Presentation

The immune system determines the status of a cell as being self or nonself, in part because cytotoxic T cells recognize peptides displayed on the cell surface as complexes with MHC class I molecules (for reviews see Yewdell and Bennink, 1992; Germain, 1994). Presented peptides are generated in the cytosol and transported into the endoplasmic reticulum where they bind MHC class I molecules (Ortmann et al., 1994; Suh et al., 1994; Grandea et al., 1995). The chaperone calnexin also binds to and retains MHC class I molecules (Jackson et al., 1994; Rajagopalan and Brenner. 1994), probably to provide additional time for proper assembly and peptide insertion prior to transport to the cell surface for display. Interestingly, the short cytosolic tail of calnexin contains a KEKE motif similar to one that we have shown can mediate protein interaction with REGα (Realini and Rechsteiner, unpublished observations). This led to speculation that a multiprotein complex of REG, proteasome, calnexin, TAP, and MHC class I molecules, may function in antigen processing and presentation (Realini et al., 1994a).

There is considerable evidence implicating REG and the proteasome in antigen processing. Proteasome inhibitors prevent production of peptide ligands for MHC class I molecules from ovalbumin (Rock et al., 1994). The cytokine INFγ, which is released during viral infection, signals production of numerous components of the antigen processing and presentation machinery, including TAP1/TAP2 transporters, MHC classI/II molecules, and β_2-microglobulin, all of which are encoded in the MHC region (Townsend and Bodmer, 1989). Stimulation of antigen presenting cells with INFγ also causes a change in proteasome subunit composition by the upregulation of two MHC-encoded proteasome subunits, LMP2 and LMP7. And the compensating downregulation of two constitutive subunits (Glynne et al., 1991;

Kelly et al., 1991; Martinez and Monaco, 1991; Yang et al., 1992; Monaco, 1992; Akiyama et al., 1994; Belich et al., 1994; Boes et al., 1994; Früh et al., 1994). We have shown that expression of REGα is also increased in reponse to INFγ (Realini et al., 1994b). Tanaka and his colleagues have extended this observation to cDNAs for REGβ and REGγ as well (Ahn et al., 1995). Proteasome preparations containing LMP2 and LMP7 have increased activity against model peptides that have Tyr or Phe N-terminal to the scissil bond (Ustrell et al., 1995a,b), and such peptides are highly represented on MHC class I molecules (Falk and Rotzschke, 1993). Furthermore, mice with a deletion of the gene encoding LMP7 have reduced levels of MHC class I cell-surface expression (Fehling et al., 1994). However, cells deficient in expression of LMP2 and LMP7 are capable of processing antigen and delivering it to the MHC class I molecules in the endoplasmic reticulum (Arnold et al., 1992; Momburg et al., 1992).

Modulation of proteasome specificity and activity by LMP2 and LMP7 is modest in comparison to the effect of REG (Boes et al., 1994; Ustrell et al., 1995b). The preference of MHC class I molecules for peptides with large hydrophobic C-terminal residues: exactly matches the sequence preference of REG activated proteasomes from cells expressing LMP2 and LMP7 (Rechsteiner et al., manuscript submitted). Proteasome-mediated cleavage after branched chain or hydrophobic amino acids is enhanced both by the presence of LMP2/LMP7 and REG. For certain substrates, these components act synergistically to increase hydrolysis after Ile, Leu, Met, Phe, and Tyr by an order of magnitude (Rechsteiner et. al., manuscript submitted).

Kloetzel and Koszinowski and their colleagues have used electrospray mass spectrometry to assay the effect of REG on the cleavage of a 25 residue peptide by proteasomes that contain low or elevated levels of LMP2 and LMP7 (Groettrup et al., 1995). This study demonstrated that REG increases the quantity and quality (i.e., variability) of peptides that can be generated in antigen presenting cells, and that this effect of REG is seen for all proteasome variants examined independently of the levels of LMP2 or LMP7 subunits. Using similar methodology this same group showed that the frequency with which peptides are presented at the cell surface by MHC class I molecules correlates with their rate of cleavage by proteasomes (Aggers et al.,1995).More recently, Groettrup et al. (1996) overexpressed REGα in mouse cells transfected with a plasmid encoding murine cytomegalovirus pp89 protein and observed marked enhancement of recognition by pp89-specific cytotoxic T cells. This *in vivo* demonstration of enhanced antigen presentation clearly points to a role for REG proteins in presentation of peptides bound to MHC class I molecules.

Which Complexes of the Proteasomal System Produce the Peptides Presented by MHC Class I Molecules?

The best cellular response towards invasion by foreign proteins is, with high probability, the production of the greatest possible variety of peptides able to bind to MHC class I molecules and eventually be recognized by pre-existing cytotoxic T lymphocytes (CTL). As mentioned above, IFNγ induces components of the proteasomal system which increase the variety and perhaps the quality of the produced peptides. It has not been shown whether there is a complete shift to one proteasome population containing LMP2, LMP7, and MECL-1 or whether a variety of proteasome populations coexist alter INFg induction *in vivo*. Apparently REGα regulates peptide production by the 20S proteasome (Groettrup et al., 1996), but the effects of REGβ and REGγ remain undefined. Inhibitors of both the 20S and 26S proteasomes lead to a decreased generation of peptides presented on MHC class I molecules (Rock et al., 1994). These inhibitors do not distinguish between the two proteases and, thus, which complex generates peptides presented is not known. The role of the 26S protease in antigen processing is unclear. It has been shown that ubiquitylation supports antigen presentation derived from ovalbumin (Michalek et al., 1993) and the initial cleavage of antigens by the 26S enzyme can be assumed. since intact proteins are mostly resistant against 20S and 20S- 11SREG complexes. Whether the 26S proteasome is able to produce peptides that can be presented by MHC class I molecules or whether it generates intermediates which must be further processed by 20S and/or 20S-11SREG complexes is not yet known. Two alternative pathways of antigen processing are possible: (1) a pathway exclusively mediated by the 26S proteasome and (2) a pathway that involves the 26S, 20S, and 20S-11S REG complexes. In the latter case initial cleavages of ubiquitylated antigens would be made by the 26S protease. The generated intermediates might remain in the lumen of the proteolytic core, the 20S proteasome, which might then be handed from the 19S RC to the 11S REG for further processing before transport of peptides into the endoplasmic reticulum.

B. 20S Proteasome Inhibitors

A number of groups have reported on inhibitors of the 20S proteasome. Etlinger and his colleagues were among the first to propose that protein inhibitors play a central role in ubiquitin-dependent proteolysis (Speiser and Etlinger, 1982). Since it is now clear that the proteasome is responsible for cleaving ubiquitylated substrates, the inhibitors presumably targeted the 26S proteasome. They suggested that ATP stimulates proteolysis in reticulocyte extract by repressing an endogenous protein inhibitor. In support of this idea, Murakami and Etlinger (1986) reported the existence of a hexameric complex of 40 kDa subunits that inhibits both calpain and the proteasome. In a follow-up paper, Li and Etlinger (1992) report that a ubiquitylated derivative of the 40 kDa inhibitor is a component of the 26S protease. Driscoll et al. (1992) similarly report that a 250 kDa ATP-stabilized inhibitor composed of 40 kDa subunits is a 26S protease component. They claim that this inhibitor corresponds to a conjugate degrading factor, CF2, identified in 1988 by Hershko et al. (1988). Both reports should be viewed with skepticism. The subunit pattern of the "26S protease" described by Li and Etlinger is not that of the 26S protease, and the

26S protease does not contain enough 40 kDa subunits for the 250 kDa inhibitor of Driscoll et al., (1992) to be present in the ATP-dependent enzyme

Etlinger and colleagues have also reported on a separate proteasome inhibitor purified from human erythrocytes (Li et al., 1991). This component is apparently a tetramer of 50 kDa subunits, and it is distinct by immunological criteria from the 40 kDa inhibitor described above. DeMartino and his colleagues have reported on a self-associating 31 kDa protein that inhibits the proteasome (Ma et al., 1992). The inhibitor, obtained from bovine erythrocytes, forms multimers under non-denaturing conditions. The protein inhibited hydrolysis of protein substrate by an activated form of the proteasome as well as fluorogenic peptides specific for the trypsin-like, chymotrypsin-like and glutamyl hydrolyzing sites. Interestingly, the inhibitor did not suppress hydrolysis of the trypsin-like substrate z-VLR-MCA when latent proteasomes were used. The properties of the 31 kDa inhibitor are similar to those of the REGγ described above (Realini et al., submitted), and it is possible that the 31 kDa inhibitor identified by DeMartino et al. is REGγ.

Three groups have shown that hsp 90 can inhibit the peptidase activity of the 20S proteasome. Tsubuki et al. (1994) purified a bovine brain protein that specifically inhibits the cleavage of the fluorogenic substrate, z-leu-leu-leu-MCA by the proteasome. Partial amino acid sequencing and immunodetection revealed that the inhibitor was hsp 90. At a molar ratio of 1:1 with the proteasome, the heat shock protein inhibited peptide hydrolysis by 50% and the observed inhibition was noncompetitive. Khan et al. (1995) reported similar findings using human erythrocyte membrane preparations. However, they found that human hsp 90 inhibited three peptidase activities: the trypsin-like, chymotrypsin-like, and the peptidyl glutamyl peptide hydrolyzing activity. Wagner and Margolis (1995) also report that hsp 90 is an inhibitor of z-leu-leu-leu-MCA cleavage by the 20S proteasome from bovine lens. In accord with Tsubuki et al. (1994), they did not observe inhibition of fluorogenic peptide cleavage by the trypsin- and chymotryspsin-like activities in the proteasome.

The finding that hsp 90 can inhibit cleavage of specific peptides by the proteasome is intriguing with regard to antigen presentation. Peptides derived from exogenous antigens are usually presented on MHC class II molecules, whereas endogenous proteins are cleaved into peptides that are presented on class I molecules (Rock, 1996) Srivastava and his colleagues have reported that gp96, an endoplasmic reticulum equivalent of hsp 90, is able to prime CTL responses to exogenous antigens (Suto and Srivastava, 1995). Presumably peptides bound to the added gp96 somehow gain access to a compartment in which they can be loaded onto MHC class I molecules. A similar situation involving cytosolic hsp 90 rather than gp96 might exist in the cytosol. REGs could activate cleavage of precursors by branch chain preferring subunits; e.g., LMP2, LMP7, and MECL-1, to generate peptides with appropriate C-termini for binding MHC class I molecules. One can imagine that after such initial cleavage, REG would dissociate, and hsp 90 would bind to the proteasome thereby inhibiting further peptide bond cleavages. It is conceivable that hsp 90 then binds the peptide and transfers it to TAPs for subsequent

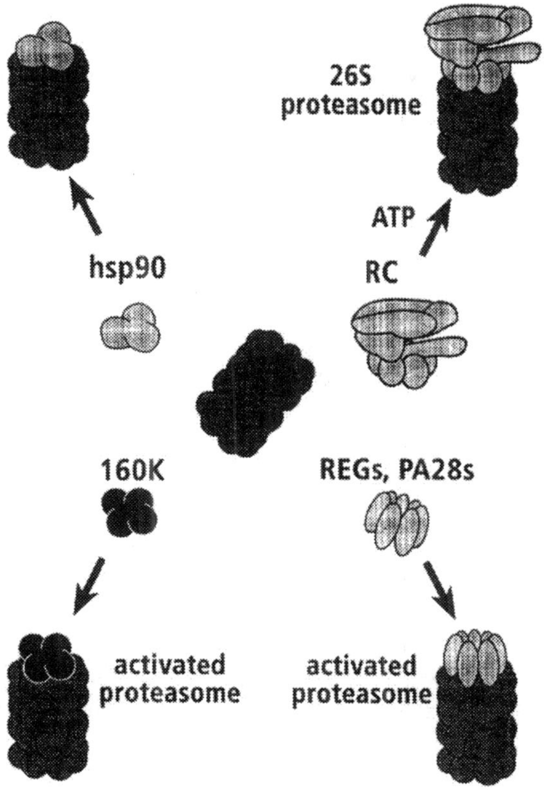

Figure 6. The proteasomal system. The system consists of a universal proteolytic core, the 20S proteasome, which associates with a number of endogeneous regulators. Three activators and one inhibitor of the 20S proteasome are shown. The presumptive 160 kDa activator has been described by Hoffman et al. (1992). The two activators, 19S regulatory complex (RC) and REGs, and the inhibitor, hsp 90, are discussed in the text.

transport into the lumen of the endoplasmic reticulum for loading onto MHC I molecules. Potential interaction of the proteasome with activators and hsp 90 is shown in Figure 6.

VII. THE PROTEASOMAL SYSTEM: A TARGET OF VIRUSES

The importance of the proteasomal system for transcription, antigen processing, and cell division makes it target for viruses. Their strategies include using the cells transcriptional/translational systems for the production of viral proteins, suppressing the cellular immune response and reproducing by replication during the cell cycle. A number of viruses are known to interfere with the conjugation of protein substrates to Ub. The African swine fever virus encodes a protein with homology to the Ub conjugating enzymes (Rodriguez et al., 1992), suggesting that viral infection may alter the specificity of the Ub pathway in the host. The viral E2 is active and ubiquitylates histones *in vitro* (Mayer et al., 1991). The two baculoviruses *Autographa california* (Guarino et al., 1995) and *Orygia pseudotsugata* (Russel and Rohrmann, 1993) encode proteins with more than 70% identity to eukaryotic Ub. Most of the residues known to be essential for Ub function have been conserved in the viral variants. The viral Ubs are closely related to each other indicating that they have evolved for a specific role in the baculovirus life cycle. Recently it has been shown for the *A. california* virus that viral Ub is not essential for productive virus infection (Reilly and Guarino, 1996). However, without Ub the infectious titers were decreased and it was suggested that viral Ub might mediate its function by regulation of the Ub pathway. Ub was also found in viral membranes suggesting that it may be involved in the assembly of viral particles or in budding of virions from infected cells (Guarino et al., 1995).

The oncogenic human papillomavirus (HPV) types 16 and 18 utilize the Ub system for targeting the tumor suppressor factor p53. The E6-associated protein (E6-AP), which belongs to the family of E3s (Ub ligases) is activated by the HPV E6 oncoprotein. The E6/E6AP complex specifically interacts with p53 leading to its ubiquitylation and degradation (Scheffner et al., 1993).

There are indications that some viral proteins interact directly with the 20S proteasome. The hepatitis B virus protein, HBx, which is believed to transactivate a wide range of viral and cellular promoters, binds to a proteasomal α subunit. It has been shown by two-hybrid assays that HBx interacts with a human proteasome α subunit called XAP C7 (Fischer et al., 1995: Huang et al., 1996). One can assume that the oncogenic potential of HBx protein in certain transgenic mice and its essential role for HBV infection is mediated by the proteasomal system, thus explaining the pleiotropic effects of the protein.

Tax, a protein encoded by human T cell leukemia virus, HTLV- 1, activates NF-κB/Rel transcription factors. By two-hybrid screening it was found that it binds to two subunits of the 20S proteasome, HsN3 and HC9 (Rousset et al., 1996) Its interaction with the proteasome accelerates the processing of the NF-κB precursor, probably by activating the 26S proteasome.

A functionally similar polypeptide encoded by the human immunodeficiency virus (HIV-1), is the Tat protein. A number of transcriptionally independent functions in determining infectivity and cytopathicity during developing AIDS have been proposed. Two Tat-binding proteins, TBP1 and TBP7, are components of the 19S RC. We have evidence that Tat also interacts with the proteasomal system and modulates its antigen processing activity (Seeger et al., in press). The Tat effect might explain the escape of HIV-1 from cytotoxic T lymphocytes.

Thus, different viruses have developed strategies for utilizing and employing the proteasomal system of the host. This knowledge of viral mechanisms may provide the basis for future therapies.

VIII. CONCLUDING REMARKS

The considerable progress achieved in recent years on the structure and function of the 26S proteasome is based on the combination of genetic and biochemical approaches. In particular, genetic studies in yeast demonstrated the requirement for the 26S proteasome in several cell cycle transitions. Yeast genetics will be a major tool for identification of new substrates and for elucidation of subunit-subunit interactions in the 19S complex. The analysis of 26S proteases with mutations in individual regulatory subunits should shed light on the functions of the various subunits within the multimeric complex. We are just beginning to understand how Ub conjugates are recognized by the enzyme. We do not know how other substrates bind to the complex or how proteins are unfolded and ultimately degraded. There is preliminary evidence that assembly and activity of the 26S proteasome is regulated by phosphorylation, but actual mechanisms are not yet clear. To date there are only a few data available on regulation of the assembly/disassembly processes and mechanisms of association of endogenous regulators with the 20S proteasome. Studies of the proteasomal system have already produced conversations among researchers in different fields such as intracellular proteolysis, cell cycle regulation, transcriptional regulation, antigen processing, viral mechanism, etc. These interactions should produce even more exciting results in the future.

REFERENCES

Aggers, M., Boes-Fabian, B., Ruppert, T., Kloetzel, P.-M., & Koszinowski, U. H. (1995). The cleavage preference of the proteasome governs the yield of antigenic peptides. J. Exp. Med. 182, 1865-1870.

Ahn, J. Y., Tanahashi, N., Akiyama, K., Hisamatsu, H., Noda, C., Tanaka, K., Chung, C. H., Shimbara, N., Willy, P. J., Mott, J. D. et al. (1995). Primary structures of two homologous subunits of PA28, a gamma-interferon-inducible protein activator of the 20S proteasome. FEBS Lett. 366, 37-42.

Aizawa, H., Kawahara, H., Tanaka, K., & Yokosawa, H. (1996). Activation of the proteasome during *Xenopus* Egg activation implies a link between proteasome activation and intracellular calcium release. Biochem. Biophys. Res. Comm. 218, 224-228.

Akiyama, K., Yokota, K., Kagawa, S., Shimbara, N., Tamura, T., Akioka, H., Nothwang, H. G., Noda, C., Tanaka, K., & Ichihara, A. (1 994). cDNA cloning and interferon gamma down-regulation of proteasomal subunits X and Y. Science 265, 1231-1234.

Akiyama, K.-y., Yokota, K.-y., Kagawa, S., Shimbara, N., DeMartino, G. N., Slaughter, C. A., Noda, C., & Tanaka, K. (1995). cDNA cloning of a new putative ATPase subunit p45 of the human 26S proteasome, a homolog of yeast transcriptional factor Sug1p. FEBS Lett. 363, 151-156.

Arnold, D., Driscoll, J., Androlewicz, M., Hughes, E., Cresswell. P., & Spies, T. (1992). Proteasome subunits encoded in the MHC are not generally required for the processing of peptides bound by MHC class I molecules. Nature 360, 171-174.

Arribas, J., & Castano, J. G. (1990). Kinetic studies of the differential effect of detergents on the peptidase activities of the multicatalytic proteinase from rat liver. J. Biol. Chem. 265, 13969- 13973.

Arrigo, A.-P., Tanaka, K., Goldberg, A. L., & Welch, W. J. (1988). Identity of the 19S 'prosome' particle with the large multifunctional protease complex of mammalian cells (the proteasome). Nature 331, 192-194.

Ball, E., Karlik, C. C., Beall, C. J., Saville, D. L., Sparrow. J. C., Bullard, B., & Fyrberg, E. A. (1987). Arthrin, a myotibrillar protein of insect flight muscle, is an actin-ubiquitin conjugate. Cell 51, 221-218.

Baumeister, W., Dahlmann, B., Hegerl, R., Kopp, F., Kuchn, L., & Pfeifer, G. (1988). Electron microscopy and image analysis of the multicatalytic proteinase. FEBS Lett. 241, 239-245.

Beal, R., Deveraux, Q., Gang, X., Rechsteiner, M., & Pickart, C. (1996). Surface hydrophobic residues essential for proteolytic targeting by ubiquitin. Proc. Natl. Acad. Sci. USA 93, 861-866.

Belich, N. P., Glynne, R. J., Senger, G., Sheer, D., & Trowsdale, J. (1994). Proteasome components with reciprocal expression to that of the MHC-encoded LMP proteins. Curr. Biol. 4, 769-776.

Beutler, B., & Cerami, A. (1989). The biology of cachectin/TNF-a primary mediator of the host response. Ann. Rev. Immunol. 7, 625-655.

Boes, B., Hengel, H., Ruppert, T., Multhaup, G., Koszinowski, U. H., & Kloetzel, P.-M. (1994). Interferon γ stimulation modulates the proteolytic activity and cleavage site preference of 20S mouse proteasomes. J. Exp. Med. 179, 901-909.

Boldin, M. P., Mett, I. L., & Wallach, D. (1995). A protein related to a proteasomal subunit binds to the intracellular domain of the p55 TNF receptor upstream to its "death domain". FEBS Lett. 367, 39-44.

Cardozo, C., Vinitsky, A., Michaud, C., & Orlowski, M. (1994). Evidence that the nature of amino acid residues in the P3 position directs substrates to distinct catalytic sites of the pituitary multicatalytic proteinase complex (proteasome). Biochemistry 33, 6483-6489.

Chau, V., Tobias, J. W., Bachmair, A., Marriott, D., Ecker, D. J., Gonda, D. K., & Varshavsky, A. (1989). A multiubiquitin chain is confined to specific lysine in a targeted short-lived protein. Science 243, 1576-1583.

Confalonieri, F., & Duguet, M. (1995). A 200-amino acid ATPase module in search of a basic function. BioEssays 17, 639-650.

Cook, W. J., Jeffrey, L. C., Kasperek, E., & Pickart, C. M. (1994). Structure of tetraubiquitin shows how multiubiquitin chains can be formed. J. Mol. Biol. 236, 601-609.

Dahlmann, B., Kopp, F., Kuchn, L., Niedel, B., Pfeiler, G., Hegerl, R., & Baumeister, W. (1989). The multicatalytic proteinase (prosome) is ubiquitous from eukaryotes to archacbactera. FEBS Lett. 251, 125-131.

DeMarini, D. J., Papa, F. R., Swaminathan, S., Ursic, D., Rasmussen, T. P., Culbertson, M. R., & Hochstrasser, M. (1995). The yeast SEN3 gene encodes a regulatory subunit of the 26S proteasome complex required for ubiquitin-dependent protein degradation *in vivo*. Mol. Cell. Biol. 15, 6311-6321.

DeMartino, G. N., Moomaw, C. R., Zagnitko, O. P., Proske, R. J., Ma, C.-P., Afendis, S. J., Swaffield, J. C., & Slaughter, C. A. (1994). PA70O, an ATP-dependent activator of the 20S proteasome, is an ATPase containing multiple members of a nucleotide-binding protein family. J. Biol. Chem. 269, 20878-20884.

DeMartino, G. N., Proske, R. J., Moomaw, C. R., Strong, A. A., Song, X., Hisamatsu, H., Tanaka, K., & Slaughter, C. A. (1996). Identification, purification, and characterization of a PA700-dependent activator of the proteasome. J. Biol. Chem. 271, 3112-3118.

Deshaies, R. J. (1995). The self-destructive personality of a cell cycle in transition. Curr. Opin. Cell Biol. 7, 781-789.

Deveraux, Q., Ustrell, V., Pickart, C., & Rechsteiner, M. (1994). A 26S protease subunit that binds ubiquitin-conjugates. J. Biol. Chem. 269, 7059-7061.

Deveraux, Q., Jensen, C., & Rechsteiner, M. (1995a). Molecular cloning and expression of a 26S protease subunit enriched in dileucine repeats. J. Biol. Chem. 270, 23726-23729.

Deveraux, Q., van Nocker, S., Mahaffey, D., Vierstra, R., & Rechsteiner, M. (1995b). Inhibition of ubiqutin-mediated proteolysis by the *Arabidopsis* 26S protease subunit S5a. J. Biol. Chem. 270, 29660-29663.

DiCola, D. (1992). Human erthrocyte contains a factor that stimulates the peptidase activities of multicatalytic proteinase complex. Ital. J. Biochem. 41, 213-224.

Driscoll, J., Frydman, J., & Goldberg, A. L. (1992). An ATP-stabilized inhibitor of the proteasome is a component of the 1500-kDa ubiquitin conjugate-degrading complex. Proc. Natl. Acad. Sci. USA 89, 4986-4990.

Dubiel, W., Pratt, G., Ferrell, K., & Rechsteiner. M. (1992a). Purification of an 11S regulator of the multicatalytic protease. J. Biol. Chem. 267, 22367-22377.

Dubiel, W., Ferrell, K., Pratt, G., & Rechsteiner, M. (1992b). Subunit 4 of the 26S protease is a member of a novel eukaryotic ATPase family. J. Biol. Chem. 267, 22699-22702.

Dubiel, W., Ferrell, K., & Rechsteiner, M. (1993). Peptide sequencing identifies MSS1, a modulator of HIV Tat-mediated transactivation, as subunit 7 of the 26S protease. FEBS Lett. 323, 276-278.

Dubiel, W., Ferrell, K., & Rechsteiner, M. (1994). Tat-binding protein 7 is a subunit of the 26S protease. Biol. Chem. Hoppe-Seyler 375, 237-240.

Dubiel, W., Ferrell, K., & Rechsteiner, M. (1995a). Subunits of the regulatory complex of the 26S protease. Mol. Biol. Rep. 21, 27-34.

Dubiel, W., Ferrell, K., Dumdey, R., Standera, S., Prehn, S., & Rechsteiner, M. (1995b). Molecular cloning and expression of subunit 12: a non-MCP and non-ATPase subunit of the 26S protease. FEBS Lett. 363, 97-100.

Eytan, E., Ganoth, D., Armon, T., & Hershko, A. (1989). ATP-dependent incorporation of 20S protease into the 26S complex that degrades proteins conjugated to ubiquitin. Proc. Natl. Acad. Sci. USA 86, 7751-7755.

Eytan, E., Armon, T., Heller, H., Beck, S., & Hershko, A. (1993). Ubiquitin C-terminal hydrolase activity associated with the 26S protease complex. J. Biol. Chem. 268, 4668-4674.

Falk, K., & Rotzschke, O. (1993). Consensus motifs and peptide ligands of MHC class I molecules. Sem. Immun. 5, 81-94.

Fehling, H. J., Swat, W., Laplace, C., Kuhn, R., Rajewsky, K., Muller, U., & Boehmer, H. V. (1994). MHC class I expression in mice lacking the proteasome subunit LMP-7. Science 265, 1234-1237.

Ferrell, K., Deveraux, Q., van Nocker, S., & Rechsteiner, M. (1996) Molecular cloning and expression of human S5a, a multiubiquitin chain binding subunit of the 26S protease. FEBS Lett. 381, 143-148.

Fischer, M., Runkel, L., & Schaller, H. (1995). HBx protein of hepatitis B virus interacts with the C-terminal portion of a novel human proteasome alpha-subunit. Virus Genes 10, 99-102.

Frentzel, S., Pesold-Hurt, B., Seelig, A., & Kloetzel, P.-M. (1994). 20S proteasomes are assembled via distinct precursor complexes. J. Mol. Biol. 236, 975-981.

Früh, K., Gossen, M., Wang, K., Bujard, H., Peterson, P. A., & Yang, Y. (1994). Displacement of housekeeping proteasome subunits by MHC-encoded LMPs: a newly discovered mechanism for modulating the multicatalytic proteinase complex. EMBO J. 13, 3236-3244.

Ganoth, D., Leshinsky, E., Eytan, E., & Hershko, A. (1988). A multicomponent system that degrades proteins conjugated to ubiquitin. J. Biol. Chem. 263, 12412-12419.

Germain, R. N. (1994). MHC-dependent antigen processing and peptide presentation: providing ligands for T lymphocytes activation. Cell 76, 287-299.

Ghislain, M., Udvardy, A., & Mann, C. (1993). *S. cerevisiae* 26S protease mutants arrest cell division in G2/metaphase. Nature 366, 358-362.

Glynne, R., Powis, S. H., Beck, S., Kelly, A., Kerr, L.-A., & Trowsdale, J. (1991). A proteasome-related gene between the two ABC transporter loci in the class II region of the human MHC. Nature 353, 357-360.

Gordon, C., McGurk, G., Dillon, P., Rosen, C., & Hastie, N. D. (1993). Defective mitosis due to a mutation in the gene for a fission yeast 26S protease subunit. Nature 366, 355-357.

Gordon, C., McGurk, G., Wallace, M., & Hastie, N. D. (1996). A conditional lethal mutant in the fission yeast 26S protease subunit mts3^{+} is defective in metaphase to anaphase transition. J. Biol. Chem. 271, 5704-5711.

Grandea, A. G. I., Androlewicz, M. J., Athwal, R. S., Geraghty, D. E., & Spies, T. (1995). Dependence of peptide binding by MCH class I molecules on their interaction with TAP. Science 270, 105-108.

Gray, C. W., Slaughter, C. A., & DeMartino, G. N. (1994). PA28 activator protein forms regulatory caps on proteasome stacked rings. J. Mol. Biol. 236, 7-15.

Groettrup, M., Ruppert, T., Kuhn, L., Seeger, M., Standera, S., Koszinowski, U., & Kloetzel, P. M. (1995). The interferon-γ-inducible 11S regulator (PA28) and the LMP2/LMP7 subunits govern the peptide production by the 20S proteasome *in vitro*. J. Biol. Chem. 270, 23808-23815.

Groettrup, M., Soza, A., Eggers, M., Kuchn, L., Dick, T. P., Schild, H., Rammensee, H.-G., Koszinowski, U. H., & Kloetzel, P.-M. (1996). A role for the proteasome regulator PA280α in antigen presentation. Nature 381, 166-168.

Guarino, L. A., Smith, G., & Dong, W. (1995). Uhiquitin is attached to membranes of the baculovirus particles by a novel type of phospholipid anchor. Cell 80, 301-309.

Haas, A., Reback, P. M., Pratt, G., & Rechsteiner, M. (1990). Ubiquitin-mediated degradation of histone H3 does not require the substrate-binding ubiquitin protein ligase, E3, or attachment of polyubiquitin chains. J. Biol. Chem. 265, 21664-21669.

Haracska, L., & Udvardy, A. (1995). Cloning and sequencing a non-ATPase subunit of the regulatory complex of the *Drosophila* 26S protease. Eur. J. Biochem. 231, 720-725.

Hartl, F. U. (1996). Molecular chaperones in cellular protein folding. Nature 381, 571-580.

Hayashi, S., & Murakami, Y. (1995). Rapid and regulated degradation of ornithine decarboxylase. Biochem. J. 306, 1-10.

Heinemeyer, W., Trondle, N., Albrecht, G., & Wolf, D. H. (1994). PRE5 and PRE6, the last missing genes encoding 20S proteasome subunits from yeast? Indication for a set of 14 different subunits in the eukaryotic proteasome core. Biochemistry 33, 12229-12237.

Henkart, P. A. (1996). ICE family proteases: Mediators of all apoptotic cell death? Immunity 4, 195-201.

Hermida-Matsumoto, M.-L., Chock, P. B., Curran, T., & Yang, D. C. H. (1996). Ubiquitinylation of transcription factors c-Jun and c-Fos using reconstituted ubiquitinylating enzymes. J. Biol. Chem. 271, 4930-4936.

Hershko, A., & Heller, H. (1985). Occurrence of a polyubiquitin structure in ubiquitin-protein conjugates. Biochem. Biophys. Res. Commun. 128, 1079-1086.

Hershko, A., Ganoth, D., Pehrson, J., Palazzo, R. E., & Cohen, L. H. (1991). Methylated ubiquitin inhibits cyclin degradation in clam embryo extracts. J. Biol. Chem. 266, 16376- 16379.

Hershko, A., & Ciechanover, A. (1992). The ubiquitin system for protein degradation. Annu. Rev. Biochem. 61, 761 -807.

Hilt, W., & Wolf, D. H. (1996). Proteasomes: Destruction as a programme. Trends in Biochem. Sci. 21, 96-102.

Hoffman, L., Pratt. G., & Rechsteiner, M. (1992). Multiple forms of the 20S multicatalytic and the 26S ubiquitin/ATP-dependent proteases from rabbit reticulocyte lysate. J. Biol. Chem. 267, 22362-22368.

Hough, R., Pratt, G., & Rechsteiner, M. (1986). Ubiquitin-lysozyme conjugates: Identification of an ATP-dependent protease from reticulocyte lysate. J. Biol. Chem. 261, 2400-2408.

Hough, R., Pratt, G., & Rechsteiner, M. (1987). Purification of two high molecular weight proteasees from rabbit reticulocyte lysate. J. Biol. Chem. 262, 8303-8311.

Hough, R., Pratt, G., & Rechsteiner, M. (1988). ATP/Ubiquitin-dependent proteases. In: *Ubiquitin* (Rechsteiner, M., ed.), pp. 101-134.

Huang, J., Kwong, J., Sun, E. C.-Y., & Liang, T. J. (1996). Proteasome complex as a potential cellular Target of hepatitis B virus X protein. J. Virol. 70, 5582-5591.

Ikai, A., Nishigai, M., Tanaka, K., & Ichihara, A. (1991). Electron microscopy of 26S complex containing 20S proteasome. FEBS Lett. 292, 21-24.

Jackson, M. R., Cohen-Doyle, M. F., Peterson, P. A., & Williams D. B. (1994). Regulation of MHC class I transport by the molecular chaperone, calnexin (p88, IP90). Science 263, 384-390.

Jariel-Encontre, I., Pariat, M., Martin, F., Carillo, S., Salvat, C., & Piechaczyk, M. (1995). Ubiquitinylation is not an absolute requirement for degradation of c-dun protein by the 26S proteasome. J. Biol. Chem. 270, 11623-11627.

Johansson, E., Lönnroth, I., Lange, S., Jonson, I., Jennische, E., & Lönnroth, Ch. (1995). Molecular cloning and expression of a pituitary gland protein modulating intestinal fluid secretion. J. Biol. Chem. 270, 20615-20620.

Johnston, S. C., Whitby, F. G., Realini, C., Rechsteiner, M., & Hill, C. P. (1997). The proteasome 11S regulator subunit REGα, (PA28a) is a heptamer. Protein Science 6, 2469-2473.

Kanayama, H.-o., Tamura, T., Ugai, S., Kagawa, S., Tanahashi, N., Yoshimura, T., Tanaka, K., & Ichihara, A. (1992). Demonstration that the human 26S proteolytic complex consists of a proteasome and multiple associated protein components and hydrolyzes ATP and ubiquitin-ligated proteins by closely linked mechanisms. Eur. J. Biochem. 206, 567-578.

Kawahara, H., & Yokosawa, H. (1994). Intracellular calcium mobilization regulates the activity of 26S proteasome during the metaphase-anaphase transition in the ascidian meiotic cell cycle. Dev. Biol. 166, 623-633.

Kelly, A., Powis, S. H., Glynne, R., Radley, E., Beck, S., & Trowsdale, J. (1991). Second proteasome-related gene in the human MHC class II region. Nature 353, 667-668.

Kessel, M., Maurizi, M. R., Kim, B., Kocsis, E., Trus, B. L., Singh, S. K., & Steven, A. C. (1995). Homology in structural organization between *E. coli* ClpAP protease and the eukaryotic 26S proteasome. J. Mol. Biol. 250, 587-594.

Khan, M. T., Wang, K., & Roufogalis, B. D. (1995). Identification and purification of a 90-kDa membrane-bound endogenous inhibitor of multicatalytic proteinase from human erythrocytes. Biochem. Biophys. Res. Commun. 214, 957-962.

Kleinschmidt, J. A., Hugle, B., Grund, C., & Franke, W. W. (1983). The 22S cylinder particles of *Xenopus* and *laevis*. I. Biochemical and electron microscopic characterization. Eur. J. Cell Biol. 32, 143-156.

Kominami, K.-i., DeMartino, G. N., Moomaw, C. R., Slaughter, C. A., Naoki, S. et al. (1995). Nin1p, a regulatory subunit of the 26S proteasome, is necessary for activation of Cdc28p kinase of *Saccharomyces cerevisiae*. EMBO J. 14, 3105-3115.

Lee, J. W., Ryan, F., Swaffield, J. C., Johnston, S. A., & Moore, D. D. (1995). Interaction of thyroid-hormone receptor with a conserved transcriptional mediator. Nature 374, 91-94.

Li, X. S., & Etlinger, J. D. (1992). Ubiquitinated proteasome inhibitor is a component of the 26S proteasome complex. Biochemistry 31, 11963-11967.

Li, X., Gu, M., & Etlinger, J. D. (1991). Isolation and characterization of a novel endogenous inhibitor of the proteasome. Biochemistry 30, 9709-9715.

Löwe, J., Stock, D., Jap, B., Zwickl, P., Baumeister, W., & Huber, R. (1995). Crystal structure of the 20S proteasome from the archaeon *T. acidophilium* at 3.4Å resolution. Science 268, 533-539.

Lu, X., Le, X., & Meissner, G. (1994). Activation of the skeletal muscle calcium release channel by a cytoplasmic loop of the dihydropyridine receptor. J. Biol. Chem. 269, 6511-6516.

Lucero, H. A., Chojnicki, E. W. T., Mandiyan, S., Nelson, H., & Nelson, N. (1995). Cloning and expression of a yeast gene encoding a protein with ATPase activity and high identity to the subunit 4 of the human 26S protease. J. Biol. Chem. 270, 9178-9184.

Ma, C.-P., Slaughter, C. A., & DeMartino, G. N. (1992a). Identification, purification and characterization of a protein activator (PA28) of the 20S proteasome (macropain). J. Biol. Chem. 267, 10515-10523.

Ma, C.-P., Slaughter, C. A., & DeMartino, G. N. (1992b). Purification and characterization of a protein inhibitor of the 20S proteasome (macropain). Biochim. et Biophys. Acta 1119, 303-311.

Ma, C.-P., Willy, P. J., Slaughter, C. A., & DeMartino, G. N. (1993). PA28, an activator of the 20S proteasome is inactivated by proteolytic modification at its carboxyl terminus. J. Biol. Chem. 268, 22514-22519.

Ma, C.-P., Vu, J. H., Proske, R. J., Slaughter, C. A., & DeMartino, G. N. (1994). Identification, purification, and characterization of a high molecular weight, ATP-dependent activator (PA700) of the 20S proteasome. J. Biol. Chem. 269, 3539-3547.

Martinez, C.K., & Monaco, J. J. (1991). Homology of proteasome subunits to a major histocompatibility complex-linked LMP gene. Nature 353, 664-667.

Mayer, J. R., Arnold, J., Laszlo, L., Landon, M., & Lowe, J. (1991). Ubiquitin in health and disease. Biochim. Biophys. Acta 1089, 141-157.

Michalek, M., Grant, E., Gramm, C., Goldberg, A., & Rock, K. (1993). A role for the ubiquitin-dependent proteolytic pathway in MHC class I-restricted antigen presentation. Nature 363, 552-554.

Momburg, F., Ortiz-Navarrete, V., Neefjes, J., Goulmy, E., van de Wal, Y., Spits, H., Powis, S. J., Butcher, G. W., Howars, J. C., Walden, P. et al. (1992). Proteasome subunits encoded in the major histocompatibility complex are not essential for antigen presentation. Nature 360, 174- 177.

Monaco, J. J. (1992). Pathways of antigen processing. Immunol. Today 13, 173-178.

Murakami, K., & Etlinger, J. D. (1986). Endogenous inhibitor of nonlysosomal high molecular weight protease and calcium-dependent protease. Proc. Natl. Acad. Sci. USA 83, 7588-7592.

Murakami, Y., Matsufuji, S., Kameji, T., Hayashi, S.-I., Igarashi, K., Tamura, T., Tanaka, K., & Ichihara, A. (1992). Ornithine decarboxylase is degraded by the 26S proteasome without ubiquitination. Nature 360, 597-599.

Nikaido, T., Shimada, K., Shibata, M., Hata, M., Sakamoto, M., Takasaki, Y., & Sato, C. (1990). Cloning and nucleotide sequence of cDNA for Ki antigen, a highly conserved nuclear protein detected with sera from patients with systemic lupus erythematosus. Clin. Exp. Immunol. 79, 209-214.

Nobel, M., Lewis, S. A., & Cowan, N. J. (1994). The microtubule binding domain of microtubule-associated protein MAP1B contains a repeated sequence motif unrelated to that of MAP2 and Tau. J. Cell. Biol. 109, 3367-3376.

Ohana, B., Moore, P. A., Ruben, S. M., Southgate, C. D., Green, M. R., & Rosen, C. A. (1993). The type I immunodeficiency virus Tat binding protein is a transcriptional activator belonging to an additional family of evolutionarily conserved genes. Proc. Natl. Acad. Sci. USA 90, 138-142.

Orlowski, M., Cardozo, C., Hidalgo, M. C., & Michaud, C. (1991). Regulation of the peptidylglutamyl-peptide hydrolyzing activity of the pituitary multicatalytic proteinase complex. Biochemistry 30, 5999-6005.

Ortmann, B., Androlewicz, M. J., & Cresswell, P. (1994). MHC classI/β2-microglobulin complexes associated with TAP transporters before peptide binding. Nature 368, 864-867.

Peters, J.-M., Harris, J. R., & Kleinschmidt, J. A. (1991). Ultrastructure of the ~26S complex containing the ~20S cylinder particle (multicatalytic proteinase/proteasome). Eur. J. Cell Biol. 56, 422-432.

Peters, J.-M., Harris, J. R., Lustig, A., Müller, S., Engel, A., Volker, S., & Franke, W. W. (1992). Ubiquitous soluble Mg^{2+}-ATPase complex. A structural study. J. Mol. Biol. 223, 557-571.

Peters, J.-M., Cejka, Z., Harris, J. R., Kleinschmidt, J. A., & Baumeister, W. (1993). Structural features of the 26S proteasome complex. J. Mol. Biol. 234, 932-937.

Rajagopalan, S., & Brenner, M. (1994). Calnexin retains unassembled major histocompatibility complex class I free heavy chains in the endoplasmic reticulum. J. Exp. Med. 180, 407-412.

Realini, C., Rogers, S. W., & Rechsteiner, M. (1994a). KEKE motifs: Proposed roles in protein-protein association and presentation of peptides by MHC Class I receptors. FEBS Lett. 348, 109-113.

Realini, C., Dubiel, W., Pratt, G., Ferrell, K., & Rechsteiner, M. (1994b). Molecular cloning and expression of a gamma-interferon inducible activator of the multicatalytic protease. J. Biol. Chem. 269, 20727-20732.

Realini, C., Jensen, C., Zhang, Z., Johnston, S., Knowlton, R., Hill, C., & Rechsteiner, M. (1997). Characterization of recombinant REGα, REGβ, and REGγ proteasome activators. J. Biol. Chem. 272, 25483-25492.

Realini, C., & Rechsteiner, M. (1995). A proteasome activator subunit binds calcium. J. Biol. Chem. 270, 29664-29667.

Rechsteiner, M. (1991). Natural substrates for the ubiquitin proteolytic pathway. Cell 66, 615-618.

Reilly, L. M., Guarino, L. A. (1996). The viral ubiquitin gene of *Autographa californica* nuclear polybedrosis virus is not essential for viral replication. Virology 218, 243-247.

Richmond, C., Gorbea, C., & Rechsteiner, M. (1997). Specific interactions between ATPase subunits of the 26S protease. J. Biol. Chem. 272, 13403-13411.

Richter-Ruoff, B., Wolf, D. H., & Hochstrasser, M. (1994). Degradation of the yeast MATα2 transcriptional regulator is mediated by the proteasome. FEBS Lett. 354, 50-52.

Rock, K. L. (1996). A new foreign policy: MHC class I molecules monitor the outside world. Immunol. Today 17, 131 -137.

Rock, K. L., Gramm, C., Rothstein, L., Clark, K., Stein, R., Dick, L., Hwang, D., & Goldberg, A. L. (1994). Inhibitors of the proteasome block the degradation of most cell proteins and the generation of peptides presented on MHC class I molecules. Cell 78, 761-771.

Rodriguez, J. M., Salas, M. L., & Vinuela, E. (1992). Genes homologous to ubiquitin-conjugating proteins and eukaryotic transcription factor SII in African swine fever virus. Virology 186, 40-52.

Rosenberg-Hasson, Y., Bercovich, Z., Ciechanover, A., & Kahana, C. (1989). Degradation of ornithine decarboxylase in mammlian cells is ATP dependent but ubiquitin independent. Eur. J. Biochem. 185, 469-474.

Rousset, R., Desbois, Ch., Bantignies, F., & Jalinot, P. (1996). Effects on NF-κB1/pl05 processing of the interaction between the HTLV-1 transactivator Tax and the proteasome. Nature 381, 328-331.

Rubin, D. M., Coux, O., Wefes, I., Hengartner, Ch., Young, R. A., Goldberg, A. L., & Finley, D. (1996). Identification of the gal4 suppressor Sug1 as a subunit of the yeast 26S proteasome. Nature 379, 655-657.

Russel, R. L., & Rohrmann, G. F. (1993). Nucleotide sequence of the ubiquitin-39K gene region from the Orgyia pseudotsugata multinucleocapsid nuclear polyhedrosis virus genome. J. Gen. Virol. 74, 1191-1195.

Sato, K., Nishikawa, T., Yokosawa, H., & Sawada, H. (1995). Phosphorylation of proteasome substrate by a protein kinase associated with the 26S proteasome is linked to the ATP-dependent proteolysis of the 26S proteasome. Biochem. Biophys. Res. Comm. 213, 7-14.

Sawada, H., Kawahara, H., Saitoh, Y., & Yokosawa, H. (1996). In: Intracellular Protein Catabolism (Suzuki, K. & Bond, J., eds.), pp. 229-232. Plenum Press, New York.

Scheffner, M., Huibregtse, J. M., Vierstra, R. D., & Howley, P. M. (1993). The HPV-16 E6 and E6-AP complex functions as a ubiquitin-protein ligase in the ubiquitination of p53. Cell 75, 495-505.

Schork, S. M., Thumm, M., & Wolf, D. H. (1995). Catabolite inactivation of fructose-1,6-bisphosphatase of *Saccharomyces cerevisiae*. Degradation occurs via the ubiquitin pathway. J. Biol. Chem. 270, 26446-26450.

Seeger, M., Gordon, C., Ferrell, K., & Dubiel, W. (1996). Characteristics of 26S proteases from fission yeast mutants, which arrest in mitosis. J. Mol. Biol. 263, 423-431.

Seemüller, E., Lupas, A., Stock, D., Lowe, J., Huber, R., & Baumeister, W. (1995). Proteasome from *Thermoplasma acidophilium*: A threonine protease. Science 268, 533-539.

Shaeffer, J. R., & Kania, M. A. (1995). Degradation of monoubiquitinated α-globin by 26S proteasomes. Biochemistry 34, 4015-4021.

Shelton, E., Kuff, E. L., Maxwell, E. S., & Harrington, J. T. (1970). Cytoplasmic particles and aminoacyl translerase I activity. J. Cell Biol. 45, 1-8.

Song, H. Y., Dunbar, J. D., Zhang, Y. X., Guo, D., & Donner, D. B. (1995). Identification of a protein with homology to hsp90 that binds the type 1 tumor necrosis factor receptor. J. Biol. Chem. 270, 3574-3581.

Speiser, S., & Etlinger, J. D. (1982). ATP stimulates proteolysis in reticulocyte extracts by repressing an endogenous protease inhibitor. Proc. Natl. Acad. Sci. USA 80, 3577-3580.

Stancovski, I., Gonen, H., Orian, A., Schwartz, A. L., & Ciechanover, A. (1995). Degradation of the proto-oncogene product c-fos by the ubiquitin proteolytic system *in vivo* and *in vitro*: Identification and characterization of the conjugating enzymes. Mol. Cell. Biol. 15, 7106-7116.

Suh, W.-K., Cohen-Doyle, M. F., Fruh, K., Wang, K., Peterson, P. A., & Williams, D. B. (1994). Interaction of MHC class I molecules with the transporter associated with antigen processing. Science 264, 1322-1326.

Suto, R., & Srivastava, P. K. (1995). A mechanism for the specific immunogenicity of heat shock protein-chaperoned peptides. Science 269, 1585-1588.

Suzuki, K., Sorimachi, H., Yoshizawa, T., Kinbara, K., & Ishiura, S. (1995). Calpain: Novel family members, activation and physiological function. Biol. Chem. Hoppe-Seyler 376, 523-529.

Swaffield, J. C., Melcher, K., & Johnston, S. A. (1995). A highly conserved ATPase protein as a mediator between acidic activation domains and the TATA-binding protein. Nature 374, 88-91.

Tanaka, K., Tamura, T., Yoshimura, T., & Ichihara, A. (1992). Proteasomes: Protein and gene structures. New Biol. 4, 173-187.

Tanaka, K (1995). Molecular biology of proteasomes. Mol. Biol. Rep. 21, 21-26.

Tartaglia, L. A., Ayres, T. M., Wong, G. H. W., & Goeddel, D. V. (1993). A novel domain within the 55 kd TNF receptor signals cell death. Cell 74, 845-853

Townsend, A. R. M., & Bodmer, H. (1989). Antigen recognition by class I-restricted T lymphocytes. Ann. Rev. Immun. 7, 601-624.

Tsubuki, S., Saito, Y., & Kawashima, S. (1994). Purification and characterization of an endogenous inhibitor specific to the Z-Leu-Leu-Leu-MCA degrading activity in proteasome and its identification as heat-shock protein 90. FEBS Lett. 344, 229-233.

Tsurumi, C., Ishida, N., Tamura, T., Kakizuka, A., Nishida, E., Okumura, E., Kishimoto, T., Inagaki, M., Okazaki, K., Sagata, N., Ichihara, A., & Tanaka, K. (1995). Degradation of c-fos by the 26S proteasome is accelerated by c-jun and multiple protein kineses. Mol. Cell. Biol. 15, 5682-5687.

Tsurumi, C., Shimizu, Y., Saeki, M., Kato, S., DeMartino, G. N., Slaughter, C. A., Fujimuro, M., Yokosawa, H., Yamasaki, M.-o, Hendil, K. B., Toh-e, A., Tanahashi, N., & Tanaka, K. (1996). cDNA cloning and functional analysis of the p97 subunit of the 26S proteasome, a polypeptide identical to the type-1 tumor-necrosis-factor-receptor-associated protein-2/55. 11. Eur. J. Biochem. 239, 912-921.

Udvardy, A. (1993). Purification and characterization of a multiprotein component of the *Drosophila* 26S (1500 kDa) proteolytic complex. J. Biol. Chem. 268, 9055-9062.

Ugai, S.-i., Tamura, T., Tanahashi, N., Takai, S., Komi, N., Chung, C. H., Tanaka. K., & Ichihara, A. (1993). Purification and characterization of the 26S proteasome complex catalyzing ATP-dependent breakdown of ubiquitin-ligated proteins from rat liver. J. Biochem. 113, 754-768.

Ustrell, V., Pratt, G., & Rechsteiner, M. (1995a). Effects of interferon gamma and major histocompatibility complex-encoded subunits on peptidase activities of human multicatalytic proteases. Proc. Natl. Acad. Sci. USA 92, 584-588.

Ustrell, V., Realini, C., Pratt, G., & Rechsteiner, M. (1995b). Human lymphoblast and erythrocyte multicatalytic proteases: Differential peptidase activities and responses to the 11S regulator. FEBS Lett. 376, 155-158.

van Nocker, S., Deveraux, Q., Rechsteiner, M., & Vierstra, R. (1996a). *Arabidopsis MBP1* gene encodes a conserved ubiquitin recognition component of the 26S proteasome. Proc. Natl. Acad. Sci. USA 93, 856-860.

van Nocker, S., Sadis, S., Rubin, D. M., Glickman, M., Fu. H., Coux, O., Wefes, I., Finley, D., & Vierstra, R. D. (1996b). The multiubiquitin-chain-binding protein Mcb1 is a component of the 26S proteasome in *Saccharomyces cerevisiae* and plays a nonessential, substrate-specific role in protein turnover. Mol. Cell. Biol. 16, 6020-6028.

Wagner, B. J., & Margolis, J. W. (1995). Age-dependent association of isolated bovine lens multicatalytic proteinase complex (proteasome) with heat-shock protein 90, an endogenous inhibitor. Arch. Biochem. Biophys. 323, 455-462.

Wenzel, T., & Baumeister, W. (1995). Conformational constraints in protein degradation by the 20S proteasome. Nature Struct. Biol. 2, 199-204.

Wilkinson, K. D., Tashayev, V. L., O'Connor, L. B., Larsen, Ch. N., Kasperek, E., & Pickart, C. M. (1995). Metabolism of the polyubiquitin degradation signal: structure, mechanism, and role of isopeptidase T. Biochemistry 34, 14535-14546.

Yang, Y., Waters, J. B., Fruh, K., & Paterson, P. A. (1992). Proteasomes are regulated by interferon gamma: implications for antigen processing. Proc. Natl. Acad. Sci. USA 89, 4928-4932.

Yang, Y., Früh, K., Ahn. K., & Peterson, P. A. (1995). *In vivo* assembly of the proteasomal complexes, implications for antigen processing. J. Biol. Chem. 270, 27687-27694.

Yewdell, J. W., & Bennink, J. R. (1992). Cell biology of antigen processing and presentation to major histocompatibility complex class I molecule-restricted T lymphocytes. Adv. Immunol. 52, 1-123.

Yokota, K.-y., Kagawa, S., Shimizu, Y., Akioka, H., Tsurumi, C., Noda, C. et al. (1996). cDNA cloning of p112, the largest regulatory subunit of the human 26S proteasome, and functional analysis of its yeast homologue, Sen3p. Mol. Biol. Cell 7, 853-870.

Yoshimura, T., Kameyama, K., Takagi, T., Ikai, A., Tokunaga, F., Koide, T., Tanahashi, N., Tamura, T., Cejka, Z., Baumesiter, W., Tanaka, K., & Ichihara, A. (1993). Molecular characterization of the "26S" proteasome complex from rat liver. J. Struct. Biol. 111, 200-211.

Yukawa, M. et al. (1991). Proteasome and its novel endogenous activator in human platelets. Biochem. Biophys. Res. Commun. 17S, 256-262.

Zwickl, P., Kleinz, J., & Baumeister, W. (1994). Critical elements in proteasome assembly. Struct. Biol. 1, 765-770.

ENDOPLASMIC RETICULUM–ASSOCIATED PROTEIN DEGRADATION: AN UNCONVENTIONAL ROUTE TO A FAMILIAR FATE

Ardythe A. McCracken, Eric D. Werner, and Jeffrey L. Brodsky

Advances in Molecular and Cell Biology
Volume 27, pages 165-200.

ISBN: 0-7623-0387-5

I. INTRODUCTION

A quality control process exists to ensure that proteins inserted into the endoplasmic reticulum (ER) are correctly folded, processed and completely assembled before they exit this compartment for further transport through the secretory pathway. Most proteins that fail to pass this checkpoint are degraded. The process by which these aberrant and unassembled secretory proteins are removed from the ER is referred to as ER-associated protein degradation, or ERAD. Although the existence of ERAD has been known for some time, the factors required for protein quality control, the proteases responsible for degradation, and even the site of proteolysis have been difficult to define (for review see Bonifacino and Klausner, 1994). In the last year, however, many facets of ERAD have emerged from divergent studies, lending new insights into how specific proteins are removed from the ER. This chapter will outline the early studies and general characteristics of ERAD, and discuss the recent advances towards investigating the molecular mechanisms of this essential cellular process. Because comprehensive reviews on the breadth of research in ERAD exist (Klausner and Sitia, 1990; Klausner et al., 1990; Bonafacino and Klausner, 1994; Brodsky and McCracken, 1997), we will focus on recent literature.

A. History

The study of ERAD can trace its roots to the 1960s and early 1970s with the discovery that rapid and selective intracellular degradation prevented the accumulation of aberrant proteins within rabbit recticulocytes (for examples see Rabinovitz and Fisher, 1964; Hendil, 1975). Because the lysosome was generally believed to be the site of intracellular protein degradation, these early experiments did not include assays to locate the sites of the proteolytic activity. However, it was soon hypothesized that another proteolytic pathway might exist since the indiscriminate nature of lysosomal proteolysis would not allow for selective degradation of aberrant species from a pool of normal proteins (Schimke, 1970). Subsequently, it was found that the degradation of ornithine decarboxylase was independent of lysosomal proteolysis (Prouty, 1976), thus supporting this premise. A candidate for a selective proteolytic pathway was revealed when a soluble, non-lysosomal proteolytic system for the degradation of aberrant proteins was uncovered (Etlinger and Goldberg 1977).

With the advancement of molecular biological techniques, it was possible to study the biogenesis of specific secretory proteins, and evidence was obtained that established two proteolytic pathways for the intracellular degradation of secretory proteins. One pathway used autophagic vacuoles and lysosomal proteolysis to degrade stored substrates such as parathyroid hormone (Habener et al., 1975), prolactin (Shenai and Wallis, 1979), insulin (Halban and Wollheim, 1980), and casein (Razooki-Hasan et al., 1982), while a second uncharacterized lysosomal protease-independent pathway disposed of aberrant and or unassembled secretory proteins (reviewed by Bienkowski, 1983). For example, studies on the degradation rates of wild-type and mutant forms of immunoglobin M (IgM) showed that the aberrant form of IgM was rapidly degraded by a lysosome protease-independent process (Sidman, 1981). When IgM was further examined in non-secreting B cells, the secreted form of IgM was rapidly degraded without transiting from the ER to the Golgi complex (Dulis et al., 1982). These findings suggested that this "novel" proteolytic pathway might function as a quality control checkpoint in an early compartment of the secretory pathway to ensure that the secretion and/or accumulation of aberrant proteins did not occur. In the late 1980s and early 1990s, ERAD received renewed interest as this pathway was uncovered in studies from diverse research areas.

B. Numerous and Varied Substrates

ERAD does not exclusively target misfolded or aberrant proteins but also regulates the activity of ER resident proteins and the expression of plasma membrane proteins, for example, the ER membrane enzyme HMG-CoA reductase (Meigs and Simoni, 1992; Hampton and Rine, 1994), the cystic fibrosis transmembrane conductance regulator, CFTR (Cheng et al., 1990), a T cell surface receptor, CD4 (Willey et al., 1992), and the heavy chain of the major histocompatibility complex (MHC) class I molecule (Wiertz et al., 1996a,b).

HMG-CoA reductase catalyzes the conversion of HMG-CoA to mevalonate, representing the rate-limiting step in the biosynthesis of sterols and non-isoprenoid sterols, and is degraded when an increased flux through the mevalonate pathway is evident (reviewed in Hampton et al., 1996a). Thus, proteolysis provides a rapid, irreversible, and efficient means to inhibit sterol and isoprenoid synthesis when compounds in the mevalonate pathway are abundant. Early studies on the regulated degradation of HMG-CoA reductase (Edwards and Gould, 1972; Faust et al., 1982; Orci et al., 1984) suggested that proteolysis occurred by a non-lysosomal pathway, and more recent investigations in permeabilized cells indicated that proteolysis of HMG-CoA reductase is not affected when brefelden A is used to inhibit transport from the ER to the Golgi (Meigs and Simoni, 1992), a result later confirmed using a reconstituted system (McGee et al., 1996). A similar result was obtained when the regulated degradation of a yeast HMG-CoA reductase isozyme was examined (Hampton and Rine, 1994), indicating that both yeast and mammalian HMG-CoA reductase are substrates for ERAD.

CFTR is an integral membrane protein that facilitates chloride transport across the apical plasma membrane of epithelial cells. The most common mutation in individuals with cystic fibrosis is a deletion of the phenylalanine at position 508 (ΔF508). Although ~75% of wild type CFTR is retained in the ER, essentially all of the ΔF508 mutant protein accumulates there. The reason for the intracellular accumulation of the mutant is that an ATP-dependent conformational change required for ER to Golgi transport is prevented (Lukacs et al., 1994). Both the immature wild-type and ΔF508 mutant forms retained in the ER are substrates for rapid degradation by ERAD (Cheng et al., 1990).

The CD4 molecule, a transmembrane glycoprotein of T cells and macrophages, acts as the receptor for the human immunodeficiency virus (HIV-1). Several mechanisms exist to down-regulate the surface expression of CD4 in HIV-1 infected cells, including the ERAD of CD4 that is induced by the viral protein Vpu (Willey et al., 1992). Vpu-stimulated degradation is highly specific for CD4 and involves a physical interaction between the cytoplasmic domains of both proteins (Bour et al., 1995; Schubert et al., 1996). Vpu is an integral membrane protein with a sequence that reveals few clues on how it targets CD4 for degradation. Because there is some homology between Vpu and the M2 protein of influenza virus, which probably functions as an ion channel (Strebel et al., 1988), Vpu might enhance CD4 degradation by altering the ionic environment of the ER lumen, by solubilizing CD4 from the ER membrane, or by radically altering its native conformation (see ERAD model below).

The recent emphasis on protein structure-function studies by genetically manipulating specific proteins has led to the expression of many mutant proteins incompetent for assembly and transport beyond the ER. In most cases, the degradation characteristics of these proteins implicates the ERAD pathway, indicating that ERAD plays an essential role in protein quality control in the ER. Greater than 30 eukaryotic proteins are substrates for ERAD (Table 1), including both aberrant and normal proteins that are either lumenal or integral membrane proteins. Indeed, it appears that protein degradation associated with the ER is a general proteolytic process, one that regulates the expression of ER and plasma membrane proteins, as well as participating in protein quality control. Hence, because of the substrate diversity of this intracellular degradation process, one is compelled to question what molecular mechanisms are involved and whether a single proteolytic pathway can maintain a substrate selectivity of this magnitude, or if ERAD encompasses multiple or branched proteolytic pathways. In this review, we will attempt to answer some of these questions.

II. DEFINITION AND THE MAJOR QUESTIONS

What characteristics define ERAD? While a large number of proteins are targeted for degradation in the cell (see other chapters), three features implicate ERAD: (1) the

Table 1. Examples of Proteins Reported to be ERAD Substrates*

Protein Substrate	*Form*	*Reference*
Acetylcholine receptor subunits	Unassembled	Blount and Merlie, 1990
Acetylcholinesterase	Normal**	Rotundo, 1988; Rotundo et al., 1989
Acid phosphatase	Mutant, unglycosylated	Riederer and Hinnen, 1991
α_1-protease inhibitor	Mutant (Z)	Le et al., 1990, 1992; McCracken and Kruse, 1993; Wu et al., 1994; McCracken et al., 1996; Qu et al., 1996; Werner et al., 1996
α-globulin-alkaline phosphatase chimera	Mislocalized	Oda et al., 1996a,b
Apolipoprotein B	Normal	Borchardt and Davis, 1987; Davis et al., 1990; Sato et al., 1990; Dixon et al., 1991; Furukawa et al., 1992; Sakata et al., 1993; Du et al., 1994
Asialoglycoprotein receptor H2a subunit	Unassembled	Amara et al., 1989; Lederkremer and Lodish, 1991; Wikström and Lodish, 1991, 1992, 1993; Yuk and Lodish, 1995
Aspartase proteinase-I	Mutant	Fukuda et al., 1996
CD4	Normal, induced by Vpu	Chen et al., 1993; Willey et al., 1992; Bour et al., 1995; Schubert et al., 1996
Carboxypeptidase *yscY* (CPY)	Mutant	Finger et al., 1993; Hiller et al., 1996; Knop et al., 1996
Chondroitin sulfate proteoglycan	Mutant	Vertel et al., 1993, 1994
Cyclic nucleotide phosphodiesterase	Induced by Ca^{2+}depletion	Coukell et al., 1992
Cystic fibrosis transmembrane conductance regulator	Normal, mutant	Cheng et al., 1990; Jensen et al., 1995; Ward et al., 1995
Cytochrome P450 (2E1)	Normal and glucagon-induced	Eliasson et al., 1992
Dipeptidyl peptidase-IV	Mutant	Erickson et al., 1992; Tsuji et al., 1992
Fcγ III receptor α subunit	Unassembled	Kurosaki et al., 1991; Lobell et al., 1993
Fibrinogen chains	Unassembled	Danishefsky et al., 1990; Roy et al., 1992
β-hexosaminidase	Mutant	Lau and Neufeld, 1989
HMG-CoA reductase	Normal	Edwards and Gould, 1972; Faust et al., 1982; Orci et al., 1984; Chin et al., 1985; Meigs and Simoni, 1992; Hampton and Rine, 1994; Hampton et al., 1996b; McGee et al., 1996
Immunoglobulin chains	Normal, unassembled, mutant	Sidman, 1981; Dulis et al., 1982; Sitia et al., 1987; Bachhawat and Pillai, 1991; Gardner et al., 1993; Shachar et al., 1994; Knittler et al., 1995

continued

Table 1. (Continued)

Protein Substrate	*Form*	*Reference*
Influenza hemagglutinin	Mutant	Doyle et al., 1986
Insulin	Misfolded	Schmitz et al., 1995
Insulin receptor	Mutant	Sawa et al., 1996
β-lactamase/α globin chimera	Mislocalized	Stoller and Shields, 1989
LDL receptors	Mutant	Esser and Russell, 1988
MHC I	Unassembled induced by US11 and US2	Raposo et al., 1995; Wiertz et al., 1996a,b
Pre-B cell Ig receptor	Unassembled	Lassoued et al., 1996
Prepro-α-factor (yeast)	Unglycosylated yeast and mammalian cells	Caplan et al., 1991; Su et al., 1993; McCracken and Brodsky, 1996; Werner et al., 1996
Preprosomatostantin/chloram-phenicol acety transferase	Chimera	Danoff et al., 1993
Proteinase *yscA* (PrA)	Mutant	Finger et al., 1993
Lysozyme	Mutant	Omura et al., 1992
Retinol binding protein	Vitamin A deficiency	Tosetti et al., 1992
Ribophorin I	Mutant	Tsao et al., 1992
Ste6p, α-factor exporter	Mutant	Loayza et al., 1996
T-cell receptor (TCR) chains	Unassembled	Hannum et al., 1987; Chen et al., 1988, Lippincott-Schwartz et al., 1988; Bonifacino et al., 1989; Wileman et al., 1991, 1990b
Tropoelastin	Normal	Davis and Mecham, 1996
Transferrin receptor	Mutant	Hoe and Hunt, 1992; Yang et al., 1993

Notes: * Revised version of Table 1, Bonifacino and Klausner, 1994.
** Normal refers to no obvious abnormality in primary structure or oligomeric assembly.

degraded substrate must reside in the ER as judged by immunological or biochemical localization methods; (2) selective proteolysis of aberrant and unassembled secreted proteins occurs under conditions in which ER to Golgi transport is blocked, as demonstrated in yeast using temperature-sensitive *sec* mutants or by cytosol or energy depletion in mammalian cells; or (3) intracellular degradation is independent of lysosomal and vacuolar hydrolases, assayed by showing that degradation is insensitive either to weak bases (which raise the pH of the lysosome and inactivate resident proteases) or to the disruption of genes encoding vacuolar proteases.

A. Substrate Specificity

How is substrate specificity achieved? ERAD is highly selective for specific proteins, while the majority of ER resident and secreted proteins are quite stable

(Lippincott-Schwartz et al., 1988; Amara et al., 1989; Wileman et al., 1990b; Wikström and Lodish, 1993; reviewed in Bonifacino and Klausner, 1994). Although the nature of this remarkable substrate selectivity is not yet known, it is likely that structural features of proteins, which are buried in the correctly folded and assembled state, function as degradation motifs.

ERAD of monomers unable to associate into protein complexes has been extensively investigated and demonstrates the substrate selectivity of ERAD (Bonifacino et al., 1990; Fra et al., 1993; reviewed in Bonifacino and Klausner, 1994). For example, the integral membrane subunits of the T cell receptor (TCR) assemble in the ER before delivery to the plasma membrane. It has been observed that three of the subunits, the monomeric forms of the α, β, and δ chains, exhibit very short half lives *in vivo* while the ε and ζ subunits are quite stable. Because the α β, and δ chains become stable during complex assembly, it was suggested that they present ERAD recognition motifs that become buried as the TCR forms (Wileman et al., 1993). Site-directed mutagenesis indicated that such motifs reside in the integral membrane domains of the α and β subunits, domains that are also required to recognize the other subunits in the TCR complex (Bonifacino et al., 1990; Wileman et al., 1990a). Thus, ERAD targets unassembled subunits and malfolded proteins that have exposed recognition motifs.

How then do assembly and folding intermediates escape degradation? To address this question, Herzog and colleagues developed a system to monitor exposed internal domains of proinsulin using tyrosine iodination to probe protein structure (Schmitz et al., 1995). They found that iodination was inhibited by the molecular chaperone BiP, indicating that BiP masks internal domains and perhaps protects these intermediates from degradation. A role for BiP and other molecular chaperones in the specificity of ERAD has been suggested from several studies (see below).

Molecular Chaperones and Substrate Specificity

The high specificity under which ERAD operates is critical, particularly as the folding and posttranslational processing of nascent polypeptides occurs in the ER. Low specificity ERAD would be devastating if secretory proteins were randomly destroyed. Therefore, the ERAD machinery must chose carefully between proteins that are en route to their native conformations and proteins that have failed to properly fold or assemble. A class of molecules that might participate in this substrate selectivity are the molecular chaperones. These factors assist protein translocation across biological membranes, facilitate protein folding, and prevent aggregation by binding to sequences consisting primarily of hydrophobic amino acids (reviewed in Gething and Sambrook, 1992; Brodsky and Schekman, 1994; Hartl, 1996). Thus, a prolonged association between lumenal chaperones and malfolded polypeptides might target these proteins for ERAD.

Indeed, molecular chaperones are known to mediate selective targeting of protein substrates for degradation in Escherichia coli (Gottesman and Maurizi, 1992;

Sherman and Goldberg, 1993; Jubete et al., 1996), in yeast (Craig et al., 1994), and in mitochondria (Wagner et al., 1994). For example, the ATP-dependent proteolytic degradation of misfolded proteins in the mitochondrial matrix by the Pim1 protease is mediated by the mitochondrial molecular chaperone mt-hsp70 and one of its partner proteins, another molecular chaperone known as Mdj1p. In the absence of Mdj1p, misfolded proteins either remain associated with mt-hsp70 or form aggregates which are no longer substrates for Pim1p (Wagner et al., 1994).

In *E. coli*, molecular chaperones and specific proteases can exist in a complex. The protease Ti (ClpAP complex) that is active in the degradation of large abnormal polypeptides (reviewed in Gottesman and Maurizi, 1992; and see chapter 1), is composed of two components, ClpA and ClpP. ClpA possesses an intrinsic ATPase activity necessary for chaperone-like functions in presenting protein substrates to ClpP, the proteolytic component (Wickner et al., 1994). Also, *E. coli* DnaK, the prokaryotic Hsp70 family member, in cooperation with GrpE and the La protease, has been implicated in the degradation of misfolded proteins (Sherman and Goldberg, 1993). More recent studies from these investigators indicated that when chaperones are unsuccessful in catalyzing proper folding of a protein, they facilitate its rapid degradation (Kandror et al., 1994).

The notion that the activities of molecular chaperones and the proteolytic machinery are interrelated in eukaryotes is supported by the observation that yeast mutants harboring defective cytosolic Hsp70s of the SSA family, *SSA1* and *SSA2*, are sensitive to the amino acid analogue canavanine (Craig et al., 1993). The incorporation of canavanine into a polypeptide prevents proper folding and the abnormal protein is rapidly degraded by the cytoplasmic multisubunit proteolytic complex known as the proteosome (reviewed in Heinemeyer et al., 1991; see below). Hsp104, another yeast heat shock protein necessary for induced thermotolerance (Sanchez and Lindquist, 1990) is homologous to *E. coli* ClpA. Defects in Hsp104 activity can be suppressed by overexpression of *SSA1*, suggesting that the Hsp104 activity may also regulate protease function by mediating substrate-protease interactions (Craig et al., 1993). More recently, Ydj1p, a yeast cytosolic DnaJ homologous chaperone, was also shown to be essential for the degradation of certain proteins (Lee et al., 1996). A temperature-sensitive *ydj1-151* mutant strain demonstrated a 4–5-fold decrease in the degradation of short-lived and abnormal proteins. Thus, the function of molecular chaperones in the degradation of abnormal cytosolic proteins appears to be conserved in prokaryotes and eukaryotes. However, a putative role for chaperones in ERAD has only recently been observed and is discussed below.

Calnexin, BiP, and ERAD Substrate Recognition

The gene *CNE1* encodes yeast calnexin (Parlati et al., 1995), a nonessential integral ER membrane protein that, like mammalian calnexin, plays a role in ER quality control by assisting nascent proteins during initial folding, and by retaining misfolded proteins and incompletely assembled protein complexes in the ER (reviewed in Bergeron et al., 1994). Unlike its mammalian homologue, Cne1p does not appear to be a calcium binding protein, lacks a cytoplasmic tail containing an ER-membrane retention signal, and is a glycoprotein (Parlati et al., 1995). We discovered that degradation of the ERAD substrate, pαF, an unglycosylated form of the yeast α factor mating pheromone, was reduced in microsomes prepared from a yeast strain with a disrupted *CNE1* gene (McCracken and Brodsky, 1996), suggesting a role for calnexin in ERAD. Furthermore, the ERAD substrate, A1PiZ (a secretion defective variant of human α-1-proteinase inhibitor), is stabilized and secreted in *CNE1* mutated strains of *Schizosaccharomyces pombe* (Parlati et al., submitted).

Mammalian calnexin also appears to play a role in ERAD. Perlmutter and colleagues found that the lag in the degradation of A1PiZ, seen in liver disease susceptible patients with α1-antitrypsin deficiency, was associated with its failure to interact with calnexin (Wu et al., 1994). Subsequently, these investigators showed that A1PiZ-calnexin complex formation plays a key role in its ER-associated degradation (Qu et al., 1996).

How might calnexin function in ERAD? Thomas and colleagues previously demonstrated that calnexin acts as a constituent of the ER quality control apparatus in *Saccharomyces cerevisiae* (Parlati et al., 1995). Perhaps the calnexin-mediated retention of misfolded substrates in the ER facilitates their targeting for rapid degradation. This hypothesis is supported by the observation that two mutant proteins known to be degraded by ERAD are among those retained in the ER by interaction with calnexin: the ΔF508 mutant of the CFTR (Pind et al., 1994), and the Z variant of human α-1 proteinase inhibitor (Le et al., 1992; McCracken and Kruse, 1993).

Together, the observations presented above implicate calnexin as a player in ERAD. Hence, it is important to consider that calnexin has been shown to serve primarily as a lectin, binding specifically to glycosylated proteins (Ou et al., 1993; Hammond et al., 1994; Hebert et al., 1995). However, calnexin also binds to proteins which lack N-linked glycans (Rajagoplan et al., 1994; Kim and Arvan 1995; Cannon et al., 1996), and there is evidence that calnexin interacts with unfolded proteins by binding to the polypeptide backbone rather than glycan determinants (Ware et al., 1995; Zhang et al., 1995) particularly if the proteins are aggregated (Cannon et al., 1996). Studies probing the interaction of calnexin with ERAD substrates are currently limited, although complexes between calnexin, misfolded proteins, and BiP/Grp78 have been observed (Hammond and Helenius, 1994). Perhaps calnexin and BiP, an ER lumenal Hsp70 family chaperone, cooperate to prevent the aggregation of aberrant proteins and mediate their degradation.

The interaction between BiP and several ERAD substrates has been investigated. Based on a correlation of the kinetics of BiP dissociation from unassembled immunoglobulin chains, with the rate of their degradation, it appears that BiP may mediate degradation of these polypeptides (Knittler et al., 1995). Recently, it was observed directly that the immunoglobulin light chain correlated with light chain degradation (Skowronek et al., 1998). In addition, *in vivo* iodination of misfolded proinsulin indicated colocalized signals for ERAD and for BiP binding (Schmitz et al., 1995). Also, results from co-immunoprecipitation analyses showed that BiP is associated with an ERAD-targeted mutant precursor of a glycosylphosphatidylinositol-anchored chimeric protein but not with the secreted form, suggesting that BiP may mediate ER-associated degradation of the mutant chimera (Oda et al., 1996a). Finally, an engaging relationship between ERAD and the association of Hsp70 and BiP with mutant insulin receptors has been observed. Tight binding of BiP and rapid ERAD correlated with mutations in the ectoplasmic (lumenal) domain of the receptor while tight binding of Hsp70 and rapid ERAD correlated with mutations in the cytoplasmic domain (Sawa et al., 1996). These results suggest a role for both cytoplasmic and ER lumenal Hsp70 chaperones in the targeting of ERAD substrates. Interestingly, Hsp70 has also been shown to be associated with the transmembrane form of apoB that is highly susceptible to ERAD (discussed below). This association between Hsp70 and apoB is transient and ATP-sensitive (Zhou et al., 1995), and thus may help to maintain nascent apoB in a translocation competent conformation, protect apoB from aggregation, or alternatively, may target apoB for ERAD. However, because the conclusions from many of these studies were arrived at by circumstantial evidence, more detailed investigation into the role for BiP and other molecular chaperones in ERAD is needed.

Toward this goal, we have recently shown that a mutant yeast strain deleted for the *cne1* gene and containing a mutation in the yeast BiP gene, *kar2-133*, exhibits a synthetic interaction, suggesting that both gene products may function in the same pathway. In addition, we find that microsomes prepared from the *kar2-133* strain and two other BiP mutants are defective for the degradation of pαF and that all three Bip mutants exhibit defective degradation of A1PiZ (Brodsky et al., submitted). Also, the degradation of another ERAD substrate, CPY*, has been shown to be defective in a *kar2* mutant strain (Plemper et al., 1997).

B. Proteolytic Activity

Where does proteolysis occur and what proteases are involved in ERAD? Both morphological and biochemical subcellular localization studies have clearly shown that ERAD substrates are present in the cisternae of the ER (reviewed in Bonifacino and Klausner, 1994). However, such studies do not rule out the possibility that ERAD substrates are transported out of the ER to a degradative compartment for rapid proteolysis. Therefore, the determination of the degradative compartment needed to be resolved by the identification and localization of the responsible protease(s). To this end, analyses of various ERAD substrates in untreated and permeabilized cells have indicated that proteases of the mammalian lysosome and the yeast vacuole, the analogous compartment in this organism, are not required for proteolysis (for example see Chen et al., 1988; Amara et al., 1989; Finger et al., 1993; McCracken and Kruse, 1993; Hampton and Rine, 1994; Yuk and Lodish, 1995; and references therein). Because of these results and those indicating that ERAD substrates are not transported beyond the ER (reviewed in Bonifacino and Klausner, 1994), and because experiments using permeabilized mammalian cells indicated initially that ERAD was cytosol and ATP-independent (Stafford and Bonifacino,1991; Meigs and Simoni 1992), ERAD-specific proteases were presumed to reside within the ER. In support of this hypothesis, Otsu et al. (1995) studied a cysteine protease purified from the ER of rat liver that specifically binds to and degrades a heterologously expressed, misfolded lysozyme mutant. Because this protease, known as ER-60 (Urade et al., 1992), also degrades stable resident ER proteins *in vitro* and has not been shown to hydrolyze other known ERAD substrates, its *in vivo* proteolytic function remains uncharacterized. In addition, this factor has recently been proposed to function primarily as a chaperone in the ER (Oliver et al., 1997).

A role in ERAD was proposed for signal peptidase (Mullins et al., 1995), an ER membrane bound endoproteolytic protein complex that removes amino-terminal signal peptides from precursor proteins during their translocation to the lumen of the ER (reviewed by Shelness, 1994). These investigators examined the degradation of a membrane bound yeast protein chimera, A189invHD, containing sequences from arganine permease, invertase, and histidinol dehydrogenase. The rapid degradation of A189invHD (95% during a 30 minute chase), was not inhibited in a *sec23* mutant in which ER transport vesicle formation is blocked. When the chimeric gene was expressed in a double mutant, *sec23/sec11*, that also possesses a temperature-sensitive defect in signal peptidase (Sec11p) activity, only 60% of A189invHD was degraded during a 30 minute chase. This result indicated that cleavage by signal peptidase was required for efficient ERAD, perhaps to generate a protein that was more effectively recognized as a substrate for ERAD. However, a role for signal peptidase in the proteolysis of unassembled or malfolded ERAD substrates remains to be demonstrated.

Nevertheless, experiments using protease inhibitors have shown that both cysteine and serine proteases may be involved in ERAD (Wileman et al., 1991; Inoue and Simoni, 1992; Tsao et al., 1992; Wikström, and Lodish, 1992; Gardner et al., 1993). In addition, inhibitors of the cytoplasmic proteasome complex limit the degradation of several ERAD substrates: ER accumulated CFTR (Jensen et al., 1995; Ward et al., 1995); immunoglobulin *k* light chain expressed in CH12 lymphoma cells in the absence of heavy chain (Gardner et al., 1993); a chimeric protein consisting of the rat α-globulin N-terminal and alkaline phosphatase C-terminal sequences when expressed in COS-1 cells (Oda et al., 1996b); the regulated degradation of HMG-CoA reductase (Hampton et al., 1996b; McGee et al., 1996); the Z variant of α1-proteinase inhibitor (Qu et al., 1996), MHC class I molecules in CMV-infected cells (Wiertz et al., 1996a,b), an immature unglycosylated precursor of the yeast α factor mating pheromone (Werner et al., 1996), tyrosinase expressed in melanoma cells (Hallaban et al., 1997), and the T-cell receptor α chain (Yu et al., 1997). These combined results indicate that the proteasome is involved in ERAD.

Proteasome-Dependent Degradation

A role for the proteasome in ER-associated degradation was first suggested by the observation that a greater number of proteasomes were associated with ER membranes in secretory cells (Rivett, 1993), and that yeast ER membranes contained the ubiquitin conjugating enzyme, Ubc6 (Sommer and Jentsch, 1993). Direct evidence that the proteasome complex supplied the proteolytic activity for ERAD was provided by studies with the yeast *S. cerevisiae* (Hiller et al., 1996; Werner et al., 1996). The results with yeast were compelling, as they were derived from *in vivo* and *in vitro* assays monitoring the degradation of both mammalian and yeast ERAD substrates. These studies are described in more detail below.

The expression of the secretion defective Z variant of human α-1 protease inhibitor (A1PiZ) has been shown to cause emphysema and liver failure because circulating levels of the protein are low and because it accumulates in the hepatic ER (Fagerhol and Cox, 1981). The secretion defect can be traced to a single amino acid substitution (McCracken et al., 1991) and much of the accumulated A1PiZ is subsequently degraded (Le et al., 1990). These results suggested that this protein would be a good candidate for investigations into the mechanisms of ERAD. Indeed, yeast expressing A1PiZ synthesize an ER form of this protein which is rapidly degraded by ERAD (McCracken and Kruse, 1993). To examine the role of the proteasome complex in ERAD, the degradation of A1PiZ was monitored in a yeast proteasome mutant, *pre 1-1 pre 2-2*. This strain is deficient in the "chymotrypsin-like" activity of the proteasome complex (Heinemeyer et al., 1991, 1993) and accumulates aberrant proteins (Seufert and Jentsch, 1992). Results from pulse chase protein radiolabeling experiments showed that degradation of A1PiZ was inhibited in the proteasome mutant strain. The half-life of A1PiZ calculated from these data was 163 minutes in the mutant, a threefold increase over the 56 minute half-life in the wild-type strain (Werner et al., 1996).

To establish that the inhibition of ERAD was specifically due to mutations in subunits of the proteasome complex and not to uncharacterized extragenic mutations, pulse chase experiments were performed with transformed *pre 1-1 pre 2-2* strains containing centromeric expression plasmids carrying the *PRE1* or *PRE2* wild-type gene. Expression of each wild-type gene restored ERAD activity in the *pre 1-1 pre 2-2* mutant, indicating that the observed ERAD inhibition was due to defective proteasomes (Werner et al., 1996). Wolf and colleagues have presented similar results investigating the degradation of a malfolded mutant form of carboxypeptidase Y (CPY*) (see chapter 2). This ERAD substrate accumulates in the ER, is rapidly degraded (Finger et al., 1993), and degradation of CPY* is limited in the *pre1-1 pre2-2* mutant strain (Hiller et al., 1996).

In vitro *Assays*

To better examine ERAD in a genetically tractable organism in which the factors required for ERAD might be more readily identified, we developed an *in vitro* system using ER-derived microsomes from the yeast *S. cerevisiae*. The ERAD substrate chosen for these studies was an unglycosylated mutant form of the yeast mating pheromone, pre-pro α factor (ppαF), that is rapidly degraded *in vivo* (Caplan et al., 1991). We found that unglycosylated, signal-sequence cleaved pro-α factor (pαF) was rapidly degraded in yeast microsomes upon addition of cytosol and ATP, while wild-type, glycosylated pαF was stable (McCracken and Brodsky, 1996). Thus, ERAD in the reconstituted system and *in vivo* exhibited identical kinetics and specificity. This *in vitro* assay was also used to show that ERAD was dependent on protein factors in the cytosol, ATP hydrolysis, and the molecular chaperone calnexin (see above; McCracken and Brodsky, 1996)

Since *in vitro* ERAD required cytosolic components, the ability of the cytosol prepared from the *pre1-1 pre2-2* proteasome mutant to support ERAD could be measured using this assay. When ERAD was examined using this cytosol, degradation of pαF was inhibited (Werner et al., 1996). Furthermore, the half-life of pαF increased threefold, an increase identical to that observed for A1PiZ *in vivo* (see above), thus confirming the conclusion that the proteasome complex was required for ERAD.

Because earlier *in vivo* studies failed to identify the protease involved in the regulated degradation of HMG-CoA reductase, Simoni and colleagues established a mammalian cell free assay to investigate the regulated degradation of HMG-CoA reductase (McGee et al., 1996). In this system neither exogenous ATP, cytosol, nor the lumenal contents of ER membranes isolated from mevalonate-treated cells was required for HMG-CoA reductase degradation once the process had been initiated by mevalonate. Furthermore, the accelerated degradation of HMG-CoA reductase was inhibited by cysteine protease inhibitors and by a newly defined specific inhibitor of the proteasome, lactacystin (Fenteany et al., 1995). Interestingly, these investigators also showed that proteasomes were stably associated with isolated ER membranes (McGee et al., 1996), an observation that correlates with the effect of lactacystin and suggests a role for the proteasome in the degradation of HMG-CoA reductase. However, as noted by these investigators, this conclusion is based on inferential data, hence, one can not rule out the possibility that unidentified lactacystin-sensitive proteases within the ER might function in the proteolysis of HMG-CoA reductase. Indeed, the regulated degradation of HMG-CoA reductase is a complex problem. It is not a quality control process requiring recognition of aberrant folding or unassembled proteins but, rather, a regulated degradation of apparently normal proteins. Thus, degradation of HMG-CoA reductase in mammalian cells appears to have all the features of the ERAD pathway but may require supplementary events prior to entry into ERAD.

However, the degradation of HMG-CoA reductase in yeast appears to be mediated by the proteasome. Hampton and colleagues (1996b) have shown that yeast strains unable to degrade an ERAD-targeted HMG-CoA reductase isozyme (Hmg2p) harbor mutations in genes encoding two novel ER membrane proteins as well as to a yeast homologue of human TRAP2/p97, a subunit of the proteasome complex (DeMartino et al., 1996; see also chapter 5). In fact, the expression of

TRAP2 cDNA in the mutant strain compliments the ERAD-deficiency (Hampton et al., 1996b).

Ubiquitin-Dependent Degradation?

Overall, the results discussed above indicate that ER-associated protein degradation, at least in yeast, is not a novel intracellular degradation process, but uses the proteasome complex, the proteolytic component of a known cytosolic protein degradation pathway. An obligatory step in the selective targeting of many short-lived cytosolic proteins to the proteasome for degradation is conjugation to ubiquitin (reviewed in Hochstrasser, 1995; Hilt and Wolf, 1996). Indeed, most known substrates of the proteasome complex are ubiquitinated, raising the question of whether ubiquitination of substrate is a prerequisite for proteasome-dependent ERAD. The answer is not yet clear since ubiquitination is required for the degradation of some ERAD substrates, but apparently not for others. Some examples are discussed below.

To define the molecular mechanism through which CFTR, a member of the ATP-binding-cassette (ABC) transporter superfamily (Riordan et al., 1989; Higgins, 1992) was degraded in the ER, Jensen et al. (1995) used protease inhibitors to selectively prevent the ATP-dependent maturation of CFTR and its degradation in cultured cells. The addition of lactacystin, a specific inhibitor of the proteasome (Fenteany et al., 1995), obstructed the degradation of the ER-retained forms of CFTR and increased the population of poly-ubiquitinated CFTR derivatives. Ward et al. (1995) definitively demonstrated that ubiquitination was required for degradation of CFTR because coexpression of a dominant negative ubiquitin mutant impeded CFTR degradation, and because cells containing a temperature-sensitive ubiquitin activating enzyme (E1) were unable to degrade CFTR (Ward et al., 1995).

Similarly, the proteasome-dependent ER-associated degradation of a mutant form of a yeast ABC transporter, Ste6p, is ubiquitin-dependent. Ste6p is a membrane protein that mediates the export of the yeast mating pheromone a-factor (Kuchler et al., 1989). Michaelis and colleagues found that the mutant Ste6-166p, accumulated in the ER, was rapidly degraded, and that proteolysis was inhibited by mutations in the ubiquitin conjugating enzymes, Ubc4 and Ubc5. In addition, a mutation in the ubiquitin isopeptidase, Doa4, known to inhibit ubiquitin-dependent proteolysis, also inhibited degradation of Ste6-166p. These results imply that, like CFTR, the ERAD of Ste6-166p was ubiquitin- and proteasome-dependent. Interestingly, these investigators found that several other *ste6* mutants (*ste6-13* and *ste6-90*) are retained in the ER and are not ERAD substrates. In fact these mutants are more stable than wild-type Ste6p (Loayza et al., submitted). The increased stability of these ER-retained forms may indicate the masking of binding sites for chaperones or other factors, such as ubiquitin conjugating enzymes, that target Ste6p for ERAD

The ubiquitin system is also involved in the degradation of a malfolded yeast pre-protein, (CPY*). Wolf and colleagues (Hiller et al., 1996) found that the mutation responsible for one of their ERAD-deficient mutants, *der2*, maps to a gene encoding Ubc7, a ubiquitin conjugating enzyme. In addition, when ERAD was inhibited, they observed ubiquitinated CPY* indicating that degradation was ubiquitin dependent. Also, Hampton and Bhakta (1997) have recently determined that HMG C0-A reductase in yeast is ubiquitinated with subsequent degradation by the proteasome complex. The interesting finding that the ER lumenal chaperone calnexin is polyubiquinated when associated with A1PiZ before proteasome-dependent degradation of the entire complex (Qu et al., 1996), further illustrates a role for ubiquitin in ERAD. Finally, an ER-associated protein known as Cue1p was recently identified as a receptor for UBc7p and is required for both retrorade transport and degradation of an ERAD substrate in yeast (Biederer et al., 1997), leading these investigators to conclude that ongoing ubiquitination might contribute mechanistically to retrograde translocation.

On the contrary, degradation of mammalian HMG-CoA reductase (McGee et al., 1996) and pαF and A1PiZ in yeast (Werner et al., 1996), do not appear to require the ubiquitin system. Simoni and colleagues investigated the regulated degradation of HMG-CoA reductase in a mutant cell line expressing a temperature sensitive form of the hamster E1 enzyme, the same cell line used to demonstrate that ERAD of CFTR was ubiquitin-dependent (see above; Kulka et al., 1988). Pulse chase analyses at both restrictive and permissive temperatures demonstrated that HMG-CoA reductase degradation was unaffected when the ubiquitin pathway was blocked (McGee et al., 1996). To confirm that the proteolysis of T-cell receptor α chains does not require ubiquitination, Yu and colleagues (1997) mutated the 11 lysine residues in this protein and still observed its degradation. Thus, these results indicate that ubiquitin is not an obligatory requirement for ERAD.

Similarly, the ubiquitin conjugating activity of the ER-localized enzyme Ubc6, was shown to be unnecessary for the ER-associated degradation of A1PiZ. Also, no effect on the degradation of pαF was observed when double deletion mutants *ubc4 ubc5* and *ubc6 ubc7* were examined *in vitro* (Werner et al., 1996). This is intriguing since *UBC4, 5, 6,* and 7, encode the known E2 ubiquitin conjugating enzymes required for degradation of aberrant proteins and all have been shown to ubiquitinate ERAD substrates (reviewed in Jentsch, 1992; Hochstrasser, 1995). Finally, the addition of ubiquitin-aldehyde, an inhibitor of the poly-ubiquitin disassembling isopeptidase known to inhibit ubiquitin-mediated proteasome-dependent proteolysis (Hershko and Rose, 1984; Dunten and Cohen, 1989), did not have a stabilizing effect on pαF in the *in vitro* assay (Werner et al., 1996). Together, these results suggest that proteolysis of certain ERAD substrates by the proteasome complex may take place by a ubiquitin-independent pathway, as observed previously for ornithine decarboxylase (Murakami et al., 1992; Elias et al., 1995) and casein (Driscoll and Goldberg, 1990). It is known that full-length proteins can be fully degraded to peptides in the absence of ubiquitin by the 20S proteasome if the proteins are first denatured (Dick et al., 1994; also see Hochstrasser, 1995). The research challenge is then to understand the role of ubiquitin in the proteasome-dependent degradation of ERAD substrates, and to discover the mechanism by which nonubiquitinated substrates are targeted to the proteasome.

Two Proteolytic Pathways?

As discussed earlier, the diversity of ERAD substrates suggests the possibility of distinct pathways for their degradation. Lodish and colleagues proposed a model invoking two proteolytic pathways for the degradation of the H2 subunit of the asialoglycoprotein receptor (Yuk and Lodish, 1995). The active human asialoglycoprotein receptor is a hetero-oligomer consisting of H1, H2a, and H2b subunits. When H1 subunits were expressed in cells in the absence of H2, mature H1 appeared on the cell surface. In contrast, all of H2a and 70% of H2b subunits expressed alone were rapidly degraded by ERAD (Shia and Lodish, 1989). During the degradation of H2a, a 35 kDa fragment transiently accumulated in the ER and the subsequent degradation of this fragment was insensitive to the serine protease inhibitors TLCK and TPCK. Furthermore, H2a mutant proteins, unable to generate the 35 kDa fragment due to alterations in the cleavage site, were degraded at the same rate as wild-type H2a by a process sensitive to TLCK and TPCK. These results indicate two proteolytic pathways for the degradation of H2a subunits. In addition, TLCK and TPCK inhibit the degradation of a vast majority of pulse-labeled H2, leading these investigators to propose that the pathway that does not target the 35 kDa intermediate is the predominant ERAD pathway (Yuk and Lodish, 1995).

The intracellular degradation of apolipoprotein B (apoB) may provide another example of multiple proteolytic pathways for ER protein degradation. ApoB is the major protein component of very low-density lipoprotein (VLDL) (reviewed in Olofsson et al., 1987; Boren et al., 1993). The secretion of apoB is regulated post-translationally and reacts rapidly to the state of lipid metabolism in the cell (reviewed in Dixon and Ginsberg, 1993). Evidence suggests that nascent apoB is maintained in a transmembrane position until adequate core lipids are available in the ER and Golgi for assembly into lipoprotein particles (Davis et al., 1990; Thrift et al., 1992; Sakata et al., 1993). In agreement with this observation, translocation of apoB into the ER of cultured cells derived from liver and intestine is stimulated by the addition of oleate (Sato et al., 1990; Dixon et al., 1991) and may be mediated by microsomal triglyceride transfer protein (Sharp et al., 1993; Du et al., 1996). A large portion of intracellular apoB is degraded and the proteolysis is sensitive to ALLN (Sakata et al., 1993; Du et al., 1994), a cysteine protease inhibitor that can also inhibit the proteasome. This ERAD of nascent apoB is reduced by oleate addition, suggesting that translocation protects this protein from degradation (Dixon et al., 1991). However, when Hep G2 cells were treated with brefeldin A (BFA), which arrests vesicular transport from the ER to the Golgi (reviewed by Klausner et al., 1990), apoB degradation continued as in control cells, and when oleate treated cells were also treated with BFA, oleate failed to protect apoB from degradation (Furukawa et al., 1992). These results suggested that both a transmembrane form and a translocated form of apoB were degraded by ERAD. However, ApoB degradation in cultured rat hepatocytes is completely blocked by BFA, suggesting the involvement of a post-ER degradative pathway (Wang et al., 1995).

Whether the degradation of nascent, partially translocated apoB, and degradation of the fully translocated species occurs by different proteolytic pathways has not been easy to determine. The degradation of apoB has been reported to occur during protein translation, translocation across the ER membrane, and vesicular transport (for review, see Yao et al., 1997); and considerable differences exist in the metabolism of apoB in human and rat hepatocytes and when apoB is expressed in Chinese hamster ovary (CHO) cells (reviewed in Dixon and Ginsberg, 1993). Indeed, the existence of a transmembrane form of apoB is controversial. Davis and co-workers have shown that ALLN induced inhibition of apoB degradation in CHO cells caused the intracellular accumulation of incompletely translocated chains of apoB, as measured by accessibility to exogenous trypsin (Bonnardel and Davis, 1995). However, Ingram and Shelness (1996) recently showed that endogenous apoB isolated from HepG2 cell membranes was stabilized by ALLN and was protected from trypsin treatment. Their results along with others (Leiper et al., 1996) indicate that apoB is efficiently translocated across the ER membrane and suggest that ERAD of apoB may be a function of its unassembled (unlipidated) form rather than result from inefficient translocation.

The premise of multiple pathways for ERAD is supported by studies of "orphaned" immunoglobulin subunits. For example when secreted and membrane bound forms of IgM were examined in a cell line that is light chain deficient, both forms were rapidly degraded by a pathway that is nonlysosomal (Taya et al., 1991) and brefeldin A insensitive (Rabinovich et al., 1993), indicating that incompletely assembled forms of IgM are retained and degraded in the ER. However, in nonsecreting B lymphocytes the membrane form of IgM is stably displayed on the cell surface, while the secreted form is rapidly degraded by an intracellular degradation process that is independent of the lysosomal pathway (Amitay et al., 1991) and inhibited by brefeldin A (Rabinovich et al., 1993). Furthermore, Bar Nun and colleagues (Winitz et al., 1996) have demonstrated that unassembled forms of IgM are degraded in the ER while the developmentally regulated degradation of secretory IgM requires vesicular transport out of the ER.

Evidence for multiple degradation sites along the secretory pathway is also revealed by the study of two carboxy-terminal truncated mutants of ribophorin I (RI), a type I ER membrane glycoprotein that is a component of the oligosaccharyl-transferase complex (Kelleher, et al., 1992). Mutant RI_{467} lacks the cytoplasmic domain of RI but is membrane anchored and is rapidly degraded with monophasic kinetics by a nonlysosomal pathway. On the other hand, mutant RI_{332}, which consists of only the luminal NH_2 terminus of RI, is degraded with biphasic kinetics by a nonlysosomal pathway. Because the second degradative phase of RI_{332} could be inhibited by carbonyl cynanide m-chlorophenylhydrazone (CCCP), monesin (both of which inhibit vesicular flow out of the ER) and brefeldin A, it was hypothesized that vesicular transport beyond the ER was required for degradation. These results imply that there are multiple nonlysosomal sites for the degradation of mutant forms of RI: the ER where the monophasic turnover of RI_{467} and the first phase degradation of RI_{332} occurs, and a post-ER pre-Golgi site where the final phase of RI_{332} degradation may occur (Tsao et al., 1992).

Finally, the degradation and maturation of CFTR (Jensen et al., 1995) provides an example of two distinct proteolytic events associated with the ER. Conversion of the CFTR precursor to the active transport-competent form is ATP-dependent and inhibited by the protease inhibitor MG-132, but not by the proteasome specific inhibitor lactacystin. Alternatively, lactacystin does inhibit the degradation of the ER-retained immature CFTR species, an event insensitive to MG-132. Thus, it appears that the maturational processing step and the rapid degradation of immature and mutant CFTR are determined by unique proteolytic processes.

Because the examples of multiple proteolytic pathways discussed here were generated by indirect evidence and in some cases were collected under nonphysiological conditions, and because the existence of a proteasome-dependent pathway for ERAD has been clearly established for several ERAD substrates (see above), further characterization of proteasome-independent pathways is certainly needed.

III. THE TOPOLOGICAL PROBLEM

In retrospect, placing a proteolytic machine in a compartment distinct from that committed to protein biogenesis makes sense; nascent proteins remain isolated from the proteasome while aberrant polypeptides are banished to the cytosol and destroyed (Brodsky and McCracken, 1997). How could ER proteins be degraded by the cytosolic proteasome complex? While integral ER membrane proteins could present external soluble domains to a cytosolic protease, a mechanism must exist to extrude or degrade the transmembrane and lumenal fragments, and the degradation of soluble proteins within the ER lumen requires that they first translocate back out of the ER. Although such events may appear improbable, several studies indicate ER to cytosol transport. Firstly, glycopeptide export from the yeast ER can occur by a cytosol and ATP-dependent mechanism (Römisch and Schekman, 1992). Secondly, free polymannose oligosaccharides that are made in the ER are transported directly into the cytosol (Moore et al., 1995). Thirdly, the ER appears to be the intracellular site for translocation of cytotoxic proteins into the cytoplasm (reviewed in Simpson et al., 1995), and a selective ATP and cytosol-dependent export of the ERAD substrate pαF from ER-derived microsomes has been demonstrated (McCracken and Brodsky, 1996). Fourthly, Hiller and colleagues (1996) demonstrated that a misfolded protein from the yeast ER was degraded in the cytosol, in vivo. Finally, a recent report on the viral-mediated degradation of the MHC class I molecule (MHCI) reveals a possible mechanism for protein export from the ER to the cytoplasm (Wiertz et al., 1996a), and is discussed below.

MHC I is an integral membrane protein that associates with β2 microglobulin (β2m) to present virally-derived peptides to cytolytic T cells. A class I-β2m-peptide complex assembles in the ER and then transits through the secretory pathway to the plasma membrane. To evade the immune system, pathways have evolved in human viruses to abolish MHCI function. In a recent report by Wiertz et al. (1996a), the product of the cytomegalovirus US11 gene, an ER membrane glycoprotein, was shown to selectively target MHCI to the cytosol where it was degraded. In contrast, β2m was secreted, demonstrating that US11 action is specific. This specificity was also apparent because tissue culture cells expressing US11 grow as well as wild-type cells, suggesting that US11-mediated protein degradation is not rampant. As was the case for the ΔF508 variant of CFTR (see above), MHCI degradation was also inhibited by lactacystin (Wiertz et al., 1996a). How US11 solubilizes an integral membrane protein from the ER is not obvious. One wonders, though, whether a nuclear encoded US11 analogue is required during protein biogenesis to remove misfolded membrane proteins, such as the immature forms of CFTR or HMG-CoA reductase. Although the sequence of US11 offers no clues to its function, future biochemical studies on this unorthodox factor will certainly be enlightening. Indeed, another cytomegalovirus gene product, US2, also targets MHCI to the cytosol for rapid degradation (Wiertz et al., 1996b). When inhibitors of the proteasome and crosslinking reagents were included in a series of co-immunoprecipitation experiments with cell lysates from *US2* transfected cells that express MHCI, stable complexes between MHCI and the proteasome were detected. Amazingly, an association between MHCI and the Sec61p complex, the protein translocation channel in the ER membrane, was also detected. Together, these results indicate that ER export and import can use the same machinery and demonstrate how proteins in the ER can be banished to the cytosol for degradation.

A. ER Export in Yeast

To test the possibility that substrates for ERAD were exported to the cytosol for proteolysis, we modified the *in vitro* ERAD assay to include a centrifugation step after the chase reaction in order to separate microsomes from cytosol. We observed that a small amount of radiolabeled unglycosylated pαF was present in the supernatant after a 25 minute chase in the absence of cytosol or hydrolyzable ATP, or in the presence of heat inactivated cytosol and ATP. However, the percent of pαF in the supernatant fraction increased approximately threefold when the chase reaction included cytosol and ATP (McCracken and Brodsky, 1996). In contrast, glycosylated pαF was not observed in the supernatant fraction in the presence or absence of cytosol and ATP, indicating that ER polypeptide export was selective for an ERAD substrate.

It was expected that when ER export was examined under conditions that reduce proteasome activity, either by mutation or protease inhibitors, an increased cytosolic pool of ERAD substrate should be evident. *In vitro* ERAD experiments comparing proteasome mutant and wild-type cytosol provided an opportunity to test this hypothesis. When the amount of pαF in the cytosol was determined using a protease protection assay, we found pαF in the supernatant after 10 minutes of the chase incubation with *pre1-1 pre 2-2* cytosol, while pαF was not detected in the wild-type cytosol until 20 minutes. Furthermore, at 20 minutes the percent of undegraded pαF in the supernatant from the *pre 1-1 pre 2-2* cytosol reaction was approximately twice that observed in the wild-type cytosol (Werner et al., 1996). In addition, when 3,4-dichlorooisocoumarin (DCI) and lactacystin, potent inhibitors of the proteasome (Fenteany et al., 1995), were added to wild-type cytosol and used in *in vitro* ERAD assays, a similar increased pool of pαF was observed in the supernatant (Werner et al., 1996).

This discovery that ERAD substrates are exported to the cytosol has further significance in that it reveals a cellular function for a retrograde protein transporter in the ER, a mechanism implicated in the delivery of cytotoxic proteins from the ER to the cytoplasm (Ogata et al., 1990; Pelham et al., 1992; Wales et al., 1993), and a transport pathway exploited by the human cytomegalovirus to mask its presence in infected cells (Wiertz et al., 1996a,b; see above).

B. ER Protein Transport, A Two-Way Street

The canon of protein biogenesis is that nascent, secreted polypeptides are translocated into the ER through an integral membrane channel or pore. As more defined systems for examining protein translocation became available, various laboratories observed that polypeptides in this channel may oscillate bi-directionally (e.g., see Nicchitta and Blobel, 1993), indicating that the translocation channel has no directional bias. Thus, it was suggested that lumenal molecular chaperones bind to increasingly longer segments of translocating proteins and drive import by acting as molecular ratchets, hydrolyzing ATP to engineer cycles of peptide binding and release (Simon et al., 1992). While this model remains controversial (Brodsky, 1996), there is now reason to suspect that the translocation channel might be required to bi-directionally transport ERAD substrates (Wiertz et al., 1996b; see discussion of MHCI above). Together, the results discussed above indicate that peptide export is required for ERAD and suggest that the protein translocation channel and the ERAD peptide exporter are identical. In fact, Pilon and colleagues (1997) have recently shown that Sec61P, a component of the yeast translocation channel, is required to export pαF from ER.

C. Genetic Approach to Study ERAD

What other ER proteins might be required to export polypeptides for degradation by the proteasome? The answer to this question may emerge from a genetic approach to study ERAD. *S. cerevisiae* mutants deficient in ER-associated degradation of mutant α1-protease inhibitor, A1PiZ (McCracken et al., 1996), malfolded CPY* (Knop et al., 1996), and a yeast HMG-CoA reductase isozyme, Hmg2p (Hampton et al., 1996b) have been isolated by elegant genetic screens for increased stability of these ERAD substrates. A gene complementing the ERAD-defective mutant *der1*, was cloned in the Wolf laboratory and encodes a polytopic ER membrane protein of unknown function (Knop et al., 1996). In addition, Hampton and colleagues isolated two genes *HRD1* and *HRD3*, required to degrade Hmg2p, that also encode novel ER membrane proteins (Hampton et al., 1996b). Such integral membrane proteins could function to export peptides to the cytosol for degradation, be required for targeting substrate (e.g., ubiquitination or chaperone activity), serve as a receptor for the proteasome, or function analogously to US11 or US2 to dislocate hydrophobic membrane spanning domains. Interestingly, the majority of the Hrd3p protein sequence is predicted to be in the lumen of the ER with a short c-terminal cytoplasmic tail, resembling the topology predicted for US11 (Hampton et al., 1996b). A third gene identified by these investigators as essential for ERAD, *HRD2*, encodes a protein homologous to the TRAP-2 activator of the 26S proteasome (see above), while the *DER2* gene isolated by Wolf and co-workers (Hiller et al., 1996) encodes a known ubiquitin conjugating enzyme, Ubc7p, supporting the role for ubiquitin and the proteasome in ERAD.

Indeed, the isolation and characterization of ERAD defective mutants has provided the genetic groundwork to understand the molecular mechanisms of ERAD. Genetic characterization of the yeast ERAD mutants has revealed seven different complementation groups for the mutants deficient in the degradation of A1PiZ (McCracken et al., 1996), four complementation groups for the mutants that fail to degrade CPY* (Knop et al., 1996), and three for the mutants deficient in the degradation of yeast HMG-CoA reductase (Hampton et al., 1996b), and none of these mutant screens were saturating. These results are not surprising since large numbers of genes have been identified in other processes associated with the secretory pathway (see Rothblatt et al., 1994). Thus, the identification of at least seven complementation groups, with the likelihood that many more genes may be involved, affords the first glimpse at the complexity of this essential cellular process.

IV. ENERGY REQUIREMENTS AND A UNIFYING MODEL FOR ERAD

Experiments to define the energy requirements for ERAD produced conflicting results. When we tabulated the results from these studies, we discovered a most intriguing correlation. The degradation of membrane proteins does not require ATP, while ATP is necessary for the proteolysis of soluble ER lumenal proteins (Table 2). The only exception is that opposite conclusions were reached on the ATP dependence for the degradation of mammalian HMG-CoA reductase. Assuming that the results of Simoni and colleagues (Meigs and Simoni, 1992; McGee et al., 1996) are valid, what model could explain such a striking relationship between an ER protein's topology and whether energy is required for its degradation?

Table 2. Energy-Dependence of ERAD

ERAD Substrate	Topology	System	Method	Reference
		ATP-Dependent		
α1-antitrypsin	Soluble	Hepatoma cell line	Metabolic poison	Le et al., 1992
Cyclic nucleotide phosphodiesterase	Soluble	Dictyostelium	Metabolic poison	Coukell et al., 1992
Ig κ light chain	Soluble	Lymphoma cell line	Metabolic poison	Gardner et al., 1993
Prepro α factor	Soluble	Yeast microsomes	ATP requiring	McCracken and Brodsky, 1996
		ATP-Independent		
Interleukin-2 receptor-TCR α subunit fusion	Membrane	Permeabilized fibroblasts	ATP depletion	Stafford and Bonifacino, 1991
HMG-CoA reductase*	Membrane	Permeabilized CHO cells isolated membranes	ATP depletion	Meigs and Simoni, 1992; McGee et al., 1996
Asialoglycoprotein receptor (H2a)**	Membrane	Permeabilized CHO cells	ATP depletion	Wikström and Lodish, 1992
TCR (α, β, δ subunits)	Membrane	Permeabilized CHO cells	ATP depletion	Young et al., 1993
CFTR	Membrane	Permeabilized CHO cells	ATP depletion	Lukacs et al., 1994

Notes: * In an earlier study, Leonard and Chen (1987) observed that degradation was ATP dependent. This disparity might arise because ATP-requiring phosphorylation events are necessary to generate a signal for degradation.

** Wikström and Lodish (1992) also demonstrated that while initial proteolytic cleavage was ATP-independent in permeabilized mouse fibroblasts, complete degradation required ATP.

Cytochrome P-450 degradation in ER microsomes has also been shown to require ATP (Eliasson et al., 1992), although the precise topology of this cytochrome is unclear.

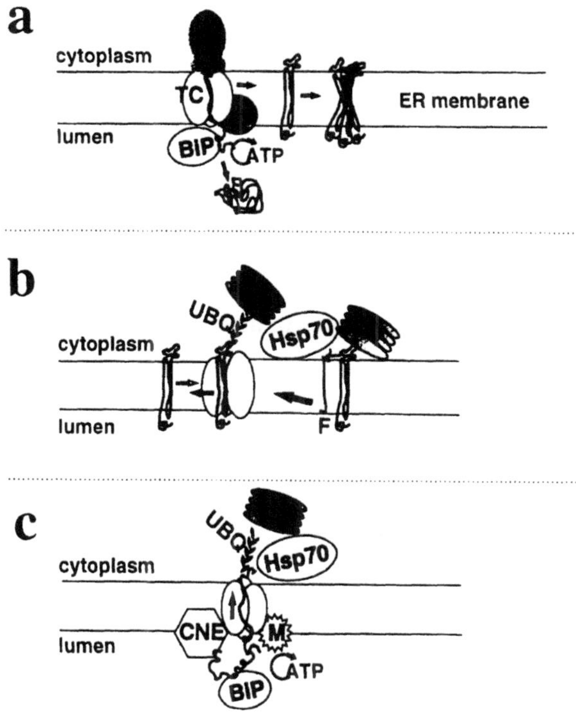

Figure 1. Working model for ERAD; Export of ERAD substrates to the cytosol. **Panel a.** Proteins enter the ER either posttranslationally or by cotranslational translocation while tethered to the ribosome (R), they traverse the membrane through the Sec61p protein translocation channel complex (TC). Soluble proteins are released into the lumen while membrane proteins diffuse laterally into the lipid bilayer (reviewed in Schatz and Dobberstein, 1996). During posttranslational translocation, lumenal Hsp70s (BiP and the "DNAJ" domain of Sec63p) facilitate translocation by coupling ATP-hydrolysis to polypeptide binding (reviewed in Brodsky, 1996). **Panel b.** Unassembled subunits of membrane multimeric protein complexes and those proteins unable to fold properly, may be exported to the cytoplasm back through the translocation channel or be solubilized from the membrane through the action of an export facilitator protein (F). Targeting of exported proteins to the proteasome (P) for degradation may be accomplished by cytosolic molecular chaperones (Hsp70), by ubiquitin tagging (UBQ), or by direct delivery to membrane associated proteasomes (P). **Panel c.** Aberrant and unassembled soluble proteins are exported to the cytoplasm by ATP-dependent retrograde translocation. Substrate selectivity could occur through the prolonged association with molecular chaperones, calnexin (CNE) and/or BiP, and movement through the Sec61p translocation channel complex could be driven by cytosolic chaperones or assisted by a protein translocation motor protein (M). As in panel b, targeting of exported proteins to the proteasome (P) for degradation may occur by the action of cytosolic molecular chaperones (Hsp70), by ubiquitin tagging (UBQ), or by direct delivery to membrane associated proteasomes.

We propose that the ATP-requirement to degrade soluble ER lumenal proteins reflects the need to export these substrates to the cytosol for proteasome-catalyzed degradation (Table 2; Brodsky and McCracken, 1997). ATP hydrolysis would be required for either molecular chaperones or an ATP-dependent transporter to drive or aid in the export of a soluble polypeptide across the membrane. Conversely, integral ER membrane proteins present cytosolic domains for degradation, and their lumenal and transmembrane segments might be exported to the cytosol through an energy-independent mechanism. Implicit in this model is that protein degradation itself is energy-independent, as shown for the removal of integral ER membrane proteins (Table 2) and for the proteolysis of unfolded polypeptides by the proteasome (reviewed in Hochstrasser, 1995).

One of the common principles of protein translocation across membranes is that transmembrane proteins can be generated by a stop transfer sequence which triggers a lateral opening of the translocation channel and allows the preprotein to diffuse out of the channel and into the lipid bilayer (reviewed in Schatz and Dobberstein, 1996). Lateral diffusion from the bilayer back into the translocation channel is probably a rare event (Figure 1a). Our model proposes that this event is inhibited by assembly of the membrane protein into a multimeric protein complex (Figure 1b), but that the probability of a protein returning to the translocation channel or entering another as yet unidentified channel can be increased dramatically by a facilitator protein (for example, US11, US2, and Vpu; see above) or by a change in the lipid composition of the membrane (e.g., increased cholesterol as occurs before the accelerated degradation of HMG-CoA reductase; Orci et al., 1984). Once in the translocation channel, the substrate protein could be released back to the cytosol for degradation by the proteasome. A requisite aspect for this model is the ATP-independent movement of the substrate protein from the translocation channel to the cytosol (Table 2). The export of substrate protein from the channel could be driven by a facilitator protein (Wiertz et al., 1996b), by molecular chaperone ATPases, or by proteasome binding and hydrolysis of the cytoplasmic tail.

Also inherent in our working model for ERAD is that soluble proteins, which are unable to assemble or acquire a native conformation in the ER lumen, are exported to the cytosol through a retrograde protein transporter or translocation channel (Figure 1c). During protein translocation into the ER, molecular chaperones (e.g., BiP) interact with the nascent polypeptides to prevent premature protein folding and may drive import (Brodsky 1996). In the absence of proper folding, incomplete translocation may delay release of the polypeptide into the ER lumen and the chaperone complex may remain associated with the nascent polypeptide that resides in the open channel. Alternatively, nascent polypeptides may be released from the channel but remain associated with molecular chaperones if unassembled into multimeric proteins. Prolonged association with BiP may recruit additional chaperones (e.g., calnexin) to the complex and eventually lead to the "decision" to expel the aberrant or orphaned polypeptide to the cytosol by retrograde translocation. The

ERAD substrates could then exit through the translocation channel (Pilon, 1997). During export, a motor protein similar in function to the ATP requiring bacterial SecA transporter (Economou and Wickner, 1994) could "push" the ERAD substrate through the channel to the cytosol, although no SecA homologue has been identified in eukaryotes. Alternatively, cytoplasmic molecular chaperones functionally similar to BiP could "pull" the substrate through the channel (reviewed in Brodsky, 1996), or ubiquitination or ongoing proteolysis might contribute to retrograde translocation (Biederer et al., 1997).

Once in the cytosol, ERAD substrates may be targeted to the proteasome either by ubiquitination or by interaction with cytosolic molecular chaperones that can maintain substrates in an unfolded state (Chirico et al., 1988; Deshaies et al., 1988) and target them to the proteasome. Alternatively, substrates emerging from translocation channels or ATP-dependent transporters may interact directly with proteasomes associated with the ER (Rivett et al., 1992; Rivett, 1993; McGee et al., 1996). Present evidence indicates that the targeting mechanism may not be universal, but could be specific to the type of ERAD substrate.

V. THE FUTURE

Many predictions from these models are easily testable. We have seen the advent of *in vitro* cell free systems to study ERAD in yeast (McCracken and Brodsky, 1996) and mammalian cells (McGee et al., 1996), and our knowledge about ER-associated protein degradation has increased considerably. Yet, as this review demonstrates, much remains to be discovered. For example, how are specific substrates targeted to the ERAD pathway? What role do the molecular chaperones play in this process? Are there other transporter/translocation systems in addition to the Sec61p complex that are required to deliver ER substrates to the cytosol? What other proteases, if any, may be involved? What targeting mechanisms are required for delivery to the proteasome? To address these questions, we now have both genetic and biochemical tools. Such a combined approach has led to the elucidation of the secretory pathway in yeast (reviewed in McCracken, 1992; and see Rothblatt et al., 1994), and we are now in a position to achieve similar goals in the molecular dissection of ERAD.

VI. SUMMARY

A protein degradation event associated with the ER can selectively remove polypeptides from the secretory pathway. Although the mechanism of ER associated protein degradation, or ERAD, has been obscure, recent reports using both yeast and mammalian cells indicate that ERAD substrates are first targeted to the cytoplasm for degradation. Proteolysis is then catalyzed in the cytosol by the proteasome. Thus, ERAD is comprised of at least three distinct events: (1) recognition of a polypeptide for degradation, (2) efflux of this substrate from the ER to the cytosol, and (3) degradation by the proteasome. Finally, while the results from many studies have permitted us to speculate on a general mechanism of ERAD, it remains possible that individual substrates may use unique proteolytic machines. The continued characterization of ERAD using *in vitro* systems and yeast mutants defective for ERAD will certainly build upon our knowledge of this essential cellular process.

ACKNOWLEDGMENTS

We would like to thank our colleagues R. Davis, R. Hampton, S. Michaelis, G. Shelness, R. Simoni, D. Thomas, and D.H. Wolf, who provided information in the form of reprints, preprints, and personal communication, J. Bonifacino for permission to use and modify Table 1, and for comments on this manuscript. We apologize to those whose relevant work is not discussed because of space limitations. The authors' research was supported by Cystic Fibrosis Foundation (G789) and National Science Foundation (MCB-951042) awards to A.A.M., National Science Foundation (MCB-9506002) and American Cancer Society (JFRA-602) awards to J.L.B., a University of Nevada, Cell and Molecular Biology Graduate Fellowship to E.D.W., and National Science Foundation award MCB-9722889 to A.A.M. and J.L.B.

REFERENCES

Amara, J.F., Lederkremer, G., & Lodish, H. F. (1989). Intracellular degradation of unassembled asialoglycoprotein receptor subunits: a pre-Golgi, non-lysosomal endoproteolytic cleavage. J. Cell Biol. 109, 3315-3324.

Amitay, R., Bar-Nun, S., Haimovich, J., Rabinovich, E., & Shachar, I. (1991). Post-translational regulation of IgM expression in B-lymphocytes. Selective nonlysosomal degradation of assembled secretory IgM is temperature-dependent and occurs prior to the trans-Golgi. J. Biol. Chem. 266, 12568-12573.

Bachhawat, A.K., & Pillai, S. (1991). Distinct intracellular fates of membrane and secretory immunoglobulin heavy chains in a pre-B cell line. J. Cell Biol. 115, 619-624.

Bergeron, J.J.M., Brenner, M.B., Thomas, D.Y., & Williams., D.B. (1994). Calnexin: a membrane-bound chaperone of the endoplasmic reticulum. Trends Biochem. Sci. 19, 124-128.

Biederer, T., Volkwein, C., and Sommer, T. (1997). Role of Cue1p in ubiquitination and degradation at the ER surface. Science 278, 1806-1809.

Bienkowski, R.S. (1983). Intracellular degradation of newly synthesized secretory proteins. Biochem. J. 214, 1-10.

Blount, P., & Merlie, J.P. (1990). Mutational analysis of muscle nicotinic acetylcholine receptor subunit assembly. J. Cell Biol. 111, 2613-2622.

Bonifacino, J.S., Yuan, L., & Sandoval, I.V. (1989). Internalization and recycling to serotonin-containing granules of the 80K integral membrane protein exposed on the surface of secreting rat basophilic leukaemia cells. J. Cell Sci. 92, 701-712.

Bonifacino, J.S., Cosson, P., & Klausner, R.D. (1990). Colocalized transmembrane determinants for ER degradation and subunit assembly explain the intracellular fate of TCR chains. Cell 63, 503-513.

Bonifacino, J.S., & Klausner, R.D. (1994). Degradation of proteins retained in the endoplasmic reticulum. In: Cellular Proteolytic Systems (Ciechanover, A.J., & Schwartz, A.L., Eds.), pp. 137-139, John Wiley, New York.

Bonnardel, J.A., & Davis, R.A. (1995). In HepG2 cells, translocation, not degradation, determines the fate of the de novo synthesized apolipoprotein B. J. Biol. Chem. 270, 28892-28896

Borchardt, R.A., & Davis, R.A. (1987). Intrahepatic assembly of very low density lipoproteins. Rate of transport out of the endoplasmic reticulum determines rate of secretion. J. Biol. Chem. 262, 16394-16402.

Boren, J., M. Wettesten, S. Rustaeus, M. Anderson, & S.O. Olofsson. (1993). The assembly and secretion of apoB-100 containing lipoproteins. Biochem. Soc. Trans. 21:4487-4493.

Bour, S., Schubert, U., & Strebel, K. (1995). The human immunodeficiency virus type 1 Vpu protein specifically binds to the cytoplasmic domain of CD4: implications for the mechanism of degradation. J. Virol. 69, 1510-1520.

Brodsky, J.L. (1996). Post-translational protein translocation: not all hsc70s are created equal. Trends Biochem. Sci. 21, 122-126.

Brodsky, J.L., & McCracken, A.A. (1997). ER-associated and proteasome-mediated protein degradation: how two topologically restricted events came together. Trends in Cell Biol. 7, 151-156.

Brodsky, J.L., & Schekman, R. (1994). Heat shock cognate proteins and polypeptide translocation across the endoplasmic reticulum membrane. In: The Biology of Heat Shock Proteins and Molecular Chaperones (Morimoto, R.I., Tissieres, A., & Georgopoulos, C., Eds.), pp. 85-110. Cold Spring Harbor Laboratory Press, Cold Spring Harbor, NY.

Brodsky, J.L., Werner, E.D., Dubas, M.E., and McCracken, A.A. Chaperone Requirements During ER-Associated Protein Degradation Indicates that Protein Import and Export are Mechanistically Distinct. (Submitted).

Cannon, K.S., Hebert, D.N., & Helenius, A. (1996). Glycan-dependent and independent association of vesicular stomatitis virus G protein with calnexin. J. Biol. Chem. 271, 14280-14284.

Caplan, S., Green, R., Rocco, J., & Kurjan, J. (1991). Glycosylation and structure of the yeast MF1 -factor precursor is important for efficient transport through the secretory pathway. J. Bacteriol. 173, 627-635.

Chen, C., Bonifacino, J.S., Yuan, L.C., & Klausner, R.D. (1988). Selective degradation of T cell antigen receptor chains retained in a pre-Golgi compartment. J. Cell Biol. 107, 2149-2161.

Chen, M.Y., Maldarelli, F., Karczewski, M.K., Willey, R.L., & Strebel, K. (1993). Human immunodeficiency virus type 1 vpu protein induces degradation of CD4 in vitro: the cytoplasmic domain of CD4 contributes to vpu sensitivity. J. Virol. 67, 3877-3884.

Cheng, S.H., Gregory, R.J., Marshall, J., Paul, S., Souza, D.W., White, G.A., O'Riordan, C.R., & Smith, A.E. (1990). Defective intracellular transport and processing of CFTR is the molecular basis of most cystic fibrosis. Cell 63, 827-834.

Chin, D.J., Gil, G., Faust, J.R., Goldstein, J.L., Brown, M.S., & Luskey, K.L. (1985). Sterols accelerate degradation of hamster 3-hydroxy-3-methylglutaryl coenzyme A reductase encoded by a constitutively expressed cDNA. Mol. Cell Biol. 5, 634-641.

Chirico, W.J., Waters, M.G., & Blobel, G. (1988). 70K heat shock related proteins stimulate protein translocation into microsomes. Nature 332, 805-809.

Coukell, M.B., Cameron, A.M., & Adames, N.R. (1992). Involvement of intracellular calcium in protein secretion in *Dictyostelium discoideum*. J. Cell Sci. 103, 371-380.

Craig, E.A., Gambill, B.D., & Nelson., R.J. (1993). Heat shock proteins: molecular chaperones of protein biogenesis. Microbiol. Rev. 57, 402-414.

Craig, E.A, Weissman, J.S., & Horwich, A.L. (1994). Heat shock proteins and molecular chaperones: mediators of protein conformation and turnover in the cell. Cell 78, 365-372.

Danishefsky, K., Hartwig, R., Banerjee, D., & Redman, C. (1990). Intracellular fate of fibrinogen B beta chain expressed in COS cells. Biochim. Biophys. Acta 1048, 202-208.

Danoff, A., Cutler, D.F., & Shields, D. (1991). Heterologous expression of preprosomatostatin. Intracellular degradation of prosomatostatin-II. J. Biol. Chem. 266, 10004-10010.

Davis, R. A., Thrift, R.N., Wu, C. C., & Howell, K.E. (1990). Apolipoprotein B is both integrated into and translocated across the endoplasmic reticulum membrane. Evidence for two functionally distinct pools. J. Biol. Chem. 265, 10005-10011.

Davis, E.C., & Mecham, R.P. (1996). Selective degradation of accumulated secretory proteins in the endoplasmic reticulum. J. Biol. Chem. 271, 3787-33794,

DeMartino, G.N., Proske, R.J., Moomaw, C.R., Strong, A.A., Song, X., Hisamatsu, H., Tanaka, K., & Slaughter, C.A. (1996). Identification, purification, and characterization of a PA700-dependent activator of the proteasome. J. Biol. Chem. 271, 3112-3118.

Deshaies, R.J, Koch, B.D., Werner-Washburne, M.E., Craig, A., & Schekman, R. (1988). A subfamily of stress proteins facilitates translocation of secretory and mitochondrial precursor polypeptides. Nature 332, 800-805.

Dick, L.R., Aldrich, C., Jameson S.D., Moomaw, C.R., Pramanik, B.C., Doyle, C.K., DeMartino, G.N., Bevan, J.J., Formen, J.M., & Slaughter, C.A. (1994). Proteolytic processing of ovalbumin and b-galactosidase by the proteasome to yield antigenic peptides. J. Immunol. 152, 3884-3394.

Dixon, J.L., Furukawa, S., & Ginsberg, H.N. (1991). Oleate stimulates secretion of apolipoprotein B-containing lipoproteins from Hep G2 cells by inhibiting early intracellular degradation of apolipoprotein B. J. Biol. Chem. 266, 5080-5086.

Dixon, J.L., & Ginsberg, H.N. (1993). Regulation of hepatic secretion of apolipoprotein B-containing lipoproteins: information obtained from cultured liver cells. J. Lipid Res. 34, 167-179.

Doyle, C., Sambrook, J., & Gething, M.J. (1986). Analysis of progressive deletions of the transmembrane and cytoplasmic domains of influenza hemagglutinin. J. Cell Biol. 103, 1193-1204.

Driscoll, J., & Goldberg, A.L. (1990). The Proteasome (multicatalytic protease) is a component of the 1500-kDa proteolytic complex which degrades ubiquitin-conjugated proteins. Proc. Natl. Acad. Sci. USA 265, 4789-4792.

Du, E.Z., Jurth, J., Wang, S-L., Humiston, P., & Davis, R.A. (1994). Proteolysis-coupled secretion of the N terminus of apolipoprotein B; characterization of a transient, translocation arrested intermediate. J. Biol. Chem. 269, 24169-24176.

Du, E.Z., Wang, S.L., Kayden, H.J., Sokol, R., Curtiss, L.K., Davis, R.A. (1996). Translocation of apolipoprotein B across the endoplasmic reticulum is blocked in abetalipoproteinemia J. Lipid Res. 37, 1309-1315.

Dulis, B.H., Kloppel, T.M., Grey, H.M., & Kubo, R.T. (1982). Regulation of catabolism of IgM heavy chains in a B lymphoma cell line. J. Biol. Chem. 257, 4369-4374.

Dunten, R.L., & Cohen, R.E. (1989). Recognition of modified forms of ribonuclease A by the ubiquitin system. J. Biol. Chem. 264, 16739-16747.

Economou, A., & Wickner, W. (1994). SecA promotes preprotein translocation by undergoing ATP-driven cycles of membrane insertion and deinsertion. Cell 78, 835-843.

Edwards, P.A., & Gould, R.G. (1972). Turnover rate of hepatic 3-hydroxy-3-methylglutaryl coenzyme A reductase as determined by use of cycloheximide. J. Biol. Chem. 247, 1520-1524.

Elias, S., Bercovich, B., Kahana, C., Coffino, P., Fischer, M., Hilt, W., Wolf, D.H., & Ciechanover, A. (1995). Degradation of orthine decarboxylase by the mammalian and yeast 26s proteasome complexes requires all the components of the protease. Eur. J. Biochem. 229, 276-283.

Eliasson, E., Mkrtchian, S., & Ingelman, S.M. (1992). Hormone- and substrate-regulated intracellular degradation of cytochrome P450 (2E1) involving MgATP-activated rapid proteolysis in the endoplasmic reticulum membranes. J. Biol.Chem. 267, 15765-15769.

Erickson, R.H., Suzuki, Y., Sedlmayer, A., & Kim, Y.S. (1992). Biosynthesis and degradation of altered immature forms of intestinal dipeptidyl peptidase IV in a rat strain lacking the enzyme. J. Biol. Chem. 267, 21623-21629.

Esser, V., & Russell, D.W. (1988). Transport-deficient mutations in the low density lipoprotein receptor. Alterations in the cysteine-rich and cysteine-poor regions of the protein block intracellular transport. J. Biol. Chem. 263, 13276-13281.

Etlinger, J.D., & Goldberg, A.L. (1977). A soluble ATP-dependent proteolytic system responsible for the degradation of abnormal proteins in recticulocytes. Proc. Natl. Acad. Sci. USA 74, 54-58.

Fagerhol, M.K., & Cox, D.W. (1981). The Pi polymorphism: genetic, biochemical and clinical aspects of human alpha 1-antitrypsin. Adv. Hum. Genet. 11, 1-62.

Faust, J.R., Luskey, K.L., Chin, D.J., Goldstein, J.L., & Brown, M.S. (1982). Regulation of synthesis and degradation of 3-hydroxy-3-methylglutaryl-coenzyme A reductase by low density lipoprotein and 25-hydroxycholesterol in UT-1 cells. Proc. Natl. Acad. Sci. USA 79, 5205-5209.

Fenteany, G., Standaert, R.F., Lane, W.S., Choi, S., Corey, E.J., & Schreiber, S.L. (1995). Inhibition of proteasome activities and subunit-specific amino-terminal threronine modification by lactacystin. Science 286, 726-731.

Finger, A., Knopp, M., & Wolf, D.H. (1993). Analysis of two mutated vacuolar proteins reveals a degradation pathway in the ER or an ER-related compartment of yeast. Eur. J. Biochem. 218, 565-574.

Fra, A.M., Fagioli, C., Finazzi, D., Sitia, R., & Alberini., C.M. (1993). Quality control of ER synthesized proteins: an exposed thiol group as a three-way switch mediating assembly, retention and degradation. EMBO J. 12, 4755-4761.

Fukuda, R., Umebayashi, J., Horiuchi, H., Ohta, A., & Takagi, M. (1996). Degradation of *Rhizopus niveus* aspartic proteinase-I with mutated prosequences occurs in the endoplasmic reticulum of *Saccharomyces cerevisiae*. J. Biol. Chem. 271, 14252-14255.

Furukawa, S., Sakata, N., Ginsberg, H.N., & Dixon, J.L. (1992). Studies of the sites of intracellular degradation of apolipoprotein B in Hep G2 cells. J. Biol. Chem. 267, 22630-22638.

Gardner, A.M., Aviel, S., & Argon, Y. (1993). Rapid degradation of an unassembled immunoglobulin light chain is mediated by a serine protease and occurs in a pre-Golgi compartment. J. Biol. Chem. 268, 25940-25947.

Gething, M.-J., & Sambrook, J. (1992). Protein folding in the cell. Nature 355, 33-45.

Gottesman, S., & Maurizi., M.R. (1992). Regulation by proteolysis: energy-dependent proteases and their targets. Microbiol. Rev. 56, 593-621.

Habener, J.F., Kemper, B., & Potts, J.T. Jr. (1975). Calcium-dependent intracellular degradation of parathyroid hormone: a possible mechanism for the regulation of hormone stores. Endocrinology 97, 431-441.

Halaban, R., Cheng, E., Zhang, Y., Moellmann, G., Hanlon, D., Michalak, M., Setaluri, V., & Hebert, D.N. (1997). Aberrant retention of tyrosinase in the endoplasmic reticulum mediates accelerated degradation of the enzyme and contributes to the dedifferentiated phenotype of amelanotic melanoma cells. Proc. Natl. Acad. Sci. USA 94, 6210-6215.

Halban, P.A., & Wollheim, C.B. (1980). Intracellular degradation of insulin stores by rat pancreatic islets *in vitro*. J. Biol. Chem. 255, 6003-6006.

Hammond, C., & Helenius, A. (1994). Folding of VSV G protein: sequential interaction with BiP and Calnexin. Science 266, 456-458

Hammond, C., Braakman, I., & Helenius., A. (1994). Role of N-linked oligosaccharide recognition, glucose trimming, and calnexin in glycoprotein folding and quality control. Proc. Natl. Acad. Sci. USA. 91, 913-917.

Hampton, R.Y., & Bhakta, H. (1997). Ubiquitin-mediated regulation of 3-hydroxy-3-methylglutaryl-CoA reductase. Proc. Natl. Acad. Sci. USA 94, 12944-12948.

Hampton, R.Y., & Rine, J. (1994). Regulated degradation of HMG-CoA reductase, an integral membrane protein of the endoplasmic reticulum, in yeast. J. Cell Biol. 125, 299-312.

Hampton, R.Y., Dimster-Denk, D., & Rine, J. (1996a). The biology of HMG-CoA reductase; the pros of contra-regulation. Trends Biochem. Sci. 21, 140-145.

Hampton, R.Y., Gardener, R., & Rine, J. (1996b). The 26s proteasome and novel proteins for the degradation of HMG-CoA reductase, an integral ER membrane protein. Mol. Biol. Cell 7, 2029-2044.

Hannum, C., Marrack, P., Kubo, R., & Kappler, J. (1987). Thymocytes with the predicted properties of pre-T cells. J. Exp. Med. 166, 874-889.

Hartl, F.U. (1996). Molecular chaperones in cellular protein folding. Nature 381, 571-580.

Hebert, D.N., Foelimer, G., & Helenius., A. (1995). Glucose trimming and reglucosylation determine glycoprotein association with calnexin in the ER. Cell 81, 425-433.

Heinemeyer, W., Kleinschmidt, J., Saidowsky, A., Saidowsky, J., Escher, C., & Wolf, D.H. (1991). Proteinase yscE, the yeast proteasome/multicatalytic-multifunctional proteinase: mutants unravel its function in stress induced proteolysis and uncover its necessity for cell survival. EMBO J. 10, 555-562.

Heinemeyer, W., Gruhler, A., Mohrle,V., Mahe, Y., & Wolfe, D.H. (1993). PRE2, highly homologous to the human major histocompatability complex-linked RING10 gene, codes for a yeast proteasome subunit necessary for chrymotryptic activity and degradation of ubiquitinated proteins. J. Biol. Chem. 268, 5115-5120.

Hendil, K.B. (1975). Degradation of abnormal proteins in HeLa Cells. J. Cell Physiol. 87, 289-296.

Hershko, A., & Rose, I.A. (1984). Ubiquitin-alldehyde: a general inhibitor of ubiquitin-recycling processes. Proc. Natl. Acad. Sci. USA 84, 1829-1833.

Hiller, M.M., Finger, A., Schweiger, M., & Wolf, D.H. (1996). ER-degradation of a misfolded luminal protein occurs via the cytosolic ubiquitin-proteasome pathway. Science 273, 1725-1728.

Hilt, W., & Wolf, D.H. (1996). Proteasomes: destruction as a program. Trends Biochem. Sci. 21, 96-102.

Hochstrasser, M. (1995). Ubiquitin, proteasomes and the regulation of intracellular degradation. Curr. Opin. Cell Biol. 7, 215-223.

Hoe, M.H., & Hunt, R.C. (1992). Loss of one asparagine-linked oligosaccharide from human transferrin receptors results in specific cleavage and association with the endoplasmic reticulum. J. Biol. Chem. 267, 4916-4923.

Ingram, M.F., & Shelness, G.S. (1996). Apolipoprotein B-100 destined for lipoprotein assembly and intracellular degradation undergoes efficient translocation across the endoplasmic reticulum membrane. J Lipid Res. 37, 2202-2214.

Inoue, S., & Simoni, R.D. (1992). 3-Hydroxy-3-methylglutaryl-coenzyme A reductase and T cell receptor alpha subunit are differentially degraded in the endoplasmic reticulum. J. Biol. Chem. 267, 9080-9086.

Jensen, T.J., Loo, M.A., Pind, S., Williams, D.B., Goldberg, A.L., & Riordan, J.R. (1995). Multiple proteolytic systems, including the proteasome, contribute to CFTR processing. Cell 83, 129-135.

Jentsch, S. (1992). The ubiquitin conjugation system. Annu. Rev. Genet. 26, 177-205.

Jubete, Y., Maurizi M.R., & Gottesman, S. (1996). Role of the heat shock protein DnaJ in the Lon-dependent degradation of naturally unstable proteins. J. Biol. Chem. 271, 30798-30803.

Kandror, O., Busconi, L., Sherman, M., & Goldberg, A.L. (1994). Rapid degradation of an abnormal protein in *E.coli* involves the chaperones GroEL and GroES. J. Biol. Chem. 269, 23575-23582.

Kelleher, D.J., Kreibich, G., & Gilmore, R. (1992). Oligosaccharyl-transferase activity is associated with a protein complex composed of ribophorins I and II and a 48 Kd protein. Cell 69, 55-65.

Kim, P.S., & Arvan, P. (1995). Calnexin and BiP act as sequential molecular chaperones during thyroglobulin folding in the endoplasmic reticulum. J. Cell Biol. 128, 29-35

Klausner, R.D., Lippincott-Schwartz, J., & Bonifacino, J.S. (1990). The T cell antigen receptor: Insights into organelle biology. Annu. Rev. Cell Biol. 6, 403-431.

Klausner, R.D., & Sitia, R. (1990). Protein degradation in the endoplasmic reticulum. Cell 62, 611-614.

Knittler, M.R., Dirks, S., & Haas, I.G. (1995). Molecular chaperones involved in protein degradation in the endoplasmic reticulum: Quantitative interaction of the heat shock cognate protein BiP with partially folded immunoglobulin light chains that are degraded in the endoplasmic reticulum. Immunology 92, 1764-1768.

Knop, M., Finger, A., Braun, T., Hellmuth, K., & Wolf, D.H. (1996). Der1, a novel protein specifically required for endoplasmic reticulum degradation in yeast. EMBO J. 15, 753-763.

Kuchler, K., Sterne, R.E., & Thorner, J. (1989). *Saccharomyces cerevisiae* STE6 gene product: a novel pathway for protein export in eukaryotic cells. EMBO J. 8, 3973-3984.

Kulka, R.G., Raboy, B., Schuster, R., Parag, H.A. Diamond, G., Ciechanover, A., & Marcus, M. (1988). A Chinese hamster cell cycle mutant arrested at G2 phase has a temperature-sensitive ubiquitin-activating enzyme. J. Biol. Chem. 263, 15726-15731.

Kurosaki, T., Gander, I., & Ravetch, J.V. (1991). A subunit common to an IgG Fc receptor and the T-cell receptor mediates assembly through different interactions. Proc. Natl. Acad. Sci. USA 88, 3837-3841.

Lassoued, K., Illges, H., Benlagha, K., & Cooper, M.D. (1996). Fate of surrogate light chains in B lineage cells. J. Exp. Med. 183, 421-429.

Lau, M.M., & Neufeld, E.F. (1989). A frameshift mutation in a patient with Tay-Sachs disease causes premature termination and defective intracellular transport of the alpha-subunit of beta-hexosaminidase. J. Biol. Chem. 264, 21376-21380.

Le, A., Graham, K.S., & Sifers, R.N. (1990). Intracellular degradation of the transport-impaired human PiZ alpha 1-antitrypsin variant. Biochemical mapping of the degradative event among compartments of the secretory pathway. J. Biol. Chem. 265, 14001-14007.

Le, A., Ferrell, G.A., Dishon, D.S., Le, Q.Q., & Sifers, R.N. (1992). Soluble aggregates of the human PiZ alpha 1-antitrypsin variant are degraded within the endoplasmic reticulum by a mechanism sensitive to inhibitors of protein synthesis. J. Biol. Chem. 267, 1072-1080.

Lee, D.H., Sherman, M.Y., & Goldberg, A.L. (1996). Involvement of the molecular chaperone Ydj1 in the ubiquitin-dependent degradation of short-lived and abnormal proteins in Saccharomyces cerevisiae. Mol. Cell. Biol. 16, 4773-4781.

Lederkremer, G.Z., & Lodish, H.F. (1991). An alternatively spliced mini exon alters the subcellular fate of the human asialoglycoprotein receptor H2 subunit. Endoplasmic reticulum retention and degradation or cell surface expression. J. Biol. Chem. 266, 1237-1244.

Leiper, J.M., Harrison, G.B., Harrison, J.D. Bayliss, J. Scott, R., & R. J. Pease. (1996). Systematic expression of the complete coding sequence of apoB-100 does not reveal transmembrane determinants. J. Lipid Res. 37, 2215-2231.

Leonard, D.A., & Chen, H.W. (1987). ATP-dependent degradation of 3-hydroxy-3-methylglutaryl coenzyme A reductase in permeabilized cells. J. Biol. Chem. 262, 7914-7919.

Lippincott-Schwartz, S.J., Bonifacino, J.S., Yuan, L.C., & Klausner, R.D. (1988). Degradation from the endoplasmic reticulum: disposing of newly synthesized proteins. Cell 54, 209-220.

Loayza, D., Tan, A., & Michaelis, S. Yeast *ste6* mutants unable to exit from the ER may define two distinct steps in ER quality control. (Submitted).

Lobell, R.B., Arm, J.P., Raizman, M.B., Austen, K.F., & Katz, H.R. (1993). Intracellular degradation of Fc gamma RIII in mouse bone marrow culture-derived progenitor mast cells prevents its surface expression and associated function. J. Biol. Chem. 268, 1207-1212.

Lukacs, G.L., Mohamed, A., Kartner, N., Chang, X.-B., Riordan, J.R., & Grinstein, S. (1994). Conformational maturation of CFTR but not its mutant counterpart (D508) occurs in the endoplasmic reticulum and requires ATP. EMBO J. 13, 6076-6086.

McCracken, A.A. (1992). Trafficking of secreted proteins, In Intracellular Trafficking of Proteins (Hannover, J., & Steer, C., eds., pp. 504-548). Cambridge University Press, Cambridge. England.

McCracken, A.A., & Brodsky, J.L. (1996). Assembly of ER-associated protein degradation in vitro: dependence on cytosol, calnexin, and ATP. J. Cell Biol. 132, 291-298.

McCracken, A.A., Karpichev, I.V., Ernaga, J.E., Werner, E.D., Dillin, A.G., & Courchesne, W.E. (1996). Yeast mutants deficient in ER-associated degradation of the Z variant of a-1-protease inhibitor. Genetics 144, 1355-1362.

McCracken, A.A., & Kruse, K.B. (1993). Selective protein degradation in the yeast exocytic pathway. Mol. Biol. Cell. 4, 729-736.

McCracken, A.A., Kruse, K.B., Valentine, J., Roberts, C., Yohannes, T.Z., & Brown, J.L. (1991). Construction and expression of alpha-1-proteinase inhibitor mutants and the effects of these mutations on secretion of the variant inhibitors. J. Biol. Chem. 226, 7578-7582.

McGee, T.P., Cheng, H.H., Humagai, H., Omura, S., & Simoni, R.D. (1996). Degradation of HMG-CoA reductase in endoplasmic reticulum membranes is accelerated as a result of increased susceptibility to proteolysis. J. Biol. Chem. 271, 25630-25638.

Meigs, T.E., & Simoni, R.D. (1992). Regulated degradation of 3-hydroxy-3-methylglutaryl-coenzyme A reductase in permeabilized cells. J. Biol. Chem. 267, 13547-13552

Moore, S.E.H., Bauvy, C., & Codogno, P. (1995). Endoplasmic reticulum-to-cytosol transport of free polymannose oligosaccharides in permeabilized HepG2 cells. EMBO J. 14, 6034-6042.

Mullins, C., Lu, Y., Campbell, A., Fang, H., & Green, N. (1995). A mutation affecting signal peptidase inhibits degradation of an abnormal membrane protein in *Saccharomyces cerevisiae.* J. Biol. Chem. 270, 17139-17147.

Murakami, Y., Matsufuji, S., Kameji, T., Hayashi, S., Igarashi, K., Tamura, T., Tanaka, K., & Ichihara, A. (1992). Orthine decarboxylase is degraded by the 26s proteasome without ubiquitination. Nature 360, 597-599.

Nicchitta, C.V., & Blobel, G. (1993). Lumenal proteins of the mammalian endoplasmic reticulum are required to complete protein translocation. Cell 73, 989-998.

Oda, K., Wada, I., Takame., N., Fujiwara, T., Misumi, Y., & Ikehara, Y. (1996a). Bip/GRP78 but not calnexin associates with a precursor of glycosylphosphatidylinositol-anchored protein. Biochem. J. 316, 623-630.

Oda, K., Ikehara, Y., & Omura, S. (1996b). Lactacystin, an inhibitor of the proteasome, blocks the degradation of a mutant precursor of glycosylphosphatidylinositol-linked protein in a pre-Golgi compartment. Biochem. Biophys. Res. Commu. 219, 800-805.

Ogata, M., Chaudhary, V.K., Pastan, I., & FitzGerald, D.J. (1990). Processing of Pseudomonas exotoxin by a cellular protease results in the generation of a 37,000-Da toxin fragment that is translocated to the Cytosol. J. Biol. Chem. 265 20678-20685.

Oliver, J.D., van der Wal, F.J., Bulleid, N.J., & High, S. (1997). Interaction of the thiol-dependent reductase Erp57 with nascent glycoproteins. Science 275, 86-88.

Olofsson, L.O., Bjursell, B., Boström, K., Carlsson P., Elovson, J., Protter, A.A., Reuben, M.A., & Bondjers, G. (1987). Apoliporpotein B: structure, biosynthesis and role in the lipoprotein assembly process. Atherosclerosis 68, 1-17.

Omura, F., Otsu, M., Yoshimori, T., Tashiro, Y., & Kikuchi, M. (1992). Non-lysosomal degradation of misfolded human lysozyme with and without an asparagine-linked glycosylation site. Eur. J. Biochem. 210, 591-599.

Orci, L., Brown, M.S., Goldstein, J.L., Garcia-Segura, L.M., & Anderson, R.G.W. (1984). Increase in membrane cholesterol: a possible trigger for degradation of HMG CoA reductase and crystalloid endoplasmic reticulum in UT-1 cells. Cell 36, 835-845.

Otsu, M., Urade, R., Kito, M., Omura, F., & Kikuchi, M. (1995). A possible role of ER-60 protease in the degradation of misfolded proteins in the endoplasmic reticulum. J. Biol. Chem. 270, 14958-14961.

Ou, W.-J., Cameron, P.H., Thomas, D.Y., & Bergeron, J.J.M. (1993). Association of folding intermediates of glycoproteins with calnexin during protein maturation. Nature 364, 771-776.

Parlati, F., Dominguez, M., Bergeron, J.J.M., & Thomas, D.Y. (1995). *Saccharomyces cerevisiae* CNE1 encodes an endoplasmic reticulum (ER) membrane protein with sequence similarity to calnexin and calreticulin and functions as a constituent of the ER quality control apparatus. J. Biol. Chem. 270, 214-253.

Parlati, F., Sifers, R., Bergeron, J.J.M., & Thomas, D.Y. Rescue of the secretion defective phenotype of mutnat α1-antitrypsin by C-terminal deletion of calnexin in *Schizosaccharomyces pombe.* (Submitted).

Pelham, H.R.B., Roberts, M.L., & Lord, J.M. (1992). Toxin entry: how reversible is the secretory pathway? Trends Cell Biol. 2, 183-185.

Pilon, M., Schekman, R., & Romisch, K. (1997). Sec61p mediates export of a misfolded secretory protein from the endoplasmic reticulum to the cytosol for degradation. EMBO J. 16, 4540-4548.

Pind, S., Riordan, J.R., & Williams, D.B. (1994). Participation of the ER chaperone calnexin (p88. IP90) in the biogenesis of the CFTR. J. Biol. Chem. 269, 12784-12788.

Plemper, R.K., Bohmler, S., Bordallo, J., Sommer, T., & Wolf, D.H. (1997). Mutant analysis links the translocon and BiP to retrograde protein transport for ER degradation. Nature 388, 891-895

Prouty, W.F. (1976). Orthine decarboxylase inactivation in Hela Cells. J. Cell. Physiol. 89, 65-76.

Qu, D. Techman, J.H., Omura, S., & Perlmutter, D.H. (1996). Degradation of a mutant secretory protein, α1-antitrypsin Z, in the endoplasmic reticulum requires proteasome activity. J. Biol. Chem. 271, 22791-22795.

Rabinovich, E., Bar-Nun, S., Amitay, R. Shachar, I., Gur, B., Taya, M., & Haimovich, J. (1993). Different assembly species of IgM are directed to distinct degradation sites along the secretory pathway. J. Biol. Chem. 268, 24145-24148.

Rabinovitz, M., & Fisher, J.M. (1964). Characteristics on the inhibition of hemoglobin synthesis in rabbit reticulocytes by *threo*-a-amino-b-chlorobutyric acid. Biochem. Biophys. Acta 91, 313-322.

Rajagopalan, S., Xu,Y., & Brenner, M.B. (1994). Retention of unassembled components of integral membrane proteins by Calnexin. Science 263, 387-390.

Raposo, G., van Santen, H.M., Leijendekker, R., Geuze, H.J., & Ploegh., H. L. (1995). Misfolded Major Histocompatibility complex class I molecules accumulate in an expanded ER-Golgi intermediate compartment. J. Cell Biol. 131, 1403-1419.

Razooki-Hasan, H., White, D.A., & Mayer, R.J. (1982). Extensive destruction of newly synthesized casein in mammary explants in organ culture. Biochem. J. 202, 133-138.

Riederer, M.A., & Hinnen, A. (1991). Removal of N-glycosylation sites of the yeast acid phosphatase severely affects protein folding. J. Bact. 173, 3539-3546.

Riordan, J.R., Rommens, J.M., Kerem, B.-S., Alon, N., Rozmahel, R., Grzelczak, Z., Zielenski, J., Lok, S., Plavsic, N., Chou, J.-L., Drumm, M.L., Iannuzzi, M.C., Collins, F.S., & Tsui, L.-C. (1989). Identification of the Cystic Fibrosis gene: Cloning and characterisation of complementary DNA. Science 245, 1066-1073.

Rivett, A.J., Palmer, A., & Knecht, E. (1992). Electron microscopic localization of the multicatalytic proteinase complex in rat liver and in cultured cells. J. Histochem. Cytochem. 40, 1165-1172.

Rivett, A.J. (1993). Proteasomes: multicatalytic proteinase complexes. Biochem. J. 291, 1-10.

Römisch, K., & Schekman, R. (1992). Distinct processes mediate glycoprotein and glycopeptide export from the endoplasmic reticulum in *S .cerevisiae*. Proc. Natl. Acad. Sci. USA 89, 7227-7231.

Rothblatt, J., Novick, P., & Stevens, T.H. (1994). Guidebook to the Secretory Pathway, Sambrook & Tooze, New York; Oxford University Press.

Rotundo, R.L. (1988). Biogenesis of acetylcholinesterase molecular forms in muscle. Evidence for a rapidly turning over, catalytically inactive precursor pool. J. Biol. Chem. 263, 19398-19406.

Rotundo, R.L., Thomas, K., Porter, J.K., Benson, R.J., Fernandez, V.C., & Fine, R.E. (1989). Intracellular transport, sorting, and turnover of acetylcholinesterase. Evidence for an endoglycosidase H-sensitive form in Golgi apparatus, sarcoplasmic reticulum, and clathrin-coated vesicles and its rapid degradation by a non-lysosomal mechanism. J. Biol. Chem. 264, 3146-3152.

Roy, S., Yu, S., Banerjee, D., Overton, O., Mukhopadhyay, G., Oddoux, C., Grieninger, G., & Redman, C. (1992). Assembly and secretion of fibrinogen. Degradation of individual chains. J. Biol. Chem. 267, 23151-23158.

Sakata, M., Wu, X., Dixon, J.L., and Ginsberg, H.N. (1993). Proteolysis and lipid-facilitated translocation are distinct but competitive processes that regulate secretion of apolipoprotein B in Hep G2 cells. J. Biol. Chem. 268, 22967-22970.

Sanchez, Y., & Lindquist, S.L. (1990). HSP104 required for induced thermotolerance. Science 248. 1112-1115.

Sato, R., Imanaka, T., Takatsuki, A., & Takano, T. (1990). Degradation of newly synthesized apolipoprotein B-100 in a pre-Golgi compartment. J. Biol. Chem. 265, 11880-11884.

Sawa, T., Imamura, T., Haruta, T., Sasaoka, T., Iskiki, M., Takata, Y., Morika, H., Ishihara, H., Usui, I., & Kobayashi, M. (1996). Hsp70 family molecular chaperones and mutant insulin receptor; differential binding specificities of BiP and Hsp70/Hsc70 determines accumulation or degradation of insulin receptor. Biochem. Biophys. Res. Commun. 218, 449-453.

Schatz, B., & Dobberstein, B. (1996). Common principles of protein translocation across membranes. Science 271, 1519-1526.

Schimke, R.T. (1970). Regulation of protein degradation in mammalian tissues. In: Mammalian Protein Metabolism (Munro, H.N., ed.) New York Academic 4 177-228.

Schmitz, A., Maintz, M., Kehle, T., & Herzog, V. (1995). *In vivo* iodination of a misfolded proinsulin reveals co-localized signals for BiP binding and for degradation in the ER. EMBO J. 14,1091-1098,

Schubert, U., Bour, S., Ferrer-Montiel, A.V., Montal, M., Maldarelli, F., & Strebel, K., (1996). The two biological activities of human immunodeficiency virus type 1 Vpu protein involve two separable structural domains. J. Virol. 70, 809-819.

Seufert, W., & Jentsch, S. (1992). In vivo function of the proteasome in the ubiquitin pathway. EMBO J. 11, 3077-3080.

Shachar, I., Rabinovich, E., Kerem, A., & Bar-Nun, S. (1994). Thiol-reducing agents and calcium perturbants alter intracellular sorting of immunoglobulin M. J. Biol. Chem. 269, 27344-27350.

Sharp, D., Blinderman, L., Combs, K.A., Kienzle, B., Ricci, B., Wager-Smith, K., Gill, C.M., Truck, C.W., Bouma, M.E., Rader, D.J., Aggerbeck, L.P., Gregg, R.E., Gordon, D.A., & Wetterau, J.R. (1993). Cloning and gene defects in microsomal triglyceride transfer protein associated with abetalipoproteinaemia. Nature 365, 65-69.

Shelness, G. (1994). Microsomal signal peptidase complex. In, Guidebook to the Secretory Pathway, Eritors, Rothblatt, J., Novick, P., & Stevens, T.H., Sambrook & Tooze Publication; Oxford University Press, New York.

Shenai, R., & Wallis, M. (1979). Biosynthesis and degradation of prolactin in the rat anterior pituitary gland. Biochem. J. 182, 735-743.

Sherman, M.Y., & Goldberg, A.L. (1993). Involvement of the chaperonin DnaK in the rapid degradation of a mutant protein in *E.coli*. EMBO J. 11, 71-77.

Shia, M.A., & Lodish, H.F. (1989). The two subunits of the human asialoglycoprotein receptor have different fates when expressed alone in fibroblasts. Proc. Natl. Acad. Sci. USA 86, 152-158.

Sidman, C. (1981). B lymphocyte differentiation and the control of IgM mu chain expression. Cell 23, 379-389.

Simon, S.M., Peskin, C.S., & Oster, G.F. (1992). What drives the translocation of proteins? Proc. Natl. Acad. Sci. USA 89, 3770-3774.

Simpson, J.C., Dascher, C., Roberts, L.M., Lord, J.M., & Balch, W.E. (1995). Ricin cytotoxicity is sensitive to recycling between the endoplasmic reticulum and the Golgi complex. J. Biol. Chem. 270, 20078-20083.

Sitia, R., Neuberger, M.S., & Milstein, C. (1987). Regulation of membrane IgM expression in secretory B cells: translational and post-translational events. EMBO J. 6, 3969-3977.

Skowronek, M.H., Hendershot, L.M., & Haas, I.G. (1998). The variable domain of nonassembled IG light chains determines both their half-life and binding to the chaperone BiP. Proc. Natl. Acad. Sci. USA 95, 1574-1578.

Sommer, T., & Jentsch, S. (1993). A protein translocation defect linked to ubiquitin conjugation at the endoplasmic reticulum. Nature 365, 176-179.

Stafford, F.J., & Bonifacino, J.S. (1991). A permeabilized cell system identifies the endoplasmic recticulum as a site of protein degradation. J. Bid. Chem. 115, 1225-1236.

Stoller, T. J., & Shields, D. (1989). The propeptide of preprosomatostatin mediates intracellular transport and secretion of a-globin from mammalian cells. J. Cell Biol. 108, 1647-1655.

Strebel, K., Klimkait, T., & Martin, M.A. (1988). A novel gene of HIV-1, vpu, and its 16 kD product. Science 241, 1221-1223.

Su, K., Stoller, T., Rocco, J., Zemsky, J., & Green, R. (1993). Pre-Golgi degradation of yeast prepro-alpha-factor expressed in a mammalian cell. J. Biol. Chem. 268, 14301-14309.

Taya, M., Rabinovich, E., & Haimovich, J. (1991). Characterization of IgM molecules in light-chain deficient variants of a B-cell tumor. Immunol. Lett. 33, 173-177.

Thrift, R.N., Drisko, J., Dueland, S., Trawick, J.D., & Davis, R.A. (1992). Translocation of apoliprotein B across the endoplasmic reticulum is blocked in a nonheptic cell line. Proc. Natl. Acad. Sci. USA 89, 9161-9165.

Tosetti, F., Ferrari, N., Pfeffer, U., Brigati, C., & Vidali, G. (1992). Regulation of plasma retinol binding protein secretion in human HepG2 cells. Exp. Cell Res. 200, 467-472.

Tsao, Y.S., Ivessa, N.E., Adesnik, M., Sabatini, D.D., & Kreibich, G. (1992). Carboxy terminally truncated forms of ribophorin I are degraded in pre-Golgi compartments by a calcium-dependent process. J. Cell Biol. 116, 57-67.

Tsuji, E., Misumi, Y., Fujiwara, T., Takami, N., Ogata, S., & Ikehara, Y. (1992). An active-site mutation (Gly633→Arg) of dipeptidyl peptidase IV causes its retention and rapid degradation in the endoplasmic reticulum. Biochemistry 31, 11921-11927.

Urade, R., Nasu, M., Moriyama, T., Wada, K., & Kito, M. (1992). Protein degradation by the phosphoinositide-specific phopholipase C-a family from rat-liver endoplasmic reticulum. J. Biol. Chem. 267, 15192-15198.

Vertel, B.M., Walthers, L.M., Grier, B, Maine, N., & Goetinck, P.F. (1993). Nonomelic chondrocytes synthesize, but fail to translocate, a truncated aggrecan precursor. Nanomelic chondrocytes, but fail to translocate, a truncated aggrecan precursor. J. Cell Sci. 104, 939-948.

Vertel, B.M., Grier, B.L., Li, H., & Schwartz, N.B. (1994). The chondrocystrophy, nanomelia: biosynthesis and processing of the defective aggrecan precursor. Biochem. J. 301, 211-216.

Wagner, I., Heike, A., van Dyke, L., Langer, T., & Neupert, W. (1994). Molecular chaperones cooperate with PIM1 protease in the degradation of misfolded proteins in mitochondria. EMBO J. 13, 5135-5145,

Wales, R., Roberts, L.M., & Lord, J.M. (1993). Addition of an endoplasmic reticulum retrieval sequence to Ricin A chain significantly increases its cytotoxicity to mammalian cells. J. Biol. Chem. 268, 23986-23990.

Wang, C-N., Hobman, T.C., and Brindley, D.N. (1995). Degradation of apolipoprotein B in cultured rat hepatocytes occurs in a post-endoplasmic reticulum compartment. J. Biol. Chem. 270, 24624-24631.

Ward, C.L., Omura, S., & Kopito, R.R. (1995). Degradation of CFTR by the ubiquitin-proteasome pathway. Cell 83, 121-127.

Ware, F.E., Vassilakos, A., Peterson, P.A., Jackson, M.R., Lehrman, M.A., & Williams, D.B. (1995). The molecular chaperone Calnexin binds GlcMan9GlcNAc2 oligosaccharide as an initial step in recognizing unfolded glycoproteins. J. Biol. Chem. 270, 4697-4704.

Werner, E.D., Brodsky, J.L., & McCracken, A.A. (1996). Proteasome-dependent ER-associated protein degradation: an unconventional route to a familiar fate. Proc. Natl. Acad. Sci. USA 93, 13797-13801.

Wickner, S., Gottesman, S., Skowyra, D., Hoskins, J. McKenney, K., & Mauruizi, M.R. (1994). A molecular chaperone, ClpA, functions like DnaK and DnaJ. Proc. Natl. Acad. Sci. USA 91, 12218-22.

Wiertz, E.J.H., Jones, T.R., Sun, L., Bogyo, M., Geuze, H.J., & Pleogh, H.L. (1996a). The human cytomegalovirus US11 gene product dislocates MHC Class I heavy chains from the endoplasmic reticulum to the cytosol. Cell 84, 769-779.

Wiertz, E.J.H., Tortorella, D., Sun, L., Bogyo, Yu, J., Mothes, W. Jones, T.R., , M., Rapoport, T.A., & Pleogh, H.L. (1996b). Sec61-mediated transfer of a membrane protein from the endoplasmic reticulum to the proteasome for destruction. Nature 384, 432-438.

Wikström, L., & Lodish, H.F. (1991). Non-lysosomal, pre-Golgi degradation of unassembled asialoglycoprotein receptor subunits: a TLCK- and TPCK-sensitive cleavage within the ER. J. Cell Biol. 113, 997-1007.

Wikström, L., & Lodish, H.F. (1992). Endoplasmic reticulum degradation of a subunit of the asialoglycoprotein receptor in vitro. Vesicular transport from endoplasmic reticulum is unnecessary. J. Biol. Chem. 267, 5-8.

Wikström, L., & Lodish, H. F. (1993). Unfolded H2b asialoglycoprotein receptor subunit polypeptides are selectively degraded within the endoplasmic reticulum. J. Biol. Chem. 268, 14412-14416.

Wileman, T., Carson, G.R., Shih, F.F., Concino, M.F., Terhorst, C. (1990a). The transmembrane anchor of the T-cell antigen receptor beta chain contains a structural determinant of pre-Golgi proteolysis. Cell Regul. 1, 907-919.

Wileman, T., Pettey, C., & Terhorst, C. (1990b). Recognition for degradation in the endoplasmic reticulum and lysosomes prevents the transport of single TCR beta and CD3 delta subunits of the T-cell antigen receptor to the surface of cells. Int. Immunol. 2, 743-754.

Wileman, T., Kane, L. P., & Terhorst, C. (1991). Degradation of T-cell receptor chains in the endoplasmic reticulum is inhibited by inhibitors of cysteine proteases. Cell Regul. 2, 753-765.

Wileman, T., Kane, L.P., Young, J., Carson, G.R., & Terhorst, C. (1993). Associations between subunit ectodomains promote T cell antigen receptor assembly and protect against degradation in the ER. J. Cell Biol. 122, 67-78.

Willey, R. L., Maldarelli, F., Martin, M. A., & Strebel, K. (1992). Human immunodeficiency virus type 1 Vpu protein induces rapid degradation of CD4. J. Virol. 66, 7193-7200.

Winitz, D., Shachar, I., Elkabetz, Y., Amitay, R., Samuelov, M., & Bar-Nun, S. (1996). Degradation of distinct assembly forms of immunoglobulin M occurs in multiple sites in permeabilized B cells. J. Biol. Chem. 271, 27645-27651.

Wu. Y, Whitman, I., Molmenti, E., Moore, K., Hippenmeyer, P., Perlmutter, D.H. (1994). A lag in intracellular degradation of mutant alpha 1-antitrypsin correlates with the liver disease phenotype in homozygous PiZZ alpha 1-antitrypsin deficiency. Proc. Natl. Acad. Sci. USA 91, 9014-9018

Yang, B., Hoe, M.H., Black, P., & Hunt, R.C. (1993). Role of oligosaccharides in the processing and function of human transferrin receptors. Effect of the loss of the three N-glycosyl oligosaccharides individually or together. J. Biol. Chem. 268, 7435-7441.

Yao, Z., Tran, K., & McLeod, R.S. (1997). Intracellular degradation of newly synthesized apolipoprotein B. J. Lipid Res. 38, 1937-1953.

Young, J., Kane, L.P., Exley, M., & Wileman, T. (1993). Regulation of selective protein degradation in the endoplasmic reticulum by redox potential. J. Biol. Chem. 268, 19810-19818.

Yu, H., Kaung, G., Kobayashi, S., & Kopito, R.R. (1997). Cytosolic degradation of T-cell receptor α chains by the proteasome. J. Biol. Chem. 272, 20800-20804.

Yuk, H.H., & Lodish, H.F. (1995). Enhanced folding and processing of a disulfide mutant of the human asialoglycoprotein receptor H2b subunit. J. Biol. Chem. 270, 20169-20176.

Zhang, Q., Tector, M., & Salter, R.D. (1995). Calnexin recognises carbohydrate and protein determinants of class I major histocompatability complex molecules. J. Biol. Chem. 270, 3944-3948.

Zhou, M., Wu, X., Huang, L-S., & Ginsberg, H.N. (1995). Apolipoprotein B100, an inefficiently translocated secretory protein, is bound to the cytosolic chaperone, heat shock protein 70. J. Biol. Chem. 270, 25220-25224.

PATHWAYS FOR THE DEGRADATION OF INTRACELLULAR PROTEINS WITHIN LYSOSOMES IN HIGHER EUKARYOTES

Erwin Knecht, José Javier Martín de Llano,
Enrique José Andreu, and Isabel Moreno Miralles

Advances in Molecular and Cell Biology
Volume 27, pages 201-234.

ISBN: 0-7623-0387-5

I. INTRODUCTION

Cell proteins are continuously being degraded by both lysosomal and non-lysosomal proteases (Doherty and Mayer, 1992; Bond and Barrett, 1993; Ciechanover and Schwartz, 1994; Bohley, 1995; Carillo et al., 1995; Suzuki and Bond, 1996). Lysosomes, which are found in almost all eukaryotic cells, participate in intracellular protein degradation by various mechanisms. This review will cover recent work on these different lysosomal pathways, focusing largely on the mammalian hepatocyte and on mammalian cultured cells, where most of the work was carried out. However, we will also discuss some findings in yeast cells that better define some of these pathways and which make them more easily accessible to genetic and molecular biological approaches. For earlier reports and references, compare Holtzman (1989), Seglen and Bohley (1992), Knop et al. (1993), Dunn (1994), Lee and Marzella (1994), Mortimore and Kadowaki (1994), Skurat and Dice (1994), and Blommart et al. (1997A).

II. GENERAL CHARACTERISTICS OF THE LYSOSOMES AND OF THEIR PROTEIN DEGRADATIVE MACHINERY

A. Lysosomal Participation in Intracellular Protein Degradation

Lysosomes (the name and concept were suggested in 1955) were discovered in December of 1949 by Christian de Duve and collaborators. These organelles were subsequently defined as particles containing, in a latent (i.e., in a membrane-bound) form, hydrolytic enzymes with optimal activity at acid pH. These organelles were also found to participate in the engulfment and degradation of extracellular (heterophagy) and intracellular (autophagy) material. Lysosomes contain, enclosed within a membrane, 40–60 different acid hydrolytic enzymes, including proteases, nucleases, glycosidases, lipases, phospholipases, phosphatases, and sulfatases. These organelles maintain an internal acidic environment (pH of about 4.5–5.0, the optimal for the destructive activity of most lysosomal hydrolases), generated by a membrane bound H^+-ATPase pump, which hydrolyses ATP to pump protons into the lysosomal lumen. The lysosomal membrane also contains several highly glycosylated proteins (the lysosome-associated membrane glycoproteins or lamps) of poorly defined functions as well as various transport systems, which either allow most of the digestive final products to exit from the lysosomal matrix for further catabolism or for reutilization in the cytosol, or transport substrates into lysosomes to aid in the breakdown processes. Thus, the cysteine-specific transport system can deliver into the lysosomes large amounts of cysteine, which help to maintain the activity of the cysteine proteases and reduce the disulfide bridges of the proteolytic substrates (Pisoni and Velilla, 1995).

It is not surprising that these organelles with a well-established proteolytic capacity (Figure 1) were soon considered strong candidates for a prominent role in intracellular protein degradation. However, the actual importance of the lysosomal pathways in this process has remained a controversial issue because: i) it is generally believed that the main lysosomal route of protein degradation (bulk nonspecific autophagy) degrades proteins at quite similar rates (see below); and ii) it is known that proteins turn over at vastly different rates (half-lives ranging from a few minutes to many days). Thus, after the discovery of an ubiquitin-dependent proteolytic pathway which required ATP, many considered that the selective protein turnover in eukaryotes occurs by this and other related pathways (see e.g., Lupas et al., 1997), whereas proteins were nonselectively degraded in lysosomes when their massive destruction was induced.

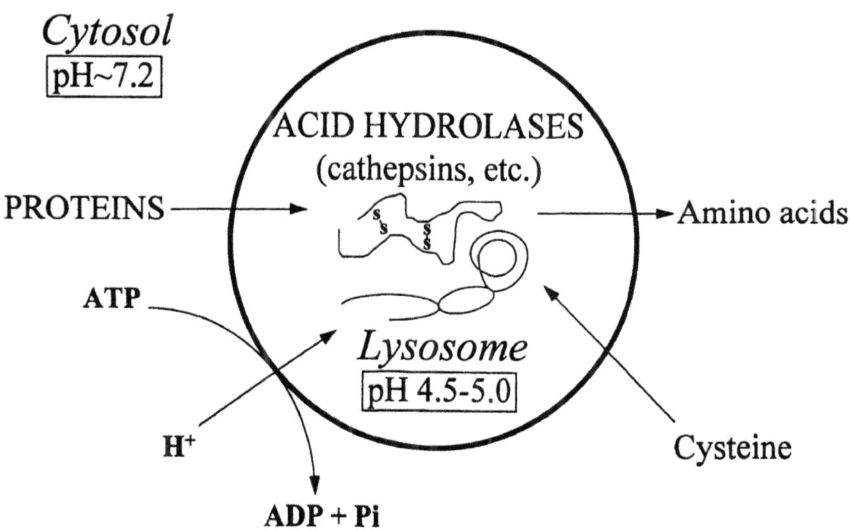

Figure 1. Lysosomal degradation of proteins. The proteins incorporated from the cytoplasm into the lysosomes are degraded in the lysosomal lumen by cathepsins, highly active at acid pH. The acid pH is maintained by an ATP-driven proton pump in the lysosomal membrane. In addition, cysteine and other metabolites may affect the conformation of proteins and aid in their breakdown. Final degradation products exit from the lysosome to the cytosol for further metabolism.

This earlier overview on the role of lysosomes in intracellular protein degradation has been extended as it has become apparent that: i) there are many, and differently regulated, lysosomal pathways to degrade proteins with various degrees of selectivity; ii) there are lysosomal-related compartments, which belong to the exocytic and endocytic routes, and which may be involved, at some stage, in proteolysis (Authier et al., 1996); iii) lysosomes and their related vacuoles are not simply a terminal proteolytic compartment, since they can also generate by limited proteolysis different bioactive protein fragments (e.g., hormones, lysosomal hydrolases, etc.) (Harding, 1994), including also antigenic peptides produced in endosomes; and iv) the lysosomal- and ubiquitin-dependent proteolytic pathways, which were considered before separately, can be functionally related at the level of the ubiquitinylation reaction (Hershko and Ciechanover, 1992; Hochstrasser, 1996; Hicke, 1997) and probably at other levels too (e.g., Laing et al., 1997).

B. Lysosomal Morphology

Lysosomes were identified in a variety of cells and defined as membrane-bound cytoplasmic organelles, most frequently with a diameter of 0.2–0.5 µm and an electron-dense appearance, but usually displaying great structural variability (Figure 2). This variability in shape, size, and other characteristics contrasts with the uniformity of other cell organelles (e.g., mitochondria) and reflects the different lysosomal functions, the wide variety of materials digested, as well as their various stages of digestion (Sahagian and Novikoff, 1994). Typical spherical lysosomes with various morphologies (Figure 2B) are usually, but not always, localized around the cell center. Tubular lysosomes, associated with the endocytic activity of the cell (Araki et al., 1993) and difficult to preserve in fixed cells (Robinson et al., 1996), occur as networks in macrophages and many other cell types, elongating radially from the perinuclear region to the cell periphery. All these structures depend on microtubules for their formation and distribution (e.g., Rajas et al., 1996). In addition, other frequently named lysosomal types are: i) multivesicular bodies, with 15 to 80 nm internal vesicles formed by vesiculation of their limiting membrane; ii) autophagic vacuoles (AVs), usually surrounded by a double or a single membrane and containing more or less recognizable cytoplasmic material for which a phagocytic origin can be ruled out; iii) dense bodies, with an electron dense matrix; iv) large (100–150 nm) and small (75–100 nm) "coated" vesicles, close to the Golgi complex or to the plasma membrane; and v) residual bodies containing myelin-like elements and other amorphous and unidentifiable digestive residues, which sometimes are exocytosed. These and other lysosomal types have been placed in the autophagic or heterophagic routes on the basis of their content, acid hydrolases activity, uptake of endocytic markers, cell localization, etc.

C. Lysosomal Proteases

Lysosomes contain in their matrix over 20 different proteases (Bohley and Seglen, 1992) with a distinct pattern of cellular expression in different tissues and

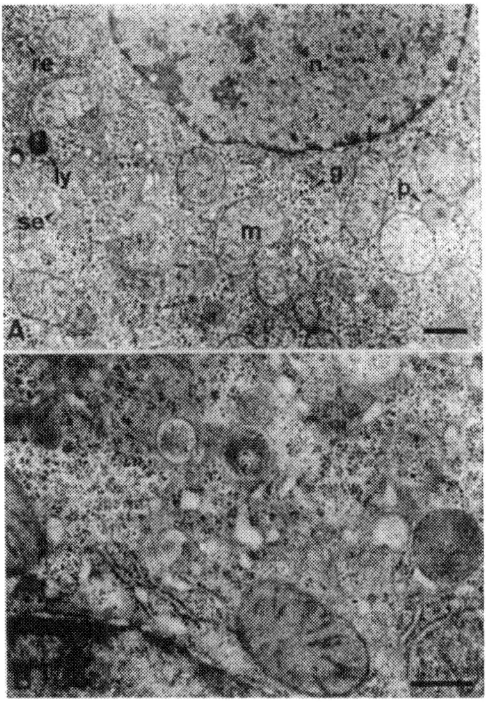

Figure 2. Electron micrographs of rat liver cells. (**A**). Cell region showing different organelles: n, nucleus; m, mitochondria; p, peroxisome; re, rough endoplasmic reticulum; se, smooth endoplasmic reticulum; ly, lysosome; g, glycogen. (**B**). High magnification view of a cell area showing various lysosomes with diverse morphology. Bars: 1 μm (**A**) and 0.5 μm (**B**).

subjected to hormonal and dietary regulation (Inubushi et al., 1996). Most, but not all, lysosomal proteases are designated with the old name "cathepsins," derived from the Greek term meaning "to digest." However, not all cathepsins are lysosomal: the aspartic proteinase cathepsin E has been found primarily in the endoplasmic reticulum (ER) (Finley and Kornfeld, 1994) and cathepsin R occurs in ribosomes (Bohley and Seglen, 1992).

Cathepsins comprise both endopeptidases (proteinases, which cleave peptide bonds localized internally in polypeptide chains) and exopeptidases (amino- and carboxy-peptidases, which cleave peptide bonds near the N- and C-termini, respectively, of polypeptide chains). Examples of endopeptidases are cathepsins D, L, and S, of exopeptidases are cathepsins A and C and carboxypeptidase B2, and of both are cathepsins B and H. Most cathepsins display an acidic optimum pH (3.5–6.0), broad substrate specificity, a relatively low molecular mass (20–42 kDa), and a monomeric and glycosylated structure. However, there are some exceptions, e.g., cathepsin M (optimum pH: 5–7), cathepsin E (molecular weight: ca 100,000), etc.

Lysosomal proteases (e.g., cathepsins B, H, L, and S) mainly belong to the group of cysteine proteases, but most lysosomes also contain high levels of cathepsin D, a very active aspartic proteinase. In contrast, very few lysosomal proteases contain a serine (e.g., cathepsin G, which is stored in the azurophil granules of the neutrophil granulocyte) or a metal in its catalytic group. Therefore, lysosomal proteolysis is most easily inhibited with specific (E-64 and its membrane permeable derivatives E64-c and E64-d, Z-Phe-Ala-CHN_2), or less specific (leupeptin, chymostatin) inhibitors of cysteine endopeptidases and also, but less efficiently, with inhibitors (pepstatin) of aspartic endopeptidases (Barrett, 1992). Although several effective cathepsin inhibitors occur in the cytosol—probably to protect against occasional escape of lysosomal proteases—they are apparently absent from the lysosomes.

The proteins sequestered within lysosomes are degraded very fast (half-life of about eight minutes) and this is possible because: i) lysosomes contain high concentrations (sometimes even in the mM range) of many cathepsins with various specificities; and ii) the low intralysosomal pH ordinarily favors processes which increase the proteolytic susceptibility of most proteins: dissociation of stabilizing ligands (associated membranes, substrates, cofactors and others), separation of subunits from oligomeric proteins, disclosure of susceptible buried sequences, etc.

Although lysosomal proteins, like any other intracellular protein, are subjected to a continuous turnover, their half-lives are quite long (days) compared to the very short half-life of the proteolytic substrates once inside the lysosomes (minutes). Therefore, lysosomal proteins should be resistant to their own strong degradative environment. Probably, the high content of acidic oligosaccharides confers to the lysosomal proteins steric protection from degradation and, indeed, the luminal side of the lysosomal membranes, where the carbohydrate moieties of the highly glycosylated lysosomal membrane proteins face, is more resistant to the hydrolytic enzymes than the cytosolic side. Also, some lysosomal proteins can probably acquire in the intralysosomal milieu a conformation which is resistant to the attack by lysosomal enzymes. For example, the stability of the heat shock cognate protein of 73 kDa (hsc73), which is found in the lumen of some lysosomal populations (see below), is higher at pH 5.4 than at pH 6.4 (Cuervo et al., 1997).

D. Biogenesis of Lysosomes

The biosynthesis and targeting of lysosomal enzymes, including cathepsins, has been amply investigated (Hasilik, 1992; Kornfeld and Sly, 1995; Mellman, 1996). They are translated in a prepro-form on the rough ER where they undergo several reactions: cleavage of the presequence, core glycosylation by high mannose oligosaccharides donated by a dolichol pyrophosphate intermediate, and deglycosylation. Further oligosaccharide processing of the proenzyme occurs in the Golgi complex, together with a very important step which takes place in the *cis* region: phosphorylation, by the concerted action of a phosphotransferase and a glycosidase, of certain mannose residues. In the *trans*-Golgi network, the mannose 6-phosphate residues, present on asparagine-linked carbohydrates, are recognized by two different mannose 6-phosphate receptors (MPRs): MPR 46 (cation-dependent MPR) and MPR 300 (cation-independent MPR, or mannose-6-phosphate/insulin-like growth factor II receptor). MPRs forward the pro-enzyme, within clathrin-coated vesicles, to a prelysosomal compartment within the endocytic route where the lysosomal proteins are released from the receptors, which recycle back to the Golgi complex. Meanwhile, the precursor matures into the active enzyme, which eventually reaches the lysosomes. Maturation of lysosomal proteases can occur by a combination of limited autoproteolysis and cleavage by other lysosomal proteinases. In some cell types, lysosomal enzymes may also be transported by a second and less predominant route, including secretion to the extracellular medium, recapture from the cell surface, and delivery to lysosomes. This pathway uses mainly MPR 300, which also, after discharging the lysosomal enzymes in endosomes, is recycled back to the plasma membrane. The importance of the MPRs is illustrated by the fact that, in the absence of the mannose-6-phosphate residue, proteases and other lysosomal hydrolases are not transported properly and are, usually, released from the cells (Kornfeld and Sly, 1995). However, certain lysosomal proteins, particularly membrane proteins, employ a mannose 6-phosphate-independent route of targeting to lysosomes, involving signals contained in their cytosolic tails (Peters and von Figura, 1994; Sandoval and Bakke, 1994; Hunziker and Geuze, 1996). Besides all these pathways, certain cathepsins (B, D, L), in response to inflammation, viral transformation or malignant transformation, can be also secreted extracellularly as active enzymes or inactive proforms (Mignatti and Rifkin, 1993; Owen and Campbell, 1995).

The yeast vacuole is considered to be the lysosomal equivalent in mammalian cells (Knop et al., 1993; see Thumm and Wolf, chapter 2) and two endopeptidases (proteinases A and B, PrA and B), as well as a set of exopeptidases, provide the activities to carry out its proteolytic roles. However, functional differences between the yeast vacuole and the mammalian lysosome exist (Lee and Goldberg, 1996). Most vacuolar hydrolases are synthesized as inactive precursors which initially follow the secretory pathway, are sorted away from the secretory proteins in a late Golgi compartment, and are transported to the vacuole, where they are activated by proteolytic cleavage produced by PrA and/or PrB (Horazdovsky et al., 1995). However, at least two vacuolar proteins, α-mannosidase (Yoshihisa and Anraku, 1990; Armstrong, 1991) and aminopeptidase I, (Klionsky et al., 1992; Seguí-Real et al., 1995) are imported into the vacuole by a different pathway which employs a direct and active transport from the cytosol. This selective pathway appears, however, to largely employ the same molecular components as macroautophagy (Harding et al., 1996; Scott et al., 1996; Straub et al., 1997; Rieder and Emr, 1997), which, under nitrogen starvation, delivers nonselectively proteins into the vacuole to provide the cell with amino acids. However, the pathway involved in the biogenesis of vacuolar resident proteins from the cytoplasm to the vacuoles is distinct from macroautophagic sequestration of substrates to be degraded in that the first employs much smaller (~150 nm diameter), double membrane-bound autophagosomes and proceeds constitutively under growing conditions (Scott et al., 1997; Baba et al., 1997).

III. MECHANISMS OF DELIVERY TO THE LYSOSOME OF INTRACELLULAR PROTEINS DESTINED FOR DEGRADATION

Lysosomes are sites where substances to be digested meet by many different routes. Classically, the most prominent of these pathways were called autophagy and heterophagy, which incorporate intra- and extracellular components, respectively, to lysosomes. However, heterophagy includes phagocytosis, pinocytosis or fluid phase endocytosis and receptor mediated endocytosis. Likewise, autophagy should be considered as a broad term which comprises the different, and most of them still largely unexplored, pathways whereby intracellular components gain access to the lysosomal matrix and are degraded. These various mechanisms will be discussed below.

A. Endocytosis

This route is followed by extracellular material which is taken up into a cell by a membrane-bound vesicle or vacuole (Berg et al., 1995; Schwartz, 1995; Mellman, 1996). These compounds are either recognized by specific receptors at the plasma membrane (a paradigmatic example is provided by the low-density lipoprotein (LDL) receptor, which binds and internalizes LDL particles to supply the cells with cholesterol, Goldstein et al., 1995), or simply trapped by nonspecific pinocytosis. Actin-dependent phagocytic uptake of bacteria, viruses, inert particles, etc. also occurs by similar mechanisms in such cells as macrophages and neutrophils (Rabinovitch, 1995).

In the main endocytic route (Figure 3), receptors plus ligands, soluble components and other material enter first into small, irregularly shaped vesicles, formed from the plasma membrane and coated by clathrin and adaptor proteins. Next, these endocytic vesicles fuse with early endosomes, where the slightly acidic environment changes the conformation of receptors and releases their ligands. From this endosomal network of interconnected tubulovesicular elements, some membrane proteins (e.g., LDL receptor, transferrin receptor, etc.) recycle back to the plasma membrane for reuse. The remaining membrane proteins, together with soluble ligands and proteins, follow via multivesicular endosomes or "endosomal carrier vesicles," into larger and rounded vesicles (late endosomes) and hydrolysis begins. Eventually, the proteins arrive to mature lysosomes, where degradation is completed. Alternatively, in another model, early endosomes mature into dense and finally very dense endosomes. Although it remains to be elucidated which of these two endocytic models (vesicle shuttle or maturation, Griffiths, 1996) is operative in the various processes, it appears that: i) the progressive decrease in the internal pH, which occurs during the transit from the plasma membrane to lysosomes, plays a

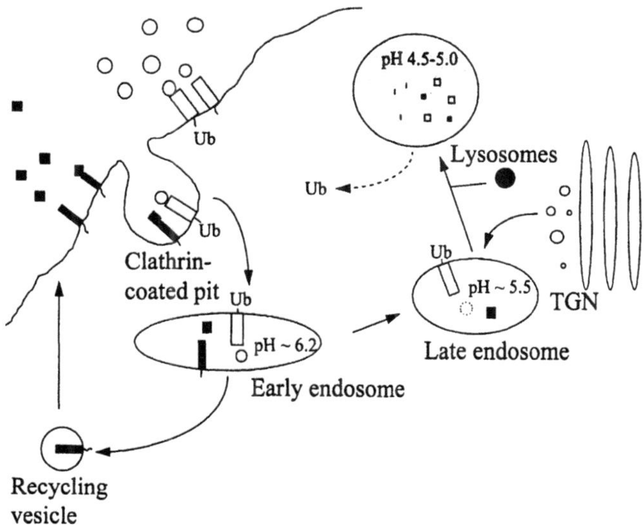

Figure 3. Endocytic internalization of plasma membrane proteins. Endocytic vesicles (not shown) are formed from noncoated regions or from clathrin-coated pits of the plasma membrane, and subsequently fuse with early endosomes. The reduced pH causes dissociation of ligands and receptors. While receptors which recycle (closed rectangles) are returned to the plasma membrane via recycling vesicles, ligands (circles, squares) and some membrane proteins (open rectangles) are transferred via vesicular transport or maturation through late endosomes to the lysosomes. The cytosolic region of plasma membrane proteins and its modifications (including ubiquitinylation, Ub) is perhaps important for the efficient transference of these membrane proteins to lysosomes. By analogy with what is known on ubiquitin-mediated degradative pathways, free ubiquitin is presumably released (-----➤) once the protein has been targeted for lysosomal proteolysis. Late endosomes acquire lysosomal proteins from the plasma membrane, from *trans*-Golgi network (TGN) vesicles and, ultimately, from other lysosomes.

notable role in the degradative events; ii) in all kind of heterophagic processes (see e.g., Authier et al., 1996; Seabright and Smith, 1996; Wattiaux et al., 1996), some degradation of the endosomal content occurs before its final delivery to lysosomes; and iii) variations in the acidification of endosomes and lysosomes are due to different activities of the same proton pumps (van Dyke, 1993).

While most receptors are recycled back, other membrane proteins, typically those involved in signal transduction and growth-factor receptors, follow the same pathway as the ligands and are degraded in lysosomes. Why some plasma membrane proteins are degraded in late endosomes/lysosomes and others not is unknown, but the cytoplasmic tails of these proteins could bear some important signal. For instance, ubiquitin could play a role in this process (see also chapter 3), since it has been found that the mammalian platelet-derived growth factor β-receptor, upon ligand binding, is multiubiquitinylated and transported to lysosomes and it has been therefore suggested that the ubiquitin modification plays a role in receptor endocytosis and turnover (Mori et al., 1992). Also, other cell surface proteins have been found to be modified by ubiquitin and, likewise, five ubiquitinylated polypeptides (most probably cell surface receptors) have been localized in multivesicular endosomes from cultured insect cells (Löw et al., 1995) and the ligand-induced endocytosis and endosomal/lysosomal degradation of the growth hormone receptor requires ubiquitin conjugation (Strous et al., 1996). In addition, specific sets of coat proteins (COPs), including clathrin and adaptor proteins, CP coatomers and probably others (Trowbridge et al., 1993; Kreis and Pepperkok, 1994; Whitney et al., 1995; Aniento et al., 1996a; Hay and Scheller, 1997) could also participate in this regulation.

In *Saccharomyces cerevisiae*, a role of ubiquitin in the endocytic pathway was already suggested by the observation of ubiquitin-protein conjugates of membranous origin in vacuoles of yeast cells lacking proteinases (Simeon et al., 1992) and also by the supression of the lethality of certain clathrin-deficient strains elicited by ubiquitin overexpression (Nelson and Lemmon, 1993). More recently, it was shown (Hicke and Riezmann, 1996) that pheromone binding to the Ste2 α-factor receptor stimulates both ubiquitinylation of the cytoplasmic tail of the receptor and its internalization, followed by its degradation in the yeast vacuole. In addition, the endocytosis and vacuolar degradation of other membrane proteins (Pdr5 multidrug transporter, Ste6 α-factor pheromone transporter and Fur4 uracil permease) seems to include similar ubiquitin modifications (Egner and Kuchler, 1996; Roth and Davis, 1996).

B. Crinophagy

By this degradative route, proteins originally destined for secretion are delivered to lysosomes when the demands for these proteins decline. Crinophagy, thus, allows the cells to adapt to their varying metabolic requirements (Schnell-Landstrom et al., 1991). It was first described in secretory granules containing prolactin from the anterior pituitary gland, following cessation of lactation (Smith and Farquhar, 1966) and, since then, it has also been recognized in liver, pancreatic acini, parathyroid cells, and so forth. Crinophagy is also operative under basal conditions since, in the liver, as much as 40% of apparently normal serum proteins are degraded via this pathway (Lenk et al., 1991). The content of the mature or immature secretory granules is digested, following their fusion either with lysosomes (Figure 4A) or with late endosomes. This fusion, which requires microtubules (Glaumann et al., 1989), gives rise to an organelle, named a crinosome or a crinolysosome, which initially shares characteristics of both secretory granules and lysosomes.

A very important question that remains unanswered is why secretory granules destined for degradation fuse with endosomes or lysosomes and not with the plasma membrane. An answer to this issue will come only after isolation and detailed structural studies of the different membranes involved. Crinophagy also raises the possibility that lysosomes could fuse with other subcellular organelles (e.g., mitochondria or peroxisomes). Although no clear evidences for such a degradative process exist, this contingency should not be ignored.

Crinophagy is not the only lysosomic way to modulate the secretory process, since this can also be achieved by at least two other mechanisms. Thus, secretory proteins can be diverted to lysosomes before entering secretory granules (see below, Figure 4B) and

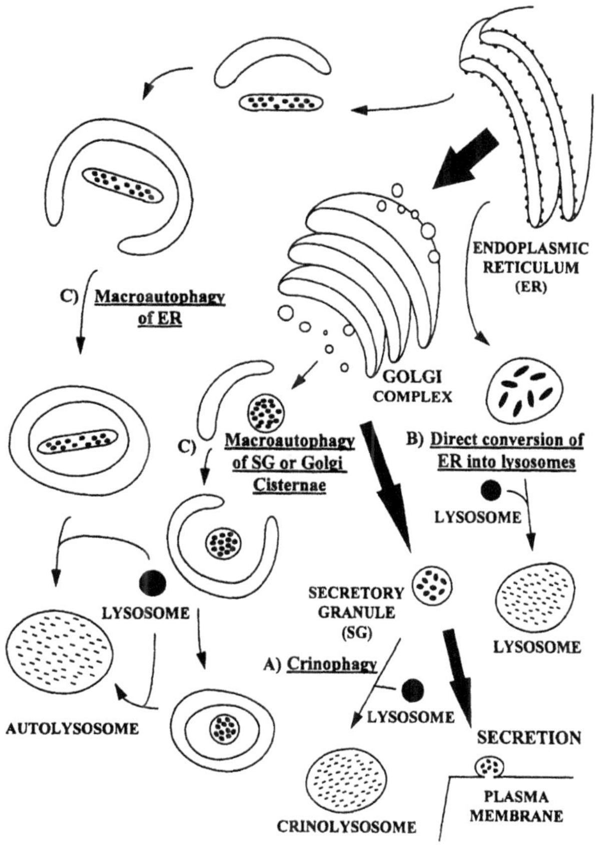

Figure 4. Schematic model for the various degradative pathways of secretory proteins. The large arrows indicate protein traffic through the secretory pathway. Secretory proteins are transported in vesicles from transitional regions of the endoplasmic reticulum (ER) to the *cis*–Golgi and then through the Golgi complex. From the *trans*-Golgi network, secretory proteins are transported to the cell surface in secretory granules (SG), which fuse with the plasma membrane. Secretory granules in excess fuse with lysosomes or late endosomes to form a crinolysosome, where the content of the secretory granules is digested (pathway A). An endoplasmic reticulum cisterna containing secretory proteins can also be directly converted into a lysosome (pathway B). In addition, secretory granules, Golgi cisternae, or endoplasmic reticulum containing secretory products can be degraded by macroautophagy in autolysosomes (pathway C, see also Figure 6).

these organelles, together with Golgi cisternae and ER containing secretory products, can be also degraded by macroautophagy (see below, Figure 4C). All these pathways to degrade secretory proteins, and probably others, can coexist, with their relative contributions changing with the cell type and its metabolic demands. For example, crinophagy appears to be the main mechanism to regulate the hormone content within peptide-secreting cells, whereas this function is typically exerted by macroautophagy in steroid-secreting cells (Yi and Tang, 1995).

C. Direct Conversion of ER Cisternae Into Lysosomes

There have been early reports of a possible differentiation of ER structures into a self-degradative compartment (e.g., Orci et al., 1984). The process was more precisely described in thyroid hormone-secreting cells of the anterior pituitary gland of thyroxine treated, thyroidectomized rats (Noda and Farquhar, 1992). Here, regions of the transitional part of the ER sequentially lose their ribosomes and ER proteins and acquire lysosomal membrane proteins by a gradual conversion process of the limiting membrane. How the ER membrane becomes a lysosomal membrane is still unknown, but these characteristics can be acquired via an early fusion with small vesicles carrying lysosomal membrane markers, as it probably occurs in macroautophagy too. Then, this converted ER saccula obtains lysosomal hydrolases by fusion with lysosomes (Figure 4B). This pathway appears to be different from macroautophagy and from crinophagy and proteolysis probably affects both membrane and lumenal resident proteins of the ER, as well as proteins in transit through certain parts of the ER. Thus, the degradation of the invariant chain, in the absence of major histocompatibility complex class II molecules (Chervonsky and Sant, 1995), or of the excess of gonadotropin β-subunits (Bassetti et al., 1995) probably occur by this process. In addition, it seems possible that a direct conversion of ER cisternae into lysosomes accounts for the differentiation-dependent degradation of N-linked glycoproteins with an altered metabolism, described in human colon adenocarcinoma cells (Houri et al., 1993, 1995; Ghidoni et al., 1996) and which appears to be controlled at the sequestration level by a specific heterotrimeric G_{i3} protein (Ogier-Denis et al., 1995, 1996, 1997). Alternatively, it is possible that AVs sequester these proteins after their exit to the cytosol or, more simply, the ER cisternae containing them (Figure 4C).

In the yeast *Schizosaccharomyces pombe* a similar degradative process has been reported (Lum and Wright, 1995). Karmellae are ER stacked membranes that partially surround the nucleus and which form when the levels of HMG-CoA reductase and of other ER membrane proteins rise. When the amounts of these proteins return to basal levels they are rapidly degraded by conversion of karmellae into a self-degradative compartment that degrades the membranes sequentially from the interior outward, forming a lipid particle as an end product. Although this mechanism also converts ER into a degradative compartment, it differs slightly from the process described by Noda and Farquhar (1992).

D. Macroautophagy

In macroautophagy or classical autophagy large areas of cytoplasm, typically containing cell organelles, are sequestered and degraded by lysosomes (Seglen and Bohley, 1992). The lysosomes implicated in this process, called AVs, are easy to distinguish in electron micrographs by their content in cell organelles and cytoplasm. However, some forms of heterophagy, where cells phagocytize cytoplasmic fragments from other cells, have a similar appearance which should not be confused with macroautophagy. Macroautophagy operates in eukaryotic cells under basal conditions at a low level. In the liver and kidney of rats, it takes place preferentially during the daytime (i.e., during the inactive phase of the animal) and decreases during the night (i.e., during the period of food intake and maximal activity). Because of the relatively few AVs which can be identified in basal proteolytic states (e.g., in rat liver, mitochondria and AVs represent approximately 20% and 0.02%, respectively, of the hepatocytic cell volume), an induced macroautophagy has been frequently employed to facilitate its study. Macroautophagic activity can be increased, among other situations and depending on cell types, by poor nutritional conditions (for example during starvation, where the above mentioned circadian rythm is no longer visible), in injured or in remodeling cells, by variations in the levels of certain hormones (e.g., glucagon administration, insulin withdrawal), and by treatment with some chemicals (e.g., microtubule depolymerizing drugs: vinblastine, etc.). In general, in most cell types, there appears to be an inverse correlation between cell growth and macroautophagy and, ordinarily, macroautophagy is depressed after short-term stimulation of anabolic processes. For example, macroautophagy is inhibited after partial hepatectomy to facilitate the compensatory growth of the regenerating liver.

The features of the macroautophagic route have been reconstructed mainly from conventional electron micrographs (Pfeifer, 1987; Seglen and Bohley, 1992), although, in some cases, either induced artificial protein complexes (Tooze et al., 1990) or immunocytochemical procedures (Rabouille et al., 1993) have been used to follow the sequence in the degradation of proteins. It is a multistep process (Figure 5), starting with sequestration of a portion of cytoplasm into an autophagosome. All subcellular components (except whole plasma membrane and nuclei, but including lysosomes and related vacuoles) have been found within AVs. Autophagosomes obtain lysosomal enzymes, via fusion with lysosomes and/or prelysosomes (late endosomes) and an autolysosome results, where degradation of the sequestered material is quickly completed.

Four differents origins have been suggested for the intracellular membranes which form the AV: *de novo* formed, or lysosomal-, Golgi- or ER-derived (see e.g., Dunn, 1990a; Furuno et al., 1990; Sakai et al., 1990; Yamamoto et al., 1990; Ueno et al., 1991; Lenk et al., 1992a; Yokota, 1993). In most cases, the enclosing membranes of the AVs correspond to ribosome-free cisternae of rough (Dunn, 1990a) or smooth (Ueno et al., 1991) ER. However, in other cases, the membranes of AVs and ER have been found to be different, and it has been proposed that the cisternae

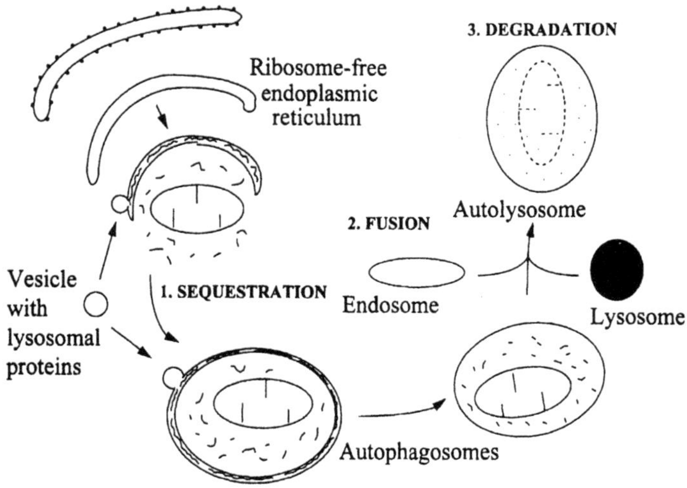

Figure 5. Macroautophagic pathway of protein degradation. Macroautophagic vacuoles probably arise from ribosome-free endoplasmic reticulum areas which undergo several modifications to constitute a segregating structure, which has been called a "phagophore" by some authors (see text), and which sequesters large areas of cytoplasm in an autophagosome. During these modifications the cisternae probably receive some lysosomal proteins (lamps, proton pump, acid phosphatase) by fusion with vesicles (probably endosomes and/or vesicles derived for the *trans*-Golgi network). Then, autophagosomes which have lost the inner membrane fuse with late endosomes or with lysosomes and acquire hydrolytic enzymes to constitute autolysosomes, which rapidly degrade its content. Finally, a residual body, with dense matrix and containing undigestible residues (not shown), results.

which will form an AV have to be considered, at some stage in their development, as independent organelles called "phagophores" (Seglen and Bohley, 1992). At any rate, it now seems widely accepted that the segregating membranes of AVs originally arise from ER cisternae despite the broad transformations which are undergone during the formation of the AV (Dunn, 1990a). Undoubtedly, the availability of purified membranes from autophagosomes (Berg et al., 1994; Seglen et al., 1996) will be advantageous to further investigate the peculiarities of the limiting membranes of AVs.

Eventually, a closed AV (called autophagosome or nascent AV) ensues. Then, the inner membrane of the vacuole disappears (probably digested) while the outer membrane thickens. Before and during these modifications, the cisternae receive certain lysosomal proteins (lamp-1 and other membranous proteins, acid phosphatase, etc.), mostly by fusion with endosomes (Furuno et al., 1990; Tooze et al., 1990; Gordon et al., 1992; Punnonen et al., 1993; Strømhaug and Seglen, 1993; Fengsrud et al., 1995; Liou et al., 1997) and probably also with vesicles developing from the *trans*-Golgi network (Dunn, 1990b). After this step, acidification of the autophagosome, containing lysosomal proteins but still deficient in acid hydrolases, occurs. Some of the acquired lysosomal proteins appear to play a role in the transition of double membrane into single membrane AVs and in the alteration of the remaining membrane to hinder its lysis by the hydrolases which will be incorporated later. This defines a population of AVs, which has been called by some authors "intermediate AVs" (AVi/d) or "acidic late autophagosomes" (Dunn, 1990b). In those cells where endosomes and autophagosomes fuse (e.g., hepatocytes), an amphisome results (Seglen and Bohley, 1992), which is equivalent to the AVi/d.

The next step is the transfer of the hydrolytic enzymes to autophagosomes, which relies on intact microtubules. Although it is possible that AVs acquire part of their hydrolytic enzymes from the lumen of the surrounding cisternae or by vesicular transport from the *trans*-Golgi network, several experiments have suggested that this could only happen at a very low scale (Dunn, 1990b; Lawrence and Brown, 1992; Punnonen et al., 1993). Therefore, obtaining lysosomal hydrolases and transformation into mature AVs occurs, almost exclusively, by fusion of autophagosomes with preexisting late endosomes or with secondary lysosomes. There may be, however, some differences in the way in which autophagosomes transform into lysosomes depending on cell type and/or experimental conditions, because in other cell types others have concluded that lysosomal proteins are acquired in one rather than in various steps (Lawrence and Brown, 1993) or that endocytic and autophagic pathways are not interconnected (Biederbick et al., 1995; Yokota et al., 1995).

The last part in the macroautophagic pathway is the degradation of the sequestered cell components. The half-life for the disappearance of the content of AVs amounts to 8–9 minutes (e.g., Pfeifer, 1987), and this value is remarkably constant in different tissues. To assure complete destruction of the incorporated organelles and fast proteolysis, high concentrations of lysosomal enzymes are probably received during these processes by multiple fusions with other secondary lysosomes. Interestingly, it has been found that the degradation of the sequestered material, during the heat stress-enhanced lysosomal degradation in cultured cells, requires protein ubiquitinylation (Gropper et al., 1991; Lenk et al., 1992b; Schwartz et al., 1992), but the meaning of this observation still remains to be elucidated.

Progress of macroautophagy is tightly linked to various cell components and functions. Intermediate filaments (Holen et al. 1992, 1993; Blankson et al., 1995), and/or microfilaments in some cell types (Aplin et al., 1992) cooperate in the initial sequestration step, whereas microtubules guide the fusion of autophagosomes with lysosomes. Although sequestration can proceed for hours in the absence of new membrane formation (Seglen and Bohley, 1992), protein synthesis appears to be necessary, at least for AV maturation. This requirement can be explained either because certain proteins needed for fusion are short-lived (Lawrence and Brown, 1993) and/or because an indirect effect due to accumulation of amino acids as a consequence of inhibition of protein synthesis occurs (Kadowaki et al., 1994). Macroautophagy strongly depends on the ATP content of the cell (relatively small changes in intracellular ATP particularly disturb early steps of the process, Schellens et al., 1990), and it is also temperature-sensitive (at 20–22 °C it is nearly fully abolished, Seglen and Bohley, 1992). In addition, Ca^{2+}, most probably from the ER, appears to be necessary, either for the sequestration or for the fusion steps (Gordon et al., 1993; Fosse et al., 1995). Also, cAMP regulates the sequestrational step in hepatocytes, most probably phosphorylating certain proteins, with dual and opposite effects: stimulating it under *in vivo* conditions, in perfused livers or in hepatocyte cultures, but inhibiting it in freshly isolated hepatocytes under most, but not all, metabolic conditions (Pfeifer, 1987; Blommaart et al., 1993; Holen et al., 1996).

Since the degradation of proteins within lysosomes is a very fast process, the rate limiting step in macroautophagy should occur earlier, during the sequestration of cellular components (Seglen and Bohley, 1992). Therefore, much of the research on macroautophagy so far has concentrated on this step. In rat liver, a group of regulatory amino acids (dominated by Leu) are considered to be the most influential suppressors of sequestration (Caro et al., 1989; Mortimore et al., 1989; Venerando et al., 1994). However, as it also happens with cAMP and with some other agents employed to investigate macroautophagy, some variations may occur depending on the experimental system employed (Bergamini et al., 1995). In addition, several hormones and growth factors influence macroautophagy. In the liver, autophagy is stimulated by glucagon, α1-adrenergic agonists and some β-agonists (isoproterenol) and is inhibited by insulin (Mortimore et al., 1989; Seglen and Bohley, 1992).

Cell swelling appears to mimic the anabolic effects produced by insulin, including the inhibition of autophagic proteolysis (Häussinger et al., 1994), and it has been proposed that the antiproteolytic effects of insulin, glutamine, and leucine are mediated by this cell alteration. In this regard, it has been shown that insulin (Mortimore et al., 1989; Seglen and Bohley, 1992) and cell swelling (Meijer et al., 1993) promote the inhibition produced by low concentrations of amino acids. Cell swelling produces a rapid and reversible pH increase in lysosomes and in other acidic prelysosomal compartments (Schreiber and Häussinger, 1995) and these pH changes, which can be produced by variations in the activity of the ATP-dependent lysosomal proton pump (Busch et al., 1996), could control the autophagic process. Although in freshly isolated hepatocytes autophagic proteolysis appears not to be controlled by changes in the lysosomal pH (Luiken et al., 1996), this is still possible in liver under different conditions (e.g., *in vivo,* perfused rat liver, cultured rat hepatocytes). In fact, the effects produced by cell swelling appear to require intact microtubular structures (vom Dahl et al., 1995), which may explain the failure to detect them in freshly isolated hepatocytes (Meijer et al., 1993), which need 6–10 hours to reorganize their microtubular structures. If this interpretation is proven, then probably some other discrepancies in the literature, when using different experimental systems, could be similarly explained.

How do the effects of leucine and of other negative regulators of macroautophagic sequestration originate? It is noteworthy that an octaleucine derivative of multiple antigen peptide (MAP), Leu_8-MAP, inhibits macroautophagy without being transported intracellularly. Therefore, it seems that leucine interacts with some binding protein located on the plasma membrane or within closely associated vesicles (Kadowaki et al., 1992; Miotto et al., 1992, 1994; Mortimore et al., 1994). This protein, which has a high molecular weight and is probably a rapidly turning over receptor, appears to be specific for leucine (Mortimore et al., 1994). Assuming the existence of this and similar receptors, activation of macroautophagic sequestration could be initiated by dissociation of amino acids, etc., from these plasma membrane sites.

In analogy with cellular responses produced by hormone and growth factor receptors to chemical signalling, interaction of amino acids and other regulators of macroautophagy with receptors may be connected to phosphorylation events of proteins involved in macroautophagic sequestration. Interestingly, okadaic acid, a potent inhibitor of serine/threonine protein phosphatases, inhibits macroautophagic sequestration, most probably because hyperphosphorylation of one or various cell components (Holen et al., 1992, 1993). The putative protein kinases responsible for the specific phosphorylation of these components have also been investigated with different inhibitors and various protein kinases (Holen et al., 1992, 1993, 1995, 1996; Gordon et al., 1995; Matsuura et al., kl1997; Blommaart et al., 1997b) could be involved in various steps of macroautophagic sequestration.

Which are the cell elements whose hyperphosphorylation inhibits macroautophagy? Okadaic acid increases the phosphorylation of intermediate filaments (cytokeratins), producing also a loss of these structures (Holen et al., 1992, 1993), and it is not difficult to imagine a role for a cytoskeletal component in guiding the movements of the sequestering membranes around areas of cytoplasm. Another possible phosphorylation target is the 31 kDa ribosomal protein S6, because its phosphorylation extent, mediated by amino acids, shows a good correlation with the degree of inhibition of autophagy (Luiken et al., 1994; Blommaart et al., 1995). Although it is unknown how S6 phosphorylation may produce an inhibitory effect on macroautophagy, this modification could promote the attachment of ribosomes to ER membranes, thus hindering the formation of autophagosomes by the ribosome-free regions of the ER (Blommaart et al., 1995). In addition, the possible role of S6 in macroautophagic degradation is especially interesting because S6 is also a component of the protein synthetic machinery and, therefore, a same shuttling event (phosphorylation of S6) could simultaneously control opposite processes in protein turnover (i.e., both synthesis and degradation). However, okadaic acid has no effect on S6 phosphorylation and the S6 kinase inhibitor rapamycin abolishes only in part the inhibition of macroautophagy produced by amino acids. Therefore, it seems that the binding of amino acids to plasma membrane receptors activates a protein kinase cascade that phosphorylates different proteins (e.g., S6, intermediate filaments and probably others), thus preventing macroautophagy.

Other steps in macroautophagy and not sequestration alone can be subjected to regulation. For example, one or more GTP-binding proteins, have been implicated in the maturation of autophagosomes (Sai and Ohkuma, 1992; Kadowaki et al., 1994), although they could play a role in the sequestration step too (Ogier-Denis et al., 1995, 1996, 1997). However and alternatively, some of the results could be differently explained (Sai et al., 1994; Kadowaki et al., 1996). The lysosomal proton pump is also a potential target to regulate, by variations in their activity, various steps in macroautophagy and in other lysosomal processes (Harada et al., 1996). In this regard, it has been proposed that amino acids (in particular, leucine) do not only affect the sequestration step but also inhibit the proton pump, increasing the lysosomal pH and affecting in that way various lysosomal functions (Luiken et al., 1996).

Cytoplasmic proteins are sequestered in the same ratio as they occur in the cytosol (Kopitz et al., 1990) and the same is believed to happen with cell organelles. However, morphometric measurements indicate that cell compartments with different turnover rates could be segregated differently in rat liver (Pfeifer, 1987; Réz et al., 1996). Furthermore, selectivity in macroautophagy, at least under certain conditions, is suggested by the rapid degradation of secretory proteins (see above) or of peroxisomes produced in excess (Heikoop et al., 1992; Luiken et al., 1992; Yokota, 1993; Yokota et al., 1994, 1995). However, except for these and probably other anomalies, macroautophagy can still be largely considered a nonspecific degradative process.

In yeast, AVs have been visualized (Takeshige et al., 1992; Baba et al., 1994) and mutants have been identified, which phenotypically are defective in autophagic protein degradation and in the maintenance of viability under nutrient starvation conditions (Takeshige et al., 1992; Tsukada and Ohsumi, 1993; Thumm et al., 1994; Kametaka et al., 1996; Straub et al., 1997). This process in yeast appears to be equivalent, but with evident peculiarities, to the macroautophagic pathway in higher cells. Thus, nitrogen starvation induces the formation of autophagosomes, composed of a portion of cytosol surrounded by a double membrane of unknown origin. Subsequently, the outer membrane of the autophagosome fuses with the vacuolar membrane and a membrane limited spherical structure, containing cytosolic components and named an autophagic body, enters into the vacuolar lumen, where it rapidly disintegrates in a process which depends on PrB and its content is degraded by the vacuolar proteases (Takeshige et al., 1992; Baba et al., 1994). As it is the case with mammalian autophagosomes, few intramembrane particles (probably corresponding to integral membrane proteins or complexes thereof) were detected by freeze-fracture on the outer membrane of the autophagosome and even less in the membrane of the autophagic body (Baba et al., 1995). Since the entry of autophagic bodies into the vacuole is inhibited by cycloheximide (Takeshige et al., 1992), some protein of short half-life on the outer membrane of the autophagosome could recognize or could be recognized by the vacuole allowing fusion. The transport of the cytoplasmic proteins into the vacuole is largely a nonselective bulk process (Egner et al., 1993). However, when cells are shifted from methylotrophic to non-methylotrophic (ethanol) conditions, peroxisomes, but not mitochondria, are selectively degraded (van der Klei et al., 1991; Tuttle et al., 1993; Titorenko et al., 1995; Tuttle and Dunn, 1995). Therefore, in yeast as in mammalian cells, macroautophagy can also be selective under certain conditions.

E. Microautophagy

This is a still poorly understood process which was first postulated on the basis that autophagy is similar to endocytosis, which comprises both macrophagocytosis and micropinocytosis. Hence, the readily recognized macroautophagic process may be the "more obvious witness of a process occurring at a much lower scale but with less easily identifiable manifestations" (de Duve and Wattiaux, 1966). In this process, called microautophagy, small bits of cytoplasm, down to the level of macromolecules (protein, glycogen), are internalized into the lysosomal matrix by various modifications in the lysosomal membrane. Usually, dense bodies containing sharply demarcated electron-lucent areas and with a high surface to volume ratio appear to be implicated in microautophagy. However, , the responsible organelle has sometimes been called a different name: multivesicular body, type A secondary lysosome, or nematolysosome (Marzella and Glaumann, 1987; Mortimore et al., 1989; Sakai et al., 1990). It appears that different forms of microautophagic sequestration (Figure 6) operate in the cells by: i) infolding of the lysosomal membrane, via an endocytosis-like

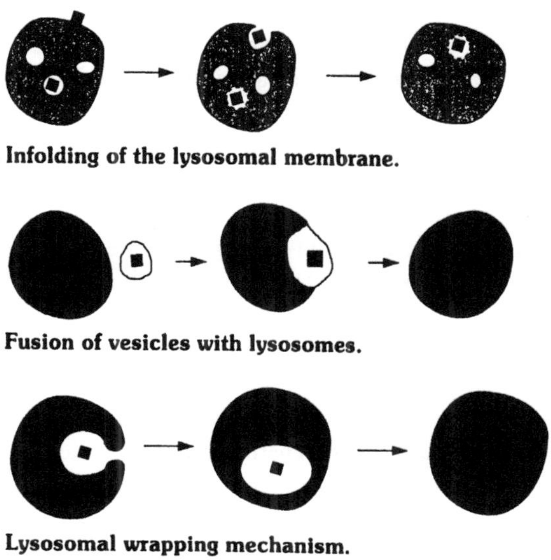

Figure 6. Microautophagic pathways of proteolysis. In path **1**, multivesicular lysosomes "micropinocytose" cytosolic proteins by infolding and pinching off of their membranes to form intralysosomal vesicles which release the protein to be degraded into the lysosomal lumen. In path **2**., small vesicles from the smooth endoplasmic reticulum, which have internalized cytosolic material, fuse with dense bodies and form a dense lysosome with an adjacent lucent area, whose content is degraded. In path **3**., flaplike extensions from lysosomal dense bodies sequester and degrade bits of cytoplasm. Some of the lysosomal types implicated in these pathways (in 1. and in 3.), have also been linked with the endocytic activity of the cells.

process, to form intralysosomal vesicles which open up or disintegrate releasing their contents; ii) fusion with lysosomal dense bodies of small single membrane vesicles (ca. 0.2 μm) from the smooth ER and which have internalized glycogen and proteins; or iii) segregation of variable portions of cytoplasm into cup-shaped or flap-like projections of dense bodies (lysosomal wrapping mechanism). In some cases (e.g., in the lysosomal wrapping mechanism) it is quite difficult to establish, on the basis of morphology alone, a sharp separation between scaled down macroautophagy and this form of microautophagy which, probably, can also internalize organelles or portions thereof. However, there are other differences (in selectivity, regulation, requirements, etc.) which may help to discriminate both processes.

Microautophagy appears to be regulated by other factors than macroautophagy, although there may be differences among the various microautophagic forms. Thus, in rat liver, microautophagy by type A secondary lysosomes remains constant during early starvation and refeeding but decreases after 1–2 days of starvation (Mortimore et al., 1989). However, the fractional volume of other forms of microautophagy may increase during nutrient starvation (de Waal et al., 1986).

In cultured cells, autophagy with serum present in the medium probably corresponds to a microautophagic process under basal conditions. In liver, microautophagy is believed to be responsible for most basal degradation of long-lived proteins in the presence of amino acids and insulin. Therefore, and to explain the different half-lives of liver proteins, microautophagy could be a selective degradation mechanism. In this regard, in experiments of *in vitro* reconstitution of the microautophagic process, methemoglobin was degraded at a five or 10 times higher rate than lysozyme or ovalbumin, respectively (Ahlberg and Glaumann, 1985). In addition, both ubiquitin and ubiquitin-conjugated proteins accumulate, by mechanisms similar to microautophagy, in lysosomes of fibroblasts treated with proteinase inhibitors (Lászlò et al., 1990), and it is well known that ubiquitinylation marks proteins for selective degradation (see Wilkinson, chapter 3; Coux et al., 1996). However, in isolated rat liver lysosomes the level of these conjugates has been shown to be comparable to that of a marker of nonspecific bulk macroautophagy (Ueno and Kominami, 1991). Therefore, it is possible that different forms of microautophagy manifest variable degrees of selectivity in the uptake of proteins.

Microautophagy has also been found in yeast, where it is possible to isolate mutants which only affect microautophagy and not macroautophagy (Tuttle and Dunn, 1995). In *Pichia pastoris*, when the cells are shifted from methanolic media to glucose, peroxisomal and cytosolic enzymes, which are present at high levels to metabolize that carbon source, are no longer needed and are selectively sequestered by microautophagy (Tuttle and Dunn, 1995). The mode of sequestration is different from macroautophagy, involving fingerlike protrusions of the vacuole and requires protein synthesis, probably because the recognition mechanism involves a short-lived membrane protein. In that organism, microautophagy also requires the α-subunit, but not the β-subunit, of phosphofructokinase, a key regulatory enzyme in the glycolytic pathway (Yuan et al., 1997).

F. Selective Transport of Cytosolic Proteins Through the Lysosomal Membrane

The selective pathway of lysosomal proteolysis was first described in serum-deprived confluent cultured fibroblasts, where it affected proteins containing peptide sequences biochemically related to Lys-Phe-Glu-Arg-Gln (KFERQ). In addition, rat proteins containing KFERQ-like sequences are preferentially depleted upon nutritional deprivation in liver, kidney, and heart but not in skeletal muscle, brain, or testis (see e.g., Dice et al., 1994; and Terlecky, 1994 for references and details of the initial studies).

Substrates of the selective pathway included RNase A and RNase S-peptide (amino acids 1–20 of RNase A), which contain the KFERQ sequence in amino acids 7–11. Antibodies against the KFERQ sequence precipitated 25–30% of the cytosolic proteins from fibroblasts and reacted with cytosolic proteins from other tissues. However, since the KFERQ motif is only found in RNase A, these antibodies as well as this lysosomal pathway may also recognize chemically related sequences (Dice, 1990). In this regard, it remains to be proven a role of such a sequence in the lysosomal uptake of glyceraldehyde 3-phosphate dehydrogenase (GAPDH), a third model protein which has been more recently used to investigate this pathway (Aniento et al., 1993; Cuervo et al., 1994, 1995a; Cuervo and Dice, 1996). Other proteins which are recognized by this selective pathway include: aspartate aminotransferase, pyruvate kinase and hemoglobin, aldolase, phosphoglyceromutase and phosphoglycerokinase (see Cuervo and Dice, 1998), the proteasome (Cuervo et al., 1995b), and c-Fos (Aniento et al., 1996b). Also it has been shown that, in general, there is a preference to import into lysosomes and degrade acidic cytosolic proteins (Aniento et al. 1997). On the contrary, RNase S-protein (amino acids 21–124 of RNase A), dihydrofolate reductase, lysozyme, ovalbumin, insulin A-chain (Terlecky et al., 1992; Cuervo et al., 1994; Cuervo adn Dice, 1998), and v-Fos (Aniento et al., 1996b) do not appear to be substrates of the pathway.

The selective pathway of lysosomal proteolysis has been investigated by a combination of *in vivo* and *in vitro* experiments. It resembles, in some respects, the import of proteins synthesized on cytosolic ribosomes into other organelles such as mitochondria. Thus, uptake of cytosolic proteins by lysosomes is stimulated by ATP and a heat-shock cognate protein, hsc73 (Chiang et al., 1989; Terlecky et al., 1992), and requires: i) recognition of peptide sequences that target proteins to the lysosomal membrane (Dice, 1990); ii) protein-containing components of the lysosomal membrane (Terlecky and Dice, 1993; Cuervo et al., 1994; Cuervo and Dice, 1996) which interact with the proteins to be degraded; and iii) hsc73 localized in the lysosomal lumen (Terlecky and Dice, 1993; Cuervo et al., 1995a), which effectively drives the protein into the lysosomal matrix (Hayes and Dice, 1996; Cuervo et al., 1997; Cuervo and Dice, 1998; Agarraberes et al., 1997). Figure 7 shows a hypothetical model for this pathway with RNase A as substrate.

A specific targeting sequence of RNase A interacts in the cytosol with hsc73 (Figure 7, #1). Although it has been recently proposed that, on a structural basis,

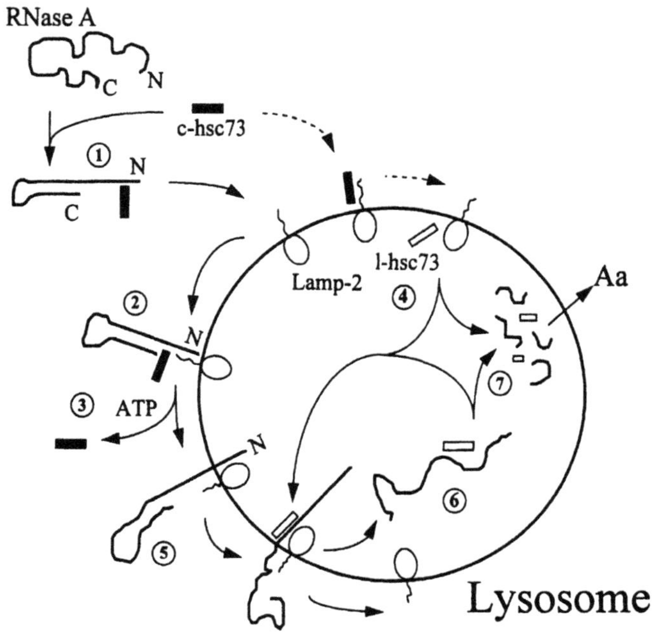

Figure 7. A plausible model for the selective transport of RNase A through the lysosomal membrane. The process has three main steps: recognition and binding of the protein to the lysosomal membrane (#1–3), transport into the lysosomal matrix (#4–6), and degradation by lysosomal proteases (#7). See detailed description in the text. l-hsc73 and c-hsc73 merely indicate the lysosomal or cytosolic localization, respectively, of the heat shock cognate protein of 73 kDa.

KFERQ is unlikely to be the true targeting signal for hsc73 recognition (Gorinsky et al., 1996), it is still possible that conformational changes in the proteins expose these motifs. Alternatively, and since the transport of proteins to lysosomes also occurs without hsc73, albeit with lower efficiencies, additional motifs and factors could be operating.

The complex hsc73-substrate protein, or the protein alone, interacts with a protein in the lysosomal membrane, most probably a receptor (Figure 7, #2). Cuervo and Dice (1996) have found in rat liver lysosomes that the lysosomal membrane glycoprotein of 96 kDa (LGP96, also called lgp110, lamp-2, or LIMP III) binds selectively to GAPDH and RNase A. This protein, as other lysosomal membrane proteins, is highly glycosylated on its lysosomal side and has a very short cytosolic tail of 12 amino acids (GLKRHHTGYEQF), which should include the sequence which recognizes and binds both substrate proteins. Although overexpression experiments have shown that lamp-2 provides a rate-limiting step for the selective pathway of lysosomal proteolysis, additional lysosomal membrane proteins are probably necessary for the transport of cytosolic proteins to the lysosomal matrix.

After binding, and before entering the protein into the lysosome, hsc73 is probably released from the complex by an ATP-dependent process (Figure 7, #3). At the same time, the N-terminal sequence of RNase A is rapidly imported, leaving outside, for a short time, approximately 2 kDa of the C-terminal portion of the protein (Cuervo et al., 1994). This intermediate is indicative of some additional step during import and it is known that organellar hsp70s are involved in driving import of cytosolic proteins into the ER, mitochondria and chloroplasts. In lysosomes, hsp70 accumulates within lysosome-related multivesicular, tubulovesicular, and dense bodies from Epstein-Barr virus-transformed lymphoblastoid cells (Mayer et al., 1994) and hsc73 was also found within lysosomes from cultured cells (Terlecky and Dice, 1993) and rat liver (Cuervo et al., 1995a). This hsc73 probably enters lysosomes (Figure 7, #4) by similar steps to those described for RNase A and GAPDH. Alternatively, but less probably, hsc73 may escort the substrate protein during its transport through the lysosomal membrane. At any rate, two different sets of observations suggest that lysosomal hsc73 is of paramount importance in the import of substrates into the lysosome. Thus, the differences found in the ability among various lysosomal subpopulations to selectively take up proteins (Terlecky and Dice, 1993; Cuervo et al., 1995a) can be explained, at least in part, by the amount of intralysosomal hsc73 present in the isolated fractions (Cuervo et al., 1997a). In addition, when intralysosomal hsc73 in cultured fibroblasts was neutralized with a specific immunoprecipitating antibody taken up via endocytosis, the enhanced lysosomal degradation produced by serum deprivation was specifically inhibited without affecting other lysosomal and non-lysosomal pathways (Agarraberes, et al., 1997). By analogy with other protein import machineries, intralysosomal hsc73 will probably act as a "translocation motor," interacting with some part of the sequence of RNase A as it emerges from the lysosomal membrane and providing the thermodynamic inward driving force for pulling the import intermediate into the lysosome, thus ensuring the irreversibility of the translocation (Figure 7, #5). In that case, this process will probably also require an interaction, regulated by ATP or by other metabolites, of the lysosomal hsc73 with some component (s) of the inner face of the lysosomal membrane. Intralysosomal hsc73 may also help in delivering substrate proteins to the lysosomal cathepsins (Figure 7, #6), promoting their rapid degradation (Figure 7, #7). Obviously, many details of this model remain to be tested and investigated, including further characterization of the sequences recognized and the identification of other components of the transport machinery.

Thus far, the identification and characterization of this pathway has relied exclusively on biochemical approaches. Molecular genetics, which could be easily carried out in yeast, provided the pathway is recognized here (Hoffman and Chiang, 1996), should complement these studies and may help to identify mutations in the translocation machinery. Using yeast protease mutants, Chiang and Schekman (1991) found that during catabolic inactivation of fructose 1,6-bisphosphatase (FBPase), the enzyme is transported and degraded in the vacuole by a process which requires a membrane protein. Entry of the substrate into the vacuole could occur either by two different forms of microautophagy or by selective transport through the vacuolar membrane mediated by heat shock proteins (Chiang et al., 1996). However, others have implicated the proteasome in the catabolic proteolysis of FBPase (Schork et al., 1994a) and the issue still remains controversial (for different points of view see Chiang and Schekman, 1994; Gamo et al., 1994; Schork et al., 1994b, 1995; Tuttle and Dunn, 1995; Chiang et al., 1996).

IV. CONCLUSIONS AND SOME QUESTIONS STILL TO BE RESOLVED

Besides non-lysosomal pathways, mammalian cells appear to have multiple lysosomal pathways for the degradation of intracellular proteins. The competitive advantage that a cell derives from having so many pathways is probably that such a variety allows all kind of proteins (cytosolic, organellar, secretory, etc.) to be degraded in quite controlled but versatile ways, at various speeds and in response to different stimuli. There are still, however, a number of questions which should be clarified before a better understanding of the lysosomal participation in protein degradation under physiological and pathological conditions can be achieved. For example, it could also be possible that, apart from the lysosomal mechanisms here described, additional lysosomal pathways of protein degradation remain to be discovered. In addition, we still know little regarding selectivity in lysosomal degradation, except that it occurs, at least under certain conditions. In this regard, we are still in need of more information on the extent and mechanisms of this selectivity. Also, except for some aspects of macroautophagy, the factors which regulate the various lysosomal mechanisms and their different steps are largely unknown. It should also be investigated to what extent the direct conversion of ER cisternae into lysosomes, as well as other lysosomal (e.g., Buscá et al., 1996) and especially proteasomal (e.g., Sommer and Wolf, 1997) mechanisms, contribute to the non-lysosomal, energy-dependent pathway which operates in or near the ER-Golgi complex (Bonifacino, 1996). Furthermore, the additional lysosomal role of ubiquitinylation, accompanying its well-characterized function in the non-lysosomal ATP-dependent degradation of proteins by the proteasome, needs to be clarified and it would be interesting to know what distinguishes a protein which follows a proteasomal pathway from another protein that follows one of the lysosomal pathways of intracellular degradation (e.g., the type of ubiquitinylation). Altogether it will eventually help to identify the signals which regulate and coordinate protein turnover (i.e., the process of protein synthesis and the various lysosomal and non-lysosomal pathways of protein degradation). Particularly important in the study of these and other issues are the needs for better methods to investigate the molecular details in the different pathways of sequestration and degradation of proteins within lysosomes, including: i) characterization of good inhibitors of these various steps, like 3-methyladenine and asparagine (e.g., Høyvik et al., 1991; Fengsrud et al.,

1995; Blommaart et al., 1997b) which have already proven their efficacy to investigate macroautophagy; ii) establishment of specific protocols to rapidly isolate the lysosomal populations involved in the different pathways of protein degradation within lysosomes; iii) development of accurate procedures to determine degradation rates of specific proteins and to differentially measure the activities of the various lysosomal pathways; and iv) complementation of these studies in yeast, using molecular genetics to identify mutations in the lysosomal machinery. Predictably all these procedures may facilitate exploring, analyzing, and unraveling the complex lysosomal mechanisms in intracellular protein degradation.

ACKNOWLEDGMENTS

The work done was supported in part by Dirección General Científica y Técnica del Ministerio de Educación y Ciencia (PB94-1281) and Fondo de Investigación Sanitaria de la Seguridad Social (96/2063). We are especially grateful to Amparo Noguera for help in the preparation of the figures.

REFERENCES

Agarraberes, F.A., Terlecky, S.R., & Dice, J.F. (1997). An intralysosomal hsp70 is required for a selective pathway of lysosomal protein degradation. J. Cell Biol. 137, 825-834.

Ahlberg, J., & Glaumann, H. (1985). Uptake-microautophagy-and degradation of exogenous proteins by isolated rat liver lysosomes. Exp. Mol. Pathol. 42, 78-88.

Aniento, F., Roche, E., Cuervo, A.M., & Knecht, E. (1993). Uptake and degradation of glyceraldehyde-3-phosphate dehydrogenase by rat liver lysosomes. J. Biol. Chem. 268, 10463-10470.

Aniento, F., Gu, F., Parton, R.G., & Gruenberg, J. (1996a). An endosomal ßCOP is involved in the pH-dependent formation of transport vesicles destined for late endosomes. J. Cell Biol. 133, 29-41.

Aniento, F., Papavassiliou, A.G., Knecht, E., & Roche, E. (1996b). Selective uptake and degradation of c-Fos and v-Fos by rat liver lysosomes. FEBS Lett. 390, 47-52.

Aniento, F., Roche, E., & Knecht, E. (1997). Acidic cytosolic proteins are preferentially imported into rat liver lysosomes. Electrophoresis 18, 2638-2644.

Aplin, A., Jasionowski, T., Tuttle, D.L., Lenk, S.E., & Dunn Jr., W.A. (1992). Cytoskeletal elements are required for the formation and maturation of autophagic vacuoles. J. Cell Physiol. 152, 458-466.

Araki, N., Ohno, J., Lee, T., Takashima, Y., & Ogawa, K. (1993). Nematolysosomes (elongate lysosomes) in rat hepatocytes: their distribution, microtubule dependence, and role in endocytic transport pathway. Exp. Cell Res. 204, 181-191.

Armstrong, J. (1991). Breaking the topological dogma. Curr. Biol. 6, 345-346.

Authier, F., Posner, B.I., & Bergeron, J.J.M. (1996). Endosomal proteolysis of internalized proteins. FEBS Lett. 389, 55-60.

Baba, M., Takeshige, K., Baba, N., & Ohsumi, Y. (1994). Ultrastructural analysis of the autophagic process in yeast: detection of autophagosomes and their characterization. J. Cell Biol. 124, 903-913.

Baba, M., Osumi, M., & Ohsumi, Y. (1995). Analysis of the membrane structures involved in autophagy in yeast by freeze-replica method. Cell Struct. Funct. 20, 465-471.

Baba, M., Osumi, M., Scott, S.V., Klinosky, D.J., & Ohsumi, Y. (1997). Two distinct pathways for targeting proteins from the cytoplasm to the vacuole/lysosome. J. Cell Biol. 139, 1687-1695.

Barrett, A.J. (1992). Cellular proteolysis. An overview. Ann. N.Y. Acad. Sci. 674, 1-15.

Bassetti, M., Pasini, D.L., & Rosa, P. (1995). Degradation of gonadotropin beta-subunits retained in the endoplasmic reticulum of the gonadotropes of castrated rats. Endocrinology 136, 1168-1176.

Berg, T.O., Strømhaug, P.E., Berg, T., & Seglen, P.O. (1994). Separation of lysosomes and autophagosomes by means of glycyl-phenylalanine-naphthylamide, a lysosome-disrupting cathepsin-C substrate. Eur. J. Biochem. 221, 595-602.

Berg, T., Gjøen, T., & Bakke, O. (1995). Physiological functions of endosomal proteolysis. Biochem. J. 307, 313-326.

Bergamini, E., Bombara, M., Del Roso, A., Gori, Z., Masiello, P., Masini, M., Pollera, M., & Vittorini, S. (1995). The regulation of liver protein degradation by amino acids *in vivo*. Effects of glutamine and leucine. Arch. Physiol. Biochem. 103, 512-515.

Biederbick, A., Kern, H.F., & Elsässer, H.P. (1995). Monodansylcadaverine (MDC) is a specific *in vivo* marker for autophagic vacuoles. Eur. J. Cell Biol. 66, 3-14.

Blankson, H., Holen, I., & Seglen, P.O. (1995). Disruption of the cytokeratin cytoskeleton and inhibition of hepatocytic autophagy by okadaic acid. Exp. Cell Res. 218, 522-530.

Blommaart, P.J.E., Zonneveld, D., Meijer, A.J., & Lamers, W.H. (1993). Effects of intracellular amino acid concentrations, cyclic AMP, and dexamethasone on lysosomal proteolysis in primary cultures of perinatal rat hepatocytes. J. Biol. Chem. 268, 1610-1617.

Blommaart, E.F.C., Krause, U., Schellens, J.P.M., Vreeling-Sindelárová, H., & Meijer, A.J. (1997b). The phosphatidylinositol 3-kinase inhibitors wortmannin and LY294002 inhibit autophagy in isolated rat hepatocytes. Eur. J. Biochem. 243, 240-246.

Blommaart, E.F.C., Luiken, J.F.P., Blommaart, P.J.E., van Woerkom, G.M., & Meijer, A.J. (1995). Phosphorylation of ribosomal protein S6 is inhibitory for autophagy in isolated rat hepatocytes. J. Biol. Chem. 270, 2320-2326.

Blommaart, E.F.C., Luiken, J.F.P., & Meijer, A.J. (1997a). Autophagic proteolysis: Control and specificity. Histochem. J. 29, 365-385.

Bohley, P. (1995). The fates of proteins in cells. The birth of a protein. Naturwissenschaften 82, 544-550.

Bohley, P., & Seglen, P.O. (1992). Proteases and proteolysis in the lysosome. Experientia 48, 151-157.

Bond, J.S., & Barrett, A.J., eds. (1993). Proteolysis and Protein Turnover. Portland Press, London.

Bonifacino, J.S. (1996). Reversal of fortune for nascent proteins. Nature 384, 405-406.

Buscá, R., Martínez, M., Vilella, E., Pognonec, P., Deeb, S., Auwerx, J., Reina, M., & Vilaró, S. (1996). The mutation $Gly^{142} \rightarrow$ Glu in human lipoprotein lipase produces a missorted protein that is diverted to lysosomes. J. Biol. Chem. 271, 2139-2146.

Busch, G.L., Lang, H.J., & Lang, F. (1996). Studies on the mechanism of swelling induced lysosomal alkalinization in vascular smooth muscle cells. Pflügers Arch. (Eur. J. Physiol.) 431, 690-696.

Carillo, S., Pariat, M., Jarielencontre, I., Steff, A.M., & Piechaczyk, M. (1995). Protein degradation: a major intracellular function. 1. Mechanisms of protein degradation. Médecine/Sciences 11, 723-734.

Caro, L.H.P., Plomp, P.J.A.M., Leverve, X.M., & Meijer, A.J. (1989). A combination of intracellular leucine with either glutamate or aspartate inhibits autophagic proteolysis in isolated rat hepatocytes. Eur. J. Biochem. 181, 717-720.

Chervonsky, A., & Sant, A.J. (1995). In the absence of major histocompatibility complex class II molecules, invariant chain is translocated to late endocytic compartments by autophagy. Eur. J. Immunol. 25, 911-918.

Chiang, H.-L., & Schekman, R. (1991). Regulated import and degradation of a cytosolic protein in the yeast vacuole. Nature (London) 350, 313-318.

Chiang, H.-L., & Schekman, R. (1994). Site of catabolite inactivation. Reply. Nature (London) 350, 284.

Chiang, H.-L., Terlecky, S.R., Plant, C.P., & Dice, J.F. (1989). A role for a 70-kilodalton heat shock protein in lysosomal degradation of intracellular proteins. Science (Washington, DC) 246, 382-385.

Chiang, H.-L., Schekman, R., & Hamamoto, S. (1996). Selective uptake of cytosolic, peroxisomal, and plasma membrane proteins into the yeast lysosome for degradation. J. Biol. Chem. 271, 9934-9941.

Ciechanover, A.J., & Schwartz, A.L., eds. (1994). Cellular Proteolytic Systems. Wiley-Liss, New York.

Coux, O., Tanaka, K., & Goldberg, A.L. (1996). Structure and functions of the 20S and 26S proteasomes. Annu. Rev. Biochem. 65, 801-847.

Cuervo, A.M., & Dice, J.F. (1996). A receptor for the selective uptake and degradation of proteins by lysosomes. Science (Washington, DC) 273, 501-503.

Cuervo, A.M., & Dice, J.F. (1998). Lysosomes, a meeting point of proteins, chaperones, and proteases. J. Mol. Med. 76, 6-12.

Cuervo, A.M., Terlecky, S.R., Dice, J.F., & Knecht, E. (1994). Selective binding and uptake of ribonuclease A and glyceraldehyde-3-phosphate dehydrogenase by isolated rat liver lysomes. J. Biol. Chem. 269, 26374-26380.

Cuervo, A.M., Knecht, E., Terlecky, S.R., & Dice, J.F. (1995a). Activation of a selective pathway of lysosomal proteolysis in rat liver by prolonged starvation. Am. J. Physiol. 269, C1200-C1208.

Cuervo, A.M., Palmer, A., Rivett, A.J., & Knecht, E. (1995b). Degradation of proteasomes by lysosomes in rat liver. Eur. J. Biochem. 227, 792-800.

Cuervo, A.M., Dice, J.F., & Knecht, E. (1997). A population of rat liver lysosomes responsible for the selective uptake and degradation of cytosolic proteins. J. Biol. Chem. 272, 5606-5615.

de Duve, C., & Wattiaux, R. (1966). Functions of lysosomes. Annu. Rev. Physiol. 28, 435-492.

de Waal, E.J., Vreeling-Sindelárová, H., Schellens, J.P.M., Houtkooper, J.M., & James, J. (1986). Quantitative changes in the lysosomal vacuolar system of the rat hepatocytes during short-term starvation. A morphometric analysis with special reference to macro-and microautophagy. Cell Tissue Res. 243, 641-648.

Dice, J.F. (1990). Peptide sequences that target cytosolic proteins for lysosomal proteolysis. Trends Biochem. Sci. 15, 305-309.

Dice, J.F., Agarraberes, F., Kirven-Brooks, M., Terlecky, L.J., & Terlecky, S.R. (1994). In: The Biology of Heat Shock Proteins and Molecular Chaperones (Morimoto, R.I., Tissieres, A., & Georgopoulos, C., eds.), pp. 137-151. Cold Spring Harbor Laboratory Press, New York.

Doherty, F.J., & Mayer, R.J. (1992). Intracellular Protein Degradation. Oxford University Press, Oxford.

Dunn Jr., W.A. (1990a). Studies on the mechanisms of autophagy: Formation of the autophagic vacuole. J. Cell Biol. 110, 1923-1933.

Dunn Jr., W.A. (1990b). Studies on the mechanisms of autophagy: maturation of the autophagic vacuole. J. Cell Biol. 110, 1935-1945.

Dunn Jr., W.A. (1994). Autophagy and related mechanisms of lysosome-mediated protein degradation. Trends Cell Biol. 4, 139-143.

Egner, R., & Kuchler, K. (1996). The yeast multidrug transporter Pdr5 of the plasma membrane is ubiquitinated prior to endocytosis and degradation in the vacuole. FEBS Lett. 378, 177-181.

Egner, R., Thumm, M., Straub, M., Simeon, A., Schüller, H.-J., & Wolf, D.H. (1993). Tracing intracellular proteolytic pathways: Proteolysis of fatty acid synthase and other cytoplasmic proteins in the yeast *Saccharomyces cerevisiae*. J. Biol. Chem. 268, 27269-27276.

Fengsrud, M., Roos, N., Berg, T., Liou, W., Slot, J.W., & Seglen, P.O. (1995). Ultrastructural and immunocytochemical characterization of autophagic vacuoles in isolated hepatocytes: effects of vinblastine and asparagine on vacuole distributions. Exp. Cell Res. 221, 504-519.

Finley, E.M., & Kornfeld, S. (1994). Subcellular localization and targeting of cathepsin E. J. Biol. Chem. 269, 31259-31266.

Fosse, M., Berg, T.O., O'Reilly, D.S.J., & Seglen, P.O (1995). Vanadate inhibition of hepatocytic autophagy. Calcium-modulated and osmolality-modulated antagonism by asparagine. Eur. J. Biochem. 230, 17-24.

Furuno, K., Ishikawa, T., Akasaki, K., Lee, S., Nishimura, Y., Tsuji, H., Himeno, M., & Kato, K. (1990). Immunocytochemical study of the surrounding envelope of autophagic vacuoles in cultured rat hepatocytes. Exp. Cell. Res. 189, 261-268.

Gamo, F.-J., Navas, M.A., Blázquez, M.A., Gancedo, C., & Gancedo, J.M. (1994). Catabolite inactivation of heterologous fructose-1,6-bisphosphatases and fructose-1,6-bisphosphatase-β-galactosidase fusion proteins in *Saccharomyces cerevisiae*. Eur. J. Biochem. 222, 879-884.

Ghidoni, R., Houri, J.-J., Giuliani, A., Ogier-Denis, E., Parolari, E., Botti, S., Bauvy, Ch., & Codogno, P. (1996). The metabolism of sphingo(glyco)lipids is correlated with the differentiation-dependent autophagic pathway in HT-29 cells. Eur. J. Biochem. 237, 454-459.

Glaumann, H., Ahlberg, J., Hultenby, K., Jansson, H., Mengarelli-Wildholm, S., & Motafeki, A.M. (1989). Isolation and characterization of crinosomes - a subclass of secondary lysosomes. Exp. Mol. Pathol. 50, 167-182.

Goldberg, A.L., & Rock, K.L. (1992). Proteolysis, proteasomes and antigen presentation. Nature (London) 357, 375-379.

Goldstein, J.L., Hobbs, H.H., & Brown, M.S. (1995). In: The Metabolic and Molecular Bases of Inherited Disease. (Scriver, C.R., Beaudet, A.L., Sly, W.S., & Valle, D., eds.), Vol. 2, pp. 1981-2030. McGraw-Hill, New York.

Gordon, P.B., Høyvik, H., & Seglen, P.O. (1992). Prelysosomal and lysosomal connections between autophagy and endocytosis. Biochem. J. 283, 361-369.

Gordon, P.B., Holen, I., Fosse, M., Rtnes, J.S., & Seglen, P.O. (1993). Dependence of hepatocytic autophagy on intracellularly sequestered calcium. J. Biol. Chem. 268, 26107-26112.

Gordon, P.B., Holen, I., & Seglen, P.O. (1995). Protection by naringin and some other flavonoids of hepatocytic autophagy and endocytosis against inhibition by okadaic acid. J. Biol. Chem. 270, 5830-5838.

Gorinsky, B., Laskowski, R.A., Lee, D.A., & Bomford, A. (1996). Conformational analysis of pentapeptide sequences matching a proposed recognition motif for lysosomal degradation. Biochim. Biophys. Acta 1293, 243-253.

Griffiths, G. (1996). On vesicles and membrane compartments. Protoplasma 195, 1-4.

Gropper, R., Brandt, R.A., Mayer, A.M., Bearer, C., Schwartz, A.L., & Ciechanover, A. (1991). The ubiquitin-activating enzyme, E1, is required for stress-induced lysosomal degradation of cellular proteins. J. Biol. Chem. 266, 3602-3610.

Harada, M., Sakisaka, S., Yoshitake, M., Kin, M., Ohishi, M., Shakado, S., Mimura, Y., Noguchi, K., Sata, M., & Tanikawa, K. (1996). Bafilomycin A1, a specific inhibitor of vacuolar-type H^+-ATPases, inhibits the receptor-mediated endocytosis of asialoglycoproteins in isolated rat hepatocytes. J. Hepatol. 24, 594-603.

Harding, C.V. (1994). In: Cellular Proteolytic Systems (Ciechanover, A.J., & Schwartz, A.L., eds.), pp. 163-182. Wiley-Liss, New York.

Harding, T.M., Hefner-Gravink, A., Thumm, M., & Klionsky, D.J. (1996). Genetic and phenotypic overlap between autophagy and the cytoplasm to vacuole protein targeting pathway. J. Biol. Chem. 30, 17621-17624.

Hasilik, A. (1992). The early and late processing of lysosomal enzymes: proteolysis and compartmentation. Experientia 48, 130-151.

Häussinger, D., Lang, F., & Gerok, W. (1994). Regulation of cell function by the cellular hydration state. Am. J. Physiol. 267, E343-E355.

Hay, J.C., & Scheller, R.H. (1997). SNAREs and NSF in targeted membrane fusion. Curr. Opin. Cell Biol. 9, 505-512.

Hayes, S.A., & Dice, J.F. (1996). Roles of molecular chaperones in protein degradation. J. Cell Biol. 132, 255-258.

Heikoop, J.C., van den Berg, M., Strijland, A., Weijers, P.J., Just, W.W., Meijer, A.J., & Tager, J.M. (1992). Turnover of peroxisomal vesicles by autophagic proteolysis in cultured fibroblasts from Zellweger patients. Eur. J. Cell Biol. 57, 165-171.

Hershko, A., & Ciechanover, A. (1992). The ubiquitin system for protein degradation. Annu. Rev. Biochem. 61, 761-807.

Hicke, L. (1997). Ubiquitin-dependent interanlization and downregulation of plasma membrane proteins. FASEB J. 11, 1215-1226.

Hicke, L., & Riezman, H. (1996). Ubiquitination of a yeast plasma membrane receptor signals its ligand-stimulated endocytosis. Cell 84, 277-287.

Hochstrasser, M. (1996). Protein degradation or regulation. Ub the judge. Cell 84, 813-815.

Hoffman, M., & Chiang, H.L. (1996). Isolation of degradation-deficient mutants defective in the targeting of fructose-1,6-bisphosphatase into the vacuole for degradation in *Saccharomyces cerevisiae*. Genetics 143, 1555-1556.

Holen, I., Gordon, P.B., & Seglen, P.O. (1992). Protein kinase-dependent effects of okadaic acid on hepatocytic autophagy and cytoskeletal integrity. Biochem. J. 284, 633-636.

Holen, I., Gordon, P.B., & Seglen, P.O. (1993). Inhibition of hepatocytic autophagy by okadaic acid and other protein phosphatase inhibitors. Eur. J. Biochem. 215, 113-122.

Holen, I., Strømhaug, P.E., Gordon, P.B., Fengsrud, M., Berg, T.O., & Seglen, P.O. (1995). Inhibition of autophagy and multiple steps in asialoglycoprotein endocytosis by inhibitors of tyrosine protein kinases (tyrphostins). J. Biol. Chem. 270, 12823-12831.

Holen, I., Gordon, P.B., Strømhaug, P.E., & Seglen, P.O. (1996). Role of cAMP in the regulation of hepatocytic autophagy. Eur. J. Biochem. 236, 163-170.

Holtzman, E. (1989). Lysosomes. Plenum Press, New York.

Horazdovsky, B.F., DeWald, D.B., & Emr, S.D. (1995). Protein transport to the yeast vacuole. Curr. Opin. Cell Biol. 7, 544-551.

Houri, J.-J., Ogier-Denis, E., Trugnan, G., & Codogno, P. (1993). Autophagic degradation of N-linked glycoproteins is downregulated in differentiated human colon adenocarcinoma cells. Biochem. Biophys. Res. Commun. 197, 805-811.

Houri, J.-J., Ogier-Denis, E., de Stefanis, D., Bauvy, Ch., Baccino, F.M., Isidoro, C., & Codogno, P. (1995). Differentiation-dependent autophagy controls the fate of newly synthesized N-linked glycoproteins in the colon adenocarcinoma HT-29 cell line. Biochem. J. 309, 521-527.

Høyvik, H., Gordon, P.B., Berg, T.O., Strømhaug, P.E., & Seglen, P.O. (1991). Inhibition of autophagic-lysosomal delivery and autophagic lactolysis by asparagine. J. Cell Biol. 113, 1305-1312.

Hunziker, W., & Geuze, H.J. (1996). Intracellular trafficking of lysosomal membrane proteins. Bioessays 18, 379-389.

Inubushi, T., Shikiji, M., Endo, K., Kakegawa, H., Kishino, Y., & Katunuma, N. (1996). Hormonal and dietary regulation of lysosomal cysteine proteinases. Biol. Chem. 377, 7-8.

Kadowaki, M., Pösö, A.R., & Mortimore, G.E. (1992). Parallel control of hepatic proteolysis by phenylalanine and phenylpyruvate through independent inhibitory sites at the plasma membrane. J. Biol. Chem. 267, 22060-22065.

Kadowaki, M., Venerando, R., Miotto, G., & Mortimore, G.E. (1994). *De novo* autophagic vacuole formation in hepatocytes permeabilized by *Staphylococcus aureus* α-toxin: Inhibition by nonhydrolyzable GTP analogs. J. Biol. Chem. 269, 3703-3710.

Kadowaki, M., Venerando, R., Miotto, G., & Mortimore, G.E. (1996). In: Intracellular Protein Catabolism (Suzuki, K., & Bond, J., eds.), pp. 113-119. Plenum Press, New York.

Kametaka, S., Matsuura, A., Wada, Y., & Ohsumi, Y. (1996). Structural and functional analyses of APG5, a gene involved in autophagy in yeast. Gene 178, 139-143.

Klionsky, D.J., Cueva, R., & Yaver, D.S. (1992). Aminopeptidase I of *Saccharomyces cerevisiae* is localized to the vacuole independent of the secretory pathway. J. Cell Biol. 119, 287-299.

Knop, M., Schiffer, H.H., Rupp, S., & Wolf, D.H. (1993). Vacuolar/lysosomal proteolysis: proteases, substrates, mechanisms. Curr. Opin. Cell Biol. 5, 990-996.

Kopitz, J., Kisen G.Ø., Gordon, P.B., Bohley, P., & Seglen, P.O. (1990). Non-selective autophagy of cytosolic enzymes by isolated rat hepatocytes. J. Cell Biol. 111, 941-953.

Kornfeld, S., & Sly, W.S. (1995). In: The Metabolic and Molecular Bases of Inherited Disease (Scriver, C.R., Beaudet, A.L., Sly, W.S., & Valle, D., eds.), Vol. 2, pp. 2495-2508. McGraw-Hill, New York.

Kreis, T.E., & Pepperkok, R. (1994). Coat proteins in intracellular membrane transport. Curr. Opin. Cell Biol. 6, 533-537.

Laing, J.G., Tadros, P.N., Westphale, E.M. & Beyer, E.C. (1997). Degradation of connexin43 gap junctions involves both the proteasome and the lysosome. Exp. cell Res. 236, 482-492.

László, L., Doherty, F.J., Osborn, N.U., & Mayer, R.J. (1990). Ubiquitinated protein conjugates are specifically enriched in the lysosomal system of fibroblasts. FEBS Lett. 261, 365-368.

Lawrence, B.P., & Brown, W.J. (1992). Autophagic vacuoles rapidly fuse with pre-existing lysosomes in cultured hepatocytes. J. Cell Sci. 102, 515-526.

Lawrence, B.P., & Brown, W.J. (1993). Inhibition of protein synthesis separates autophagic sequestration from the delivery of lysosomal enzymes. J. Cell Sci. 105, 473-480.

Lee, H.K., & Marzella, L. (1994). Regulation of intracellular protein degradation with special reference to lysosomes: role in cell physiology and pathology. Int. Rev. Exp. Pathol. 35, 39-147.

Lee, D.H., & Goldberg, A.L. (1996). Selective inhibitors of the proteasome-dependent and vacuolar pathways of protein degradation in *Saccharomyces cerevisiae*. J. Biol. Chem. 271, 27280-27284.

Lenk, S.E., Fisher, D.L., & Dunn Jr., W.A. (1991). Regulation of protein secretion by crinophagy in perfused rat liver. Eur. J. Cell Biol. 56, 201-209.

Lenk, S.E., Bhat, D., Blakeney, W., & Dunn Jr., W.A. (1992a). Effects of streptozotocin-induced diabetes on rough endoplasmic reticulum and lysosomes of rat liver. Am. J. Physiol. 263, E856-E862.

Lenk, S.E., Dunn Jr., W.A., Trausch, J.S., Ciechanover, A., & Schwartz, A.L. (1992b). Ubiquitin-activating enzyme, E1, is associated with maturation of autophagic vacuoles. J. Cell Biol. 118, 301-308.

Liou, W., Geuze, H.J. Greelen, M.J.H., & Slot, J.W. (1997). The autophagic and endocytic pathways converge at the nascent autophagic vacuoles. J. Cell Biol. 136, 61-70.

Löw, P., Doherty, F.J., Fellinger, E., Sass, M., Mayer, R.J., & László, L. (1995). Related organelles of the endosome-lysosome system contain a different repertoire of ubiquitinated proteins in Sf9 insect cells. FEBS Lett. 368, 125-131.

Luiken, J.J.F.P., van den Berg, M., Heikoop, J.C., & Meijer, A.J. (1992). Autophagic degradation of peroxisomes in isolated rat hepatocytes. FEBS Lett. 304, 93-97.

Luiken, J.J.F.P., Blommaart, E.F.C., Boon, L., van Woerkom, G.M., & Meijer, A.J. (1994). Cell swelling and the control of autophagic proteolysis in hepatocytes: involvement of phosphorylation of ribosomal protein S6? Biochem. Soc. Trans. 22, 508-511.

Luiken, J.J.F.P., Aerts, J.M.F.G., & Meijer, A.J. (1996). The role of the intralysosomal pH in the control of autophagic proteolytic flux in rat hepatocytes. Eur. J. Biochem. 235, 564-573.

Lum, P.Y., & Wright, R. (1995). Degradation of HMG-CoA reductase-induced membranes in the fission yeast, *Schizosaccharomyces pombe*. J. Cell Biol. 131, 81-94.

Lupas, A., Flanagan, J.M., Tamura, T., & Baumeister, W. (1997). Self-compartmentalizing proteases. Trends Biochem. Sci. 22, 399-404.

Marzella, L., & Glaumann, H. (1987). In: Lysosomes: Their Role in Protein Breakdown (Glaumann, H., & Ballard, F.J., eds.), pp. 319-367. Academic Press, London.

Mayer, T.J., Tipler, C., László, L., Arnold, J., Lowe, J., & Landon, M. (1994). Endosome-lysosomes and neurodegeneration. Biomed. Pharmacother. 48, 282-286.

Matsuura, A., Tsukada, M., Wada, Y., & Ohsumi, Y. (1997). Apg1p, a novel protein kinase required for the autophagic process in *Saccharomyces cerevisiae*. Gene 192, 245-250.

Meijer, A.J., Gustafson, L.A., Luiken, J.J.F.P., Blommaart, P.J.E., Caro, L.H.P., van Woerkom, G.M., Spronk, C., & Boon, L. (1993). Cell swelling and the sensitivity of autophagic proteolysis to inhibition by amino acids in isolated rat hepatocytes. Eur. J. Biochem. 215, 449-454.

Mellman, I. (1996). Endocytosis and molecular sorting. Annu. Rev. Cell Dev. Biol. 12, 575-625.

Mignatti, P., & Rifkin, D.B. (1993). Biology and biochemistry of proteinases in tumor invasion. Physiol. Rev. 73, 161-195.

Miotto, G., Venerando, R., Khurana, K.K., Siliprandi, N., & Mortimore G.E. (1992). Control of hepatic proteolysis by leucine and isovaleryl-L-carnitine through a common locus. J. Biol. Chem. 267, 22066-22072.

Miotto, G., Venerando, R., Marin, O., Siliprandi, N., & Mortimore, G.E. (1994). Inhibition of macroautophagy and proteolysis in the isolated rat hepatocyte by a nontransportable derivative of the multiple antigen peptide Leu_8-Lys_4-Lys_2-Lys-βAla. J. Biol. Chem. 269, 25348-25353.

Mori, S., Heldin, C.-H., & Claesson-Welsh, L. (1992). Ligand-induced polyubiquitination of the platelet-derived growth factor β-receptor. J. Biol. Chem. 267, 6429-6434.

Mortimore, G.E., & Kadowaki, M. (1994). In: Cellular Proteolytic Systems (Ciechanover, A.J., & Schwartz, A.L., eds.), pp. 65-87. Wiley-Liss, New York.

Mortimore, G.E., Pösö, A.R., & Lardeux, B.R. (1989). Mechanism and regulation of protein degradation in liver. Diabetes Metab. Rev. 5, 49-70.

Mortimore, G.E., Wert, J.J., Miotto, G., Venerando, R., & Kadowaki, M. (1994). Leucine-specific binding of photoreactive Leu_7-Map to a high molecular weight protein on the plasma membrane of the isolated rat hepatocyte. Biochem. Biophys. Res. Commun. 203, 200-208.

Nelson, K.K., & Lemmon S.K. (1993). Suppressors of clathrin deficiency: Overexpression of ubiquitin rescues lethal strains of clathrin-deficient *Saccharomyces cerevisiae*. Mol. Cell. Biol. 13, 521-532.

Noda, T., & Farquhar, M.G. (1992). A non-autophagic pathway for diversion of ER secretory proteins to lysosomes. J. Cell Biol. 119, 85-97.

Ogier-Denis, E., Couvineau, A., Maoret, J.J., Houri, J.J., Bauvy, C., De Stefanis, D., Isidoro, C., Laburthe, M., & Codogno, P. (1995). A heterotrimeric G_{i3}-protein controls autophagic sequestration in the human colon cancer cell line HT-29. J. Biol. Chem. 270, 13-16.

Ogier-Denis, E., Houri, J.J., Bauvy, C., & Codogno, P. (1996). Guanine nucleotide exchange on heterotrimeric G_{i3} protein controls autophagic sequestration in HT-29 cells. J. Biol. Chem. 271, 28593-28600.

Ogier-Denis, E., Petiot, A., Bauvy, C., & Codogno, P. (1997). Control of the expression and activity of the G_α-interacting protein (GAIP) in human intestinal cells. J. Biol. Chem. 272, 24599-24603.

Orci, L., Brown, M.S., Goldstein, J.L., García-Segura, L.M., & Anderson, R.G.W. (1984). Increase in membrane cholesterol: a possible trigger for degradation of HMG-CoA reductase and crystalloid endoplasmic reticulum in UT-1 Cells. Cell 36, 835-845.

Owen C.A., & Campbell E.J. (1995). Neutrophil proteinases and matrix degradation. The cell biology of pericellular proteolysis. Semin. Cell Biol. 6, 367-376.

Peters, C., & von Figura, K. (1994). Biogenesis of lysosomal membranes. FEBS Lett. 346, 108-114.

Pfeifer, U. (1987). In: Lysosomes: Their Role in Protein Breakdown (Glaumann, H., & Ballard F.J., eds.), pp. 3-59. Academic Press, London.

Pisoni, R.L., & Velilla, V.Q. (1995). Evidence for an essential histidine residue located in the binding site of the cysteine-specific lysosomal transport protein. Biochim. Biophys. Acta 1236, 23-30.

Punnonen, E.-L., Autio, S., Kaija, H., & Reunanen, H. (1993). Autophagic vacuoles fuse with the prelysosomal compartment in cultured rat fibroblasts. Eur. J. Cell Biol. 61, 54-66.

Rabinovitch, M. (1995). Professional and non-professional phagocytes: an introduction. Trends Cell Biol. 5, 85-87.

Rabouille, C., Strous, G.J., Crapo, J.D., Geuze, H.J., & Slot, J.W. (1993). The differential degradation of two cytosolic proteins as a tool to monitor autophagy in hepatocytes by immunocytochemistry. J. Cell Biol. 120, 897-908.

Rajas, F., Gire, V., & Rousset, B. (1996). Involvement of a membrane-bound form of glutamate dehydrogenase in the association of lysosomes to microtubules. J. Biol. Chem. 271, 29882-29890.

Réz, G., Csák, J., Fellinger, E., László, L., Kovács, A.L., Oliva, O., & Kovács, J. (1996). Time cours of vinblastine-induced autophagocytosis and changes in the endoplasmic reticulum in murine pancreatic acinar cells: a morphometric and biochemical study. Eur. J. Cell Biol. 71, 341-350.

Rieder, S.E., & Emr, S.D. (1997). A novel RING finger protein complex essential for a late step in protein transport to the yeast vacuole. Mol. Biol. Cell 8, 2307-2327.

Robinson, J.M., Chiplonkar, J., & Luo, Z. (1996). A method for co-localization of tubular lysosomes and microtubules in macrophages : fluorescence microscopy of individual cells. J. Histochem Cytochem. 44, 1109-1114.

Roth, A.F., & Davis, N.G. (1996). Ubiquitination of the yeast a-factor receptor. J. Cell Biol. 134, 661-674.

Sahagian, G.G., & Novikoff, P.M. (1994). In: Liver. Biology and Pathobiology, 3rd edn. (Arias, J.M., ed.), pp. 275-291. Raven Press, New York.

Sai, Y., & Ohkuma, S. (1992). Small GTP-binding proteins on rat liver lysosomal membranes. Cell Struct. Funct. 17, 363-369.

Sai, Y., Arai, K., & Ohkuma, S. (1994). Cytosol treated with GTPgammaS disintegrates lysosomes *in vitro*. Biochem. Biophys. Res. Commun. 198, 869-877.

Sakai, M., Araki, N., & Ogawa, K. (1990). Regulatory factors of intracellular lysosomal activity associated with heterophagy and autophagy. Acta Histochem. Cytochem. 23, 19-30.

Sandoval, I.V., & Bakke, O. (1994). Targeting of membrane proteins to endosomes and lysosomes. Trends Cell Biol. 4, 292-297.

Schellens, J.P.M., Vreeling-Sindelárová, H., Plomp, P.J.A.M., & Meijer, A.J. (1990). Cytochemical and morphometric analysis of autophagy in energy depleted rat hepatocytes. Cell Biol. Int. Rep. 14, 805-814.

Schnell-Landstrom, A.H., Andersson, A., & Borg, L.A.H. (1991). Lysosomes and pancreatic islet function: Adaptation of β-cell lysosomes to various metabolic demands. Metabolism 40, 399-405.

Schork, S.M., Bee, G., Thumm, M., & Wolf, D.H. (1994a). Site of catabolite inactivation. Nature (London) 369, 283-284.

Schork, S.M., Bee, G., Thumm, M., & Wolf, D.H. (1994b). Catabolite inactivation of fructose-1,6-bisphosphatase in yeast is mediated by the proteasome. FEBS Lett. 349, 270-274.

Schork, S.M., Thumm, M., & Wolf, D.H. (1995). Catabolite inactivation of fructose-1,6-bisphosphatase *of Saccharomyces cerevisiae*. Degradation occurs via the ubiquitin pathway. J. Biol. Chem. 270, 26446-26450.

Schreiber, R., & Häussinger, D. (1995). Characterization of the swelling-induced alkalinization of endocytic vesicles in fluorescein isothiocyanate-dextran-loaded rat hepatocytes. Biochem. J. 309, 19-24.

Schwartz, A.L. (1995). Receptor cell biology: receptor-mediated endocytosis. Pediat. Res. 38, 835-843.

Schwartz, A.L., Brandt, R., Geuze, H., & Ciechanover, A. (1992). Stress-induced alterations in autophagic pathway: relationship to ubiquitin system. Am. J. Physiol. 262, C1031-C1038.

Scott, S.V., Baba, M., Ohsumi, Y., & Klionsky, D.J. (1997). Aminopeptidase I is targeted to the vacuole by a nonclassical vesicular mechanism. J. Cell Biol. 138, 37-44.

Scott, S.V., Hefner-Gravink, A., Morano, K.A., Noda, T., Ohsumi, Y., & Klionsky, D.J. (1996). Cytoplasm-to-vacuole targeting and autophagy employ the same machinery to deliver proteins to the yeast vacuole. Proc. Natl. Acad. Sci. USA 93, 12304-12308.

Seabright, P.J., & Smith, G.D. (1996). The characterization of endosomal insulin degradation. Biochem. J. 320, 947-956.

Seglen, P.O., & Bohley, P. (1992). Autophagy and other vacuolar protein degradation mechanisms. Experientia 48, 158-172.

Seglen, P.O., Berg, T.O., Blankson, H., Fengsrud, M., Holen, I., & Strømhaug, P.E. (1996). In: Intracellular Protein Catabolism (Suzuki, K., & Bond, J., eds.), pp. 103-111. Plenum Press, New York.

Seguí-Real, B., Martínez, M., & Sandoval, I.V. (1995). Yeast aminopeptidase I is post-translationally sorted from the cytosol to the vacuole by a mechanism mediated by its bipartite N-terminal extension. EMBO J. 14, 5476-5484.

Simeon, A., van der Klei, I.J., Veenhuis, M., & Wolf, D.H. (1992). Ubiquitin, a central component of selective cytoplasmic proteolysis, is linked to proteins residing at the locus of non-selective proteolysis, the vacuole. FEBS Lett. 301, 231-235.

Skurat, R.M., & Dice, J.F. (1994). In: Liver. Biology and Pathobiology 3rd edn. (Arias, J.M., ed.), pp. 229-238. Raven Press, New York.

Smith, R.E., & Farquhar, M.G. (1966). Lysosome function in the regulation of the secretory process in cells of the anterior pituitary gland. J. Cell Biol. 31, 319-347.

Sommer, T.., & Wolf, D.H. (1997). Endoplasmic reticulum degradation: Reverse protein flow of no return. FASEB J. 11, 1227-1233.

Straub, M., Bredschneider, M., & Thumm, M. (1997). AUT3, a serine/threonine kinase gene, is essential for autophagocytosis in *Saccharomyces cerevisiae*. J. Bacteriol. 179, 3875-3883.

Strømhaug, P.E., & Seglen, P.O. (1993). Evidence for acidity of prelysosomal autophagic/endocytic vacuoles (amphisomes). Biochem. J. 291, 115-121.

Strous, G.J., van Kerkhof, P., Govers, R., Ciechanover, A., & Schwartz, A.L. (1996). The ubiquitin conjugation system is required for ligand-induced endocytosis and degradation of the growth hormone receptor. EMBO J. 15, 3806-3812.

Suzuki, K., & Bond, J., eds. (1996). Intracellular Protein Catabolism. Plenum Press, New York.

Takeshige, K., Baba, M., Tsuboi, S., Noda, T., & Ohsumi, Y. (1992). Autophagy in yeast demonstrated with proteinase-deficient mutants and conditions for its induction. J. Cell Biol. 119, 301-311.

Terlecky, S.R. (1994). Hsp70s and lysosomal proteolysis. Experientia 50, 1021-1025.

Terlecky, S.R., & Dice J.F. (1993). Polypeptide import and degradation by isolated lysosomes. J. Biol. Chem. 268, 23490-23495.

Terlecky, S.R., Chiang H.-L., Olson, T.S., & Dice, J.F. (1992). Protein and peptide binding and stimulation of *in vitro* lysosomal proteolysis by the 73-kDa heat shock cognate protein. J. Biol. Chem. 267, 9202-9209.

Thumm, M., Egner, R., Koch, B., Schlumpberger, M., Straub, M., Veenhuis, M., & Wolf, D.H. (1994). Isolation of autophagocytosis mutants of *Saccharomyces cerevisiae*. FEBS Lett. 349, 275-280.

Titorenko, V.I., Keizer, I., Harder, W., & Veenhuis, M. (1995). Isolation and characterization of mutants impaired in the selective degradation of peroxisomes in the yeast *Hansenula polymorpha*. J. Bacteriol. 177, 357-363.

Tooze, J., Hollinshead, M., Ludwig, T., Howell, K., Hoflack, B., & Kern, H. (1990). In exocrine pancreas, the basolateral pathway converges with the autophagic pathway immediately after the early endosome. J. Cell Biol. 111, 329-345.

Trowbridge, I.S., Collawn, J.F., & Hopkins, C.R. (1993). Signal-dependent membrane protein trafficking in the endocytic pathway. Annu. Rev. Cell Biol. 9, 129-161.

Tsukada, M., & Ohsumi, Y. (1993). Isolation and characterization of autophagy-defective mutants of *Saccharomyces cerevisiae*. FEBS Lett. 333, 169-174.

Tuttle, D.L., & Dunn Jr., W.A. (1995). Divergent modes of autophagy in the methylotrophic yeast *Pichia pastoris*. J. Cell Sci. 108, 25-35.

Tuttle, D.L., Lewin, A.S., & Dunn Jr., W.A. (1993). Selective autophagy of peroxisomes in methylotrophic yeast. Eur. J. Cell Biol. 60, 283-290.

Ueno, T., & Kominami, E. (1991). Mechanism and regulation of lysosomal sequestration and proteolysis. Biomed. Biochim. Acta 4-6, 365-371.

Ueno, T., Muno, D., & Kominami, E. (1991). Membrane markers of endoplasmic reticulum preserved in autophagic vacuolar membranes isolated from leupeptin-administered rat liver. J. Biol. Chem. 266, 18995-18999.

van der Klei, I.J., Harder, W., & Veenhuis, M. (1991). Selective inactivation of alcohol oxidase in two peroxisome-deficient mutants of the yeast *Hansenula polymorpha*. Yeast 7, 813-821.

van Dyke, R.W. (1993). Acidification of rat liver lysosomes. Quantitation and comparison with endosomes. Am. J. Physiol. 265, C901-C917.

Venerando, R., Miotto, G., Kadowaki, M., Siliprandi, N., & Mortimore, G.E. (1994). Multiphasic control of proteolysis by leucine and alanine in the isolated rat hepatocyte. Am. J. Physiol. 266, C455-C461.

vom Dahl, S., Stoll, B., Gerok, W., & Häussinger, D. (1995). Inhibition of proteolysis by cell swelling in the liver requires intact microtubular structures. Biochem. J. 308, 529-536.

Wattiaux, R., Jadot, M., Dubois, F., & Wattiaux-De Coninck, S. (1996). Phagocytosis by rat liver: relationship between phagosomes and lysosomes. Biochem. Biophys. Res. Commun. 220, 569-574.

Whitney, J.A., Gómez, M., Sheff, D., Kreis, T.E., & Mellman, I. (1995). Cytoplasmic coat proteins involved in endosome function. Cell 83, 703-713.

Wolf, P.R., & Ploegh, H.L. (1995). How MHC class II molecules acquire peptide cargo: Biosynthesis and trafficking through the endocytic pathway. Annu. Rev. Cell. Biol. 11, 267-306.

Yamamoto, A., Masaki, R., & Tashiro, Y. (1990). Characterization of the isolation membranes and the limiting membranes of autophagosomes in rat hepatocytes by lectin cytochemistry. J. Histochem. Cytochem. 38, 573-580.

Yi, J., & Tang, X.M. (1995). Functional implication of autophagy in steroid-secreting cells of the rat. Anat. Rec. 242, 137-146.

Yokota, S. (1993). Formation of autophagosomes during degradation of excess peroxisomes induced by administration of dioctyl phthalate. Eur. J. Cell Biol. 61, 67-80.

Yokota, S., Himeno, M., & Kato, K. (1994). Degradation of excess peroxisomes by cellular autophagy: Immunocytochemical and biochemical analysis. Acta Histochem. Cytochem. 27, 573-579.

Yokota, S., Himeno, M., & Kato, K. (1995). Formation of autophagosomes during degradation of excess peroxisomes induced by di-(2-ethylhexyl)-phthalate treatment. III. Fusion of early autophagosomes with lysosomal compartments. Eur. J. Cell Biol. 66, 15-24.

Yoshihisa, T., & Anraku, Y. (1990). A novel pathway of import of α-mannosidase, a marker enzyme of vacuolar membrane, in *Saccharomyces cerevisiae*. J. Biol. Chem. 265, 22418-22425.

Yuan, W., Tuttle, D.L., Shi, Y-J., Ralph, G.S., & Dunn Jr., W.A. (1997). Gluocose-induced microautophagy in *Pichia pastoris* requires the α-subunit of phosphofructokinase. J. Cell Sci. 110, 1935-1945.

THE CRITICAL ROLE OF THE UBIQUITIN-PROTEASOME PATHWAY IN MUSCLE WASTING IN COMPARISON TO LYSOSOMAL AND Ca^{2+}-DEPENDENT SYSTEMS

Didier Attaix and Daniel Taillandier

Advances in Molecular and Cell Biology
Volume 27, pages 235-266.

ISBN: 0-7623-0387-5

I. INTRODUCTION

A. Background and Significance of Muscle Wasting

Proteins in skeletal muscle, as in other mammalian tissues, undergo a continuous process of synthesis and degradation (Waterlow et al., 1978). Rates of protein turnover regulate the levels of specific proteins, as well as the overall muscle protein mass during growth and atrophy. For example, increased proteolysis contributes to the muscle wasting seen with fasting and many pathological states (e.g., sepsis, cancer, chronic renal failure, and trauma). In such states, enhanced protein breakdown in skeletal muscle, the major protein reservoir in the body, is a key metabolic adaptation providing the organism with free amino acids for energy metabolism by increased gluconeogenesis and direct oxidation. In some instances, for example in sepsis, net mobilization of muscle protein also provides free amino acids for acute phase protein synthesis in the liver. However, despite its importance in growth and human disease, the regulation of muscle proteolysis by nutritional, hormonal, and mechanical factors remains poorly understood (Kettelhut et al., 1988; Sugden and Fuller, 1991). In this review, we will discuss the methods available to study skeletal muscle protein breakdown, and our current understanding of the role of three major proteolytic pathways so far well-characterized in this tissue (e.g., the lysosomal, Ca^{2+}-activated and ubiquitin-proteasome-dependent processes). A particular emphasis is placed on the regulation of the ubiquitin-proteasome pathway that is presumably involved in the breakdown of contractile components (Attaix et al., 1994; Mitch and Goldberg, 1996).

B. Methods for Measuring Skeletal Muscle Protein Breakdown

Studying skeletal muscle proteolysis is difficult because myofibrillar proteins represent approximately 60–70% of the proteins and turn over rather slowly under normal conditions (Waterlow et al., 1978). Even in severe muscle wasting conditions, measured changes in protein turnover are difficult to quantitate reliably, and affect muscle protein mass relatively slowly. In addition, methodological problems associated with the reliable measurements of muscle proteolysis impede the understanding of its regulation (Sugden and Fuller, 1991).

In Vivo

Most *in vivo* measurements of skeletal muscle protein breakdown are indirect and provide little if any information on intracellular regulatory mechanisms. The average rate of muscle protein breakdown is often calculated as the difference between the rate of protein synthesis and the rate of protein deposition in growing animals (Waterlow et al., 1978). Conversely, the rate of protein breakdown is the sum of protein synthesis and protein loss upon a catabolic state. Such estimations are very imprecise because the rate of protein deposition or loss must be estimated over several days, the rate of protein synthesis being measured over a few minutes or hours after the administration of a labeled amino acid. However, rates of protein deposition and synthesis are not necessarily constant over the period of measurements.

A second method is based on the loss of radioactivity from prelabeled proteins. Amino acid reutilization results in artificially low rates of protein breakdown, and the data critically depend on the time period over which the measurements are made. Furthermore, this technique provides only an estimated rate of protein breakdown for a group of animals killed at different time points, and not individual rates (Waterlow et al., 1978).

A third widely used technique is the measurement of urinary 3-methylhistidine excretion. 3-methylhistidine is formed by a posttranslational methylation of specific histidine residues in actin (and also in myosin heavy chains of fast-twitch glycolytic skeletal muscles). This amino acid is neither reutilized for protein synthesis nor significantly metabolized, and is quantitatively excreted in urine of some (e.g., rat and man) but not all species (Young and Munro, 1978). Therefore, the urinary excretion of 3-methylhistidine can provide an index of myofibrillar protein breakdown. Unfortunately, the visceral smooth muscles of the gastrointestinal tract and other tissues like skin contain significant amounts of actin. These tissues contribute disproportionately for their size to 3-methylhistidine excretion, because of their high rates of protein turnover (Wassner and Li, 1982). In addition, changes in renal clearance of 3-methylhistidine may affect the interpretation of such experiments. Consequently, the use of 3-methylhistidine excretion as a valid and specific index of skeletal muscle protein breakdown is controversial (Millward et al., 1980; Wassner and Li, 1982).

A fourth method involves selective catheterization of an artery and a vein draining a muscle bed, and is based on the isotopic dilution of infused [^{3}H]phenylalanine, an amino acid which is not significantly degraded or synthesized in muscle. An index of both the rates of protein breakdown and synthesis is calculated by measuring the phenylalanine concentration and specific radioactivity in arterial and venous blood, and the blood flow (Barrett and Gelfand, 1989). Although very useful in man, this method provides only a minimal estimate of the rate of protein breakdown because of the reutilization of the label.

Botbol and Scornick (1997) have recently proposed a new method for studying acute changes in skeletal muscle protein breakdown in live mice. The procedure is based on the accumulation of bestatin-induced peptides. Bestatin is an aminopeptidase inhibitor, which permits the breakdown of intracellular proteins by either the autophagic or cytosolic pathways (see below) to di- and tri-peptides but interferes with the further degradation of these peptides to amino acids. Because the accumulation of bestatin-induced peptides is measured in 10 minutes, this technique provides a unique tool in the study of rapid and transient changes in proteolysis such as those provided by nutrients, hormones, or exercise.

Finally, another approach based on the measurements of mRNA levels for proteinases or cofactors involved in proteolysis has been developed in the past few years (Attaix et al., 1994). These mRNA levels increased in skeletal muscle in many catabolic conditions, and may provide information about specific proteinases or proteolytic pathways activated. However, the significance of variation in mRNA levels for almost all proteinases is still unclear. For example, Ilian and Forsberg (1992) reported increased mRNA levels for μ- and m-calpains in muscles from fasted rabbits without any variation in proteinase activities. Discrepancies between levels of proteinase mRNA and actual proteinase activity may result from changes in either the rate of proteinase turnover or in endogenous proteinase inhibitor levels or activity. Thus, this approach must be combined with other techniques (i.e., *in vitro* methods and/or measurements of proteinases activities) because of its theoretical limitations.

In Vitro

In vitro muscle preparations have been widely used to study protein breakdown. Most studies employ very thin incubated muscles to avoid the formation of an hypoxic core. The slow-twitch oxidative soleus, and the fast-twitch glycolytic extensor digitorum longus muscle from young rats (< 80 g) or mice have been extensively used. Alternatively, the flat diaphragm and epithrochlearis muscles from adult and old rats are used. Hindlimbs and hinquarters of rats have also been perfused (see Kettelhut et al., 1988 for a review). Net tyrosine or phenylalanine release in the incubation media or the perfusate reflects the absolute rate of protein breakdown when protein synthesis is blocked by cycloheximide, since neither amino acid is significantly metabolized in muscle (Tischler et al., 1982). Tyrosine release is very easily measured fluorometrically. Alternatively, absolute rates of protein breakdown can be obtained from simultaneous measurements of net tyrosine release and of protein synthesis within the same preparation. The rate of protein synthesis is measured by following the incorporation of [^{14}C]phenylalanine;

this rate of phenylalanine incorporation is multiplied by the molar ratio of tyrosine to phenylalanine in muscle proteins to obtain tyrosine equivalents. The absolute rate of protein breakdown is then calculated by adding the net tyrosine release to the rate of tyrosine incorporation into proteins (Tischler et al., 1982). Rates of tyrosine release reflect the breakdown of all classes of proteins. Alternatively, the rate of 3-methylhistidine release gives an estimate of the breakdown of myofibrillar proteins (e.g., actin and some myosins) (Lowell et al., 1986; Tiao et al., 1994).

It should be pointed out that muscle preparations are always in a catabolic state. Therefore, observations *in vitro* may not necessarily reflect the *in vivo* situation, and caution must be exercised in interpreting such data. However, if incubated muscles are catabolic, this is in large part due to abnormally low rates of protein synthesis, and not only to increased proteolysis. Furthermore, muscle preparations are highly responsive to nutrient supply and hormones, as occurs *in vivo*. Many groups using isolated muscles incubated in the least possible negative conditions of nitrogen balance (with oxygen supply, tension, insulin and amino acids, especially the branched chain residues; see Kettelhut et al., 1988) have reported variations in both rates of protein synthesis and breakdown that closely mimic observations reported in intact animals. Finally, this approach is particularly important, since incubated muscles are the only system where selective inhibition may be used to identify the contribution of different proteolytic pathways to overall skeletal muscle protein breakdown (see below).

Cultured myoblasts (Wing and Bedard, 1996) and myotubes (Hong and Forsberg, 1995; Ebisui et al., 1995) have also been used to determine the mechanisms regulating muscle protein breakdown. Again, caution must be exercised in extrapolating such data to the *in vivo* situation. For example, insulin downregulated the expression of the 14-kDa ubiquitin conjugating enzyme (14 kDa E_2) *in vitro* (Wing and Banville, 1994), but had no effect *in vivo* (Larbaud et al., 1996).

C. Major Proteolytic Pathways in Muscle

Like other mammalian tissues, skeletal muscle contains multiple proteolytic systems that are presumably responsible for the breakdown of specific proteins. Three major proteolytic pathways operate in skeletal muscle, although such a simple division is a gross oversimplification.

The Lysosomal System

The best known proteolytic system is the lysosomal pathway (for review see Knecht et al., 1998, in this volume). Lysosomes are particularly abundant in liver, and two lysosomal vacuolar processes (macro and microautophagy) play a major role in the degradation of hepatic long-lived proteins (Mortimore et al., 1992). By contrast, skeletal muscle contains few lysosomes. The major lysosomal proteinases (cathepsins B, H, L, and D) do not contribute significantly to overall protein breakdown in muscles incubated under optimal conditions (e.g., with tension and in the presence of insulin and amino acids) (Furuno et al., 1990; Wing and Goldberg, 1993; Temparis et al., 1994; Tiao et al., 1994; Baracos et al., 1995; Taillandier et al., 1996; Voisin et al., 1996). In addition, lysosomes are not involved in the degradation of myofibrillar proteins (Lowell et al., 1986; Furuno et al., 1990; Tiao et al., 1994). This pathway is presumably responsible for the breakdown of some long-lived, membrane and endocytosed proteins (Mitch and Goldberg, 1996). Increased cathepsin activities and mRNA levels have been reported in some, but not all, instances of muscle wasting (Table 1). Lysosomal proteolysis is widely believed to be nonselective, but proteins containing KFERQ (Lys, Phe, Glu, Arg, Gln) sequences that specifically target proteins for lysosomal degradation (see Knecht et al., 1998, in this volume) exist in skeletal muscle (Wing et al., 1991). However, proteins containing KFERQ sequences were selectively depleted in the liver and heart but not skeletal muscle of fasted rats (Wing et al., 1991). These data suggest again that lysosomal proteolysis plays a minor role in muscle protein breakdown.

Ca^{2+}-Activated Proteinases

The best characterized nonlysosomal proteinases are two Ca^{2+}-dependent enzymes, μ-calpain and m-calpain, that differ in their affinities for Ca^{2+} (see Croall and DeMartino, 1991, and Sorimachi et al., 1997, for reviews). Calpains are heterodimeric molecules consisting of distinct large subunits associated with a common small subunit, all of which are encoded by different genes. The activity of these ubiquitous cysteine proteinases is regulated not only by Ca^{2+}, but also by their endogenous inhibitor, calpastatin (rewieved by Johnson, 1990). A third calpain, p94, that is expressed abundantly only in skeletal muscle, has been characterized (Sorimachi et al., 1989). Unfortunately, its precise role and physiological importance is still unknown (reviewed by Sorimachi et al., 1993). The Ca^{2+}-dependent process does not contribute significantly to overall proteolysis in muscles from control and catabolic animals (Wing and Goldberg, 1993; Mitch et al., 1994; Temparis et al., 1994; Tiao et al., 1994; Baracos et al., 1995; Taillandier et al., 1996; Voisin et al., 1996), except in muscular dystrophy (Turner et al., 1988; MacLennan et al., 1991). In muscles from mdx mice, the animal model that reproduces Duchenne muscular dystrophy (DMD), there was an increase in total calpain content, due to an increase in m-calpain concentration (Spencer et al., 1995). This increased m-calpain content correlated with a nonsignificant (Spencer et al., 1995) or significant (Combaret et al., 1996) increase in m-calpain mRNA levels. By contrast, mutations in p94 cause limb-girdle muscular dystrophy type 2A (Richard et al., 1995). Although the mRNA levels for μ- and m-calpain, and p94 did not change in muscle biopsies from DMD patients (Sakamoto et al., 1994; Combaret et al., 1996), there is some evidence that p94 was expressed in proportion of dystrophin, whose absence causes DMD in such patients (Sakamoto et al., 1994). In fact, p94 content decreased in mdx muscle (Spencer et al., 1995). Interestingly, the expression of p94 was downregulated in the atrophying muscles from interleukin-6 (IL-6) transgenic mice (Tsujinaka et al., 1996).

Table 1. Changes in Lysosomal Proteolysis in Different Muscle Wasting Conditions

Wasting Condition	*Species*	*Measurements*	*Reference*
Short-term fasting	Rat	⇑ lysosomal proteolysis	Wing and Goldberg, 1993
		⇔ or ⇓ cathepsin B, H, and B+L activities	Belkhou et al., 1994
		⇔ mRNA for cathepsin L and D	Medina et al., 1995
Fasting	Rabbit	⇑ mRNA for cathepsin D	Ilian and Forsberg, 1992
Denervation atrophy	Rat	⇑ cathepsin B activity	Furuno et al., 1990
Simulated weightlessness	Rat	⇑ cathepsin B and D activities	Goldspink et al.,1986
		⇑ mRNA for cathepsins B, L and D	Taillandier et al., 1996
		⇑ cathepsin B and B+ L activities	
Cancer	Rat	⇔ mRNA and activity of cathepsin B	Temparis et al., 1994
		⇓ cathepsin B, and B+L activities	Llovera et al., 1994
Acidosis	Rat	⇔ mRNA for cathepsin B	Mitch et al., 1994
Muscular dystrophy	mdx mice	⇔ mRNA for cathepsin D	Combaret et al., 1996
Sepsis	Rat	⇔ or ⇑ mRNA* for cathepsin B	Voisin et al., 1996
Diabetes	Rat	⇔ lysosomal proteolysis	Pepato et al., 1996
Glucocorticoid treatment	Rat	⇑ mRNA for cathepsin B and D	Dardevet et al., 1995
IL-6 transgenic animals	Mice	⇑ mRNA for cathepsin B and L	Tsujinaka et al., 1996
		⇑ cathepsin B and B+L activities	
Head trauma	Man	⇑ mRNA for cathepsin D	Mansoor et al., 1996

Note: * In acute or chronis sepsis, respectively.

Table 2. Changes in Ca^{2+}-Dependent Proteolysis in Different Muscle Wasting Conditions

Wasting Condition	*Species*	*Measurements*	*Reference*
Short-term fasting	Rat	⇔ Ca^{2+}-dependent proteolysis	Wing and Goldberg, 1993
Fasting	Rabbit	⇑ mRNA for μ-,m-calpain, and calpastatin	Ilian and Forsberg, 1992
		⇔ μ- and m-calpain activities	
Simulated weightlessness	Rat	⇑ Ca^{2+}-dependent proteolysis	Tischler et al., 1990
		⇑ mRNA and activity of m-calpain	Taillandier et al., 1996
Cancer	Rat	⇑ mRNA for m-calpain	Temparis et al., 1994
Muscular dystrophy	Duchenne muscular dystrophy patients	⇔ mRNA for μ-, m-calpain, and p94	Sakamoto et al.,1994
		⇔ mRNA for m-calpain	Combaret et al., 1996
	mdx mice	⇑ Ca^{2+}-dependent proteolysis	Turner et al., 1988
		⇑ mRNA for m-calpain	Combaret et al., 1996
		⇑ mRNA* and content of m-calpain	Spencer et al., 1995
		⇓ p94 content	
Sepsis	Rat	⇔ or ⇑ mRNA† for m-calpain	Voisin et al., 1996
Diabetes	Rat	⇑ Ca^{2+}-dependent proteolysis	Pepato et al., 1996
Glucocorticoid treatment	Rat	⇑ mRNA for m-calpain	Dardevet et al., 1995
IL-6 transgenic animals	Mice	⇑ mRNA for m-calpain	Tsujinaka et al., 1996
		⇓ mRNA for p94	
Head trauma	Man	⇑ mRNA for m-calpain	Mansoor et al., 1996

Notes: * Not significant.
† In acute or chronic sepsis, respectively.

Thus, muscle atrophy seems to be associated with decreased p94 content and mRNA levels. Otherwise, there are few biological models in which significant changes in calpain activities occur (Ilian and Forsberg, 1992), and evidence that Ca^{2+}-dependent proteinases are activated in any muscle wasting condition is lacking (Table 2). Goll et al. (1992) suggested that calpains may initiate myofibrillar protein breakdown. However, inhibitors of calpains do not affect myofibrillar protein breakdown *in vitro* (Lowell et al., 1986; Tiao et al., 1994). In fact, there is now growing evidence that calpains do not play a major role in general turnover of intracellular protein, but are involved in limited proteolysis of some specific target proteins, e.g., processing of protein kinase C (Johnson, 1990; Melloni and Pontremoli, 1991). The Ca^{2+}-dependent proteinases may be qualitatively important for the degradation of specific but quantitatively minor muscle proteins. They are involved in rapid removal of Z-disks and N_2 lines (Goll et al., 1992; Taylor et al., 1995), and degrade nebulin, titin, troponin, tropomyosin, but not myosin and actin (Goll et al., 1992), and probably membrane proteins (Combaret et al., 1996). Many short-lived proteins contain PEST (Pro, Glu, Ser, Thr) sequences (Rogers et al., 1986). Such sequences (very negatively charged) may bind Ca^{2+}, thus creating a microenvironment of higher Ca^{2+} concentration favorable to calpain-mediated proteolysis. In support of this hypothesis, Smith et al. (1993) reported that the insulin receptor substrate (IRS-1) was a PEST protein susceptible to calpain degradation *in vitro*. Unfortunately, evidence that PEST motifs did not influence substrate susceptibility to calpain proteolysis was also recently provided (Molinari et al., 1995; Carillo et al., 1996).

The Ubiquitin-Proteasome System

Fagan et al. (1987) demonstrated that skeletal muscle also contains a soluble ATP-ubiquitin-dependent proteolytic system. The first step in this pathway is the covalent attachment of multiple ubiquitin molecules to the targeted protein (reviewed by Wilkinson, 1998, in this volume). In brief, ubiquitylation of proteins always requires ATP, the ubiquitin activating enzyme, E_1, and one of the ubiquitin conjugating enzymes, E_2s, to generate the first attachment of a single ubiquitin moiety to a substrate. Some conjugation reactions also require a third enzyme, known as ubiquitin protein ligases, E_3s, for substrate recognition. Multiubiquitylation of the substrate is always observed in E_3-dependent reactions, but also occurs in some E_3-independent reactions. E_1 is barely expressed in human skeletal muscle (Handley et al. 1991), but its mRNA level was regulated in biopsies from acidotic patients with renal failure (Roberts et al., 1996), and in atrophying muscles from hindlimb suspended rats (Taillandier and Attaix, unpublished observations). Various E_2s are found in rodent skeletal muscle, including the 14 kDa (Wing and Banville, 1994), the 17 kDa (Wing and Jain, 1995), and the 20 kDa (Wefes et al., 1995) species. Among these species, only the 14 kDa E_2 was found to play a significant role in muscle wasting (Wing, Taillandier, and Attaix, unpublished observations). The 14 kDa E_2 is one of the major mammalian E_2s that best supports E_3-dependent conjugate formation and protein breakdown (Hersko and Ciechanover, 1992); its

1.2 kb transcript was reported to be tightly upregulated in various instances of muscle atrophy both in rats (Temparis et al., 1994; Wing and Banville, 1994; Dardevet et al., 1995; Taillandier et al., 1996; Voisin et al. 1996) and humans (Mansoor et al., 1996). Very little is known about E_3s in skeletal muscle. A very large ubiquitin-protein-ligase, E_3L, was recently demonstrated to be involved in the degradation of muscle proteins such as actin, troponin T, and MyoD (Gonen et al., 1996). The degradation of these ubiquitylated proteins appeared to be E_2-F1-dependent, a previously identified "non-N-end-rule" ubiquitin conjugating enzyme, whereas the 14 kDa E_2 is a "N-end rule" enzyme. Interestingly, Gonen et al. (1996) also provided evidence that proteins can be conjugated and degraded by two distinct ubiquitin-dependent pathways. Lysozyme was conjugated either by 14 kDa E_2 and $E_3\alpha$ (i.e., recognized via its basic N-terminal residue, Lys, according to the N-end rule), or by the non-N-end rule pathway consisting of E_2-F1 and E_3L.

The second step in the ubiquitin-proteasome pathway is the degradation of ubiquitin conjugates by the 26S proteasome complex that requires ATP for activation and substrate hydrolysis (Coux et al., 1996; Baumeister et al., 1998). The 26S complex contains the 20S proteasome with proteolytic function, and the 19S regulatory complex (reviewed by Tanaka and Tanahashi, 1998, and Dubiel and Rechsteiner, 1998, in this volume). Incidentally, skeletal muscle was one of the first tissues in which the 20S proteasome was isolated. So far, the enzyme has been purified from muscle of several mammalian and fish species, as well as lobster (reviewed by Dahlmann and Kuehn, 1995). A single report indicated that the 20S proteasome from rabbit skeletal muscle can degrade proteins in an ATP-dependent and ubiquitin-independent fashion (Driscoll and Goldberg, 1989). However, these data are in contrast with several reports in which ATP had no, or even an inhibitory, effect on the activities of muscle 20S proteasome (Dahlmann and Kuehn, 1995). Lobster muscle 20S proteasome has been extensively studied, and contains five peptidase activities (reviewed by Mykles, 1993), which have been described in proteasomes from other tissues and species (Rivett, 1993). *In vitro* studies have sown that the branched chain amino acid-preferring (BrAAP), the peptidylglutamyl peptide hydrolase (PGPH), and trypsin-like activities were involved in the degradation of contractile proteins. The BrAAP activity appears to be the most important under certain conditions, and may degrade myosin heavy and light chains, paramyosin, tropomyosin, and troponin-T, -I, and –C, and actin depolymerizing factor (Mykles and Haire, 1991, 1995). The PGPH activity apparently degraded myosin heavy chain, and the trypsin-like activity may mediate a limited cleavage of troponin-T_2 (Mykles, 1993). However, properties of muscle proteasome could be species-specific, and adapted to specific biological requirements (i.e., molt-induced claw muscle atrophy in lobster). In contrast with the 20S proteasome, very little is known about the 26S proteasome complex in muscle. Dahlmann et al. (1995) demonstrated that ATP stimulated some activities of the 26S proteasome from rat skeletal muscle. Extensive changes in the peptidases activities and ATPase amounts of the 26S proteasome were observed in abdominal intersegmental muscles of the tobacco horn moth *Manduca sexta* during programmed cell death (Dawson et al., 1995). This process is ubiquitin-dependent (Haas et al., 1995). By contrast, to our knowledge, whether or not the mammalian 26S proteasome degrades ubiquitylated and/or nonubiquitylated substrates in skeletal muscle is not documented.

Contribution of Specific Proteolytic Pathways to Overall Protein Breakdown

Rates of protein degradation attributable to different proteinases or proteolytic systems must be measured in an unambiguous way. Incubated muscles offer unique and convenient advantages to study the regulation of a given proteolytic pathway. Inhibitors of either lysosomal function (e.g., E64 and leupeptin that inhibit the cysteine proteinases cathepsins B, H, and L) or acidification (e.g., methylamine, chloroquine, propylamine) have been used to estimate the relative contribution of the lysosomal process to overall rates of protein breakdown (Lowell et al., 1986; Furuno et al., 1990; Wing and Goldberg, 1993; Mitch et al., 1994; Temparis et al., 1994). However, E64 and leupeptin are also inhibitors of the calpains, as is a Ca^{2+}-free medium. A nonlysosomal, Ca^{2+}-independent, but ATP-dependent proteolytic process has also been quantified in incubated muscles (Kettelhut et al., 1988; Tawa et al., 1992; Wing and Goldberg 1993; Mitch et al., 1994; Temparis et al., 1994). ATP depletion (up to 95%) is performed by adding to the incubation media 2-deoxyglucose and dinitrophenol that block glycolysis, and oxidative phosphorylation, respectively. The quantitation of the ATP-dependent process must be performed following the inhibition of Ca^{2+}-dependent proteinases (by E64 and using a Ca^{2+}-free medium), because ATP depletion actually stimulates Ca^{2+}-dependent proteolysis (Kettelhut et al., 1988; Fagan et al., 1992). Since E64 inhibits cysteine lysosomal proteinases, methylamine is added to the medium to maximally suppress lysosomal proteolysis. Such procedure is, however, not specific because skeletal muscle contains ATP-dependent but ubiquitin-independent proteinases (Fagan and Waxman, 1989). Furthermore, ATP-depleted muscles are dying. More specific approaches have been developed with specific inhibitors of the proteasome (e.g., peptidyl aldehydes and/or lactacystin). MG 132 (CBZ-leu-leu-leucinal, a potent inhibitor of the BrAAP, PGPH, and chymotryptic activities; see Coux et al., 1996) was reported to block increased skeletal muscle proteolysis in acidosis (Bailey et al., 1996), diabetes (Price et al., 1996), denervation, atrophy, sepsis, and following the administratin of thyroid hormones (Tawa et al., 1997). Furthermore, lactacystin (to date the most specific inhibitor of the proteasome) also blunted increased total and myofibrillar protein breakdown in septic rats (Hobler et al., 1998).

Another approach is based on the stimulation of the lysosomal process (in the absence of insulin and amino acids that block the formation of autophagic vacuoles) or of Ca^{2+}-dependent proteolysis (in the presence of Ca^{2+} ionophores in muscles incubated without tension) (Kettelhut et al., 1988). These experiments give an estimation of the maximal capacity of the muscles to carry out either lysosomal or Ca^{2+}-dependent proteolysis in a particular condition.

In conclusion, *in vitro* preparations are the only system where selective inhibition/activation may be used to identify the contribution of the different proteolytic systems. Since most available inhibitors have limitations, other indications of levels or activity of proteolytic systems are a useful adjunct to inhibitor studies. However, measures of proteinase activity require specific substrates and are confounded by the presence of endogenous inhibitors. In addition, most proteolytic assays utilize artificial and not physiological radiolabeled or peptide substrates. The hydrolysis of such substrates by a given proteinase may not reflect the *in vivo* situation. Proteinase activity is also regulated by rates of proteinase synthesis (to which proteinase mRNA levels may be related) and proteinase degradation. Thus, whenever possible, it appears highly advisable to combine multiple approaches to elucidate the precise role of any proteolytic process.

II. UBIQUITIN-PROTEASOME PROTEOLYSIS IN EXPERIMENTAL MODELS OF MUSCLE ATROPHY

A. Fasting

Short-term fasting leads to a mobilization of muscle protein, resulting from a decrease in protein synthesis and from an increase in non-myofibrillar and myofibrillar protein breakdown (Li and Goldberg, 1976; Li et al., 1979; Lowell et al., 1986). Increased protein breakdown of non-myofibrillar proteins is presumably mediated by the lysosomal pathway (Lowell et al., 1986; Wing and Goldberg, 1993). The low availability of amino acids and the low level of circulating insulin may contribute to increased proteolysis. Both insulin and amino acids inhibit the formation of autophagic vacuoles in liver (Mortimore et al., 1992) and protein breakdown in skeletal muscle (Kettelhut et al., 1988). However, Lowell et al. (1986) reported that the large increase in myofibrillar protein breakdown observed in perfused hindquarters of fasted rats does not occur by the lysosomal or the Ca^{2+}-dependent pathways. In these experiments, agents which block lysosomal acidification (chloroquine, NH_4Cl) or the autophagic pathway (insulin, amino acids), and an inhibitor of cathepsins B, H and L and calpains (leupeptin) do not affect 3-methylhistidine release. The first evidence that the skeletal muscle ATP-ubiquitin-dependent proteolytic pathway is activated upon starvation was obtained by Goldberg and co-workers (Medina et al., 1991). Most of the increased proteolysis observed in the muscles from fasted rats was not affected by inhibitors of the lysosomal and Ca^{2+}-dependent pathways. By contrast, incubation of these muscles under conditions where intracellular ATP was depleted almost totally suppressed increased proteolysis due to fasting (Medina et al., 1991, 1992; Wing and Goldberg, 1993). The rise in ATP-dependent proteolysis observed during starvation was associated with increased mRNA levels for ubiquitin (Medina et al., 1991, 1995; Wing and Goldberg 1993) and the 14 kDa E_2 that mediates the formation of ubiquitin-protein conjugates (Wing and Banville, 1994), accumulation of ubiquitylated proteins (Wing et al., 1995), and increased expression of subunits of the 20S proteasome (Medina et al., 1995). Upon refeeding total and ATP-dependent proteolysis, levels of ubiquitin-protein conjugates, and expression of ubiquitin (Medina et al., 1995), 14 kDa E_2 (Wing and Banville, 1994), or proteasome subunits (Medina et al., 1995) returned to normal. ATP-ubiquitin-stimulated proteolysis was also enhanced in the soluble extracts of skeletal muscle from fasted rabbits (Medina et al., 1992). Accordingly, Ilian and Forsberg (1992) reported an increased mRNA level for the C2 20S proteasome subunit in similar conditions.

Interestingly, ubiquitin mRNA did not increase in a variety of other tissues during fasting, including heart, liver, kidney, and fat, and it has been suggested that this adaptation is specific to striated muscle (Medina et al., 1995). In fact, Samuels et al. (1996) reported increases in mRNA levels for ubiquitin, 14 kDa E_2 or 20S proteasome subunits in the intestines of fasted rats. The serosal layer of the intestines contains myofibrillar proteins, which are depleted during fasting, suggesting the increase in these mRNA levels may in part stem from proteolytic activity in this part of the gut.

B. Protein Deficiency

Both rates of muscle protein synthesis and breakdown decrease in young animals fed an inadequate dietary intake (Tawa and Goldberg, 1994). In contrast with starvation, these changes in protein turnover preserve muscle mass and protein content. Protein breakdown is only impaired in growing animals fed a deficient protein diet. Adult mice actually exhibit an increase in muscle proteolysis following protein deprivation (Botbol and Scornick, 1997). Tawa et al. (1992) reported that the suppression of skeletal muscle protein breakdown observed in growing rats fed a protein-deficient diet was accompanied by reduced lysosomal proteolysis. Protein deficiency lowered the levels of several lysosomal enzymes (cathepsins B, H, C, and carboxypeptidases A and C) without any effect on cathepsin D. However, following the inhibition of the lysosomal proteolytic process, proteolysis remained lower in protein deficient animals than in controls rats. Indeed, the reduction in protein breakdown observed in these animals was due to a marked reduction in ATP-dependent proteolysis that explained the fall in overall proteolysis. Accordingly, the muscle content of proteasome fell significantly in animals fed a protein-deficient diet (Tawa and Goldberg, 1994). Therefore, the stimulation or inhibition of the ubiquitin-proteasome pathway following nutritional manipulation is clearly correlated with either muscle protein loss or preservation, in starvation and protein deprivation, respectively.

C. Disuse

Muscle mass and protein content are extremely sensitive to disuse. In rats, muscle atrophy seen after section of the sciatic nerve results primarily from enhanced proteolysis (Furuno et al., 1990). Treatments that maximally activate either the lysosomal or the Ca^{2+}-dependent pathway were more effective in stimulating overall protein degradation in the denervated muscles than in controls. However, inhibitors of these processes did not affect the breakdown of myofibrillar components (Furuno et al., 1990). By contrast, ATP depletion in denervated muscles inhibited 3-methylhistidine release by 56%, and suppressed the overall increase in total proteolysis (Medina et al., 1991). The enhancement in ATP-dependent proteolysis observed in atrophying muscles from denervated animals was associated with a rise in mRNA levels for ubiquitin and 20S proteasome subunits (Medina et al., 1995) and an increase in ubiquitin conjugates (Wing et al., 1995).

Another example of disuse is real or simulated weightlessness. In such conditions, increased protein breakdown is the major determinant of muscle wasting in antigravity muscles, i.e., soleus (Thomason and Booth, 1990). Experiments by Goldspink et al. (1986) suggested an activation of cathepsin D. By contrast, Tischler et al. (1990) reported that enhanced proteolysis in such conditions occurred by a nonlysosomal but Ca^{2+}-dependent mechanism. Taillandier et al. (1996) have recently found that the lysosomal and Ca^{2+}-dependent proteolysis was enhanced (254%) in the soleus muscle of nonweight bearing rats. Accordingly, large increases in mRNA levels for lysosomal (cathepsins B, L, and D), and Ca^{2+}-dependent (m-calpain) proteinases were observed in the atrophying soleus muscle. The enhanced expression of cathepsin B or L and m-calpain was associated with increased enzyme activities, suggesting a transcriptional regulation. However, the lysosomal and Ca^{2+}-dependent proteolysis did not exceed 18% of total proteolysis in the atrophying muscles, and the inhibition of both processes did not suppress increased proteolysis. Haas and Riley (1988) previously reported significant alterations in the ubiquitin pools in unweighted muscles. Accordingly, we found large increases in mRNA levels for both ubiquitin and the 14 kDa E_2 in the soleus muscle of nonweight bearing rats, as well as increased expression of C2 and C9 20S proteasome subunits (Taillandier et al., 1996). Furthermore, the increased abundance of mRNA for the C9 proteasome subunit underwent active translation in unweighted muscles.

Therefore, the major proteolytic systems are coordinately activated in skeletal muscle during disuse. However, a nonlysosomal Ca^{2+}-independent, but presumably ubiquitin-proteasome-dependent proteolytic process appears to be primarily responsible for muscle wasting.

D. Pathological Conditions

Many studies have provided evidence that muscle wasting in sepsis results mainly from increased protein breakdown, although depressed protein synthesis also contributes to the catabolic response (reviewed by Hasselgren, 1993). Tiao et al. (1994) have shown that sepsis stimulated an energy-dependent pathway that accounted for the breakdown of myofibrillar proteins. Moreover, these authors reported increased ubiquitin mRNA levels in the muscles from septic rats, suggesting that the ATP-dependent proteolytic pathway activated was the ubiquitin system. However, sepsis was induced in these experiments by cecal ligation and puncture.

This model is characterized by a very high mortality rate within a short period of illness (so that muscle wasting is not apparent), and is representative of an acute sepsis. Voisin et al. (1996) carried out experiments to elucidate the possible role of the ubiquitin-proteasome pathway using a long lasting septic rat model that reproduces a sustained and reversible catabolic state, as observed in humans. Septic rats were injected intravenously with live *Escherichia coli* and were studied in an acute septic phase (day two postinfection), in a chronic septic phase (day six), and in a late septic phase (day 10). Protein breakdown was elevated during both the acute and chronic septic phases (when significant muscle wasting occurred) and returned to control values in the late septic phase (when wasting was stopped). A nonlysosomal and Ca^{2+}-independent process accounted for the enhanced proteolysis, and only mRNA levels for ubiquitin and subunits of the 20S proteasome paralleled the increased and decreased rates of proteolysis throughout. However, increased mRNA levels for the 14 kDa E_2, cathepsin B, and m-calpain were also observed in chronic sepsis. These observations clearly supported a major role for ubiquitin-proteasome proteolysis during sepsis, but also suggested that the activation of the lysosomal and Ca^{2+}-activated systems may be important in the chronic phase.

Muscle wasting is also commonly observed in cancer patients and tumor-bearing animals, and results from impaired protein synthesis and/or enhanced proteolysis (Pisters and Brennan, 1990). In rats bearing a Yoshida sarcoma, muscles close to the tumor did atrophy before generalized muscle wasting was apparent (Temparis et al., 1994). This local atrophy resulted from increased protein breakdown, without any significant change in protein synthesis. Incubating the atrophying muscles from tumor-bearing rats in the presence of inhibitors of the lysosomal and Ca^{2+}-dependent pathways did not reduce the enhanced proteolysis. In contrast, an additional ATP depletion suppressed increased protein breakdown. In addition, increased expression of ubiquitin, 14 kDa E_2, and C8 or C9 proteasome subunits was observed in the atrophying muscles from cancer rats. Similar observations were obtained when generalized muscle wasting was apparent, nine days after Yoshida sarcoma implantation (Attaix et al., 1997). Baracos et al. (1995) and Llovera et al. (1995) also showed that accelerated proteolysis in muscles of cachectic rats bearing the fast-growing AH-130 Yoshida ascites hepatoma resulted primarily from activation of ATP-ubiquitin-dependent proteolysis. An accumulation of ubiquitin-protein conjugates was observed in skeletal muscles of such cachectic animals (Llovera et al., 1994; Baracos et al., 1995).

Evidence was also provided that the enhanced skeletal muscle proteolysis seen in acidotic rats (a model of chronic renal failure) resulted from the activation of a nonlysososomal, Ca^{2+}-independent, but ATP-dependent, pathway (Mitch et al., 1994; Bailey et al., 1996). Increased ubiquitin mRNA levels observed in acidotic muscles returned to control levels within 24 hours after cessation of acidosis (Mitch et al., 1994). Using nuclear run-on experiments, Bailey et al. (1996) demonstrated that the increased expression of ubiquitin and 20S proteasome subunits seen in muscles from acidotic rats resulted from increased transcription of their mRNAs.

Furthermore, the proteasome inhibitor MG 132 suppressed increased protein breakdown in the atrophying incubated muscles, as did ATP depletion. Thus, this elegant study clearly establishes a role for the proteasome in muscle wasting.

Muscle wasting also prevails in diabetes, and results from a decrease in protein synthesis, and an increase in protein breakdown. The pattern of these changes is complex and time-dependent in streptozotocin-treated animals (Pepato et al., 1996). In such animals, there was no change in lysosomal proteolysis, but an increased activity of Ca^{2+}-activated and ATP-dependent proteinases, which was suppressed by insulin treatments. Other experiments clearly demonstrated that the increased protein breakdown was attributable to the ubiquitin-proteasome pathway: an inhibitor of the proteasome (MG 132) totally blocked the enhanced proteolysis. Furthermore, increased transcription of the ubiquitin gene, and elevated mRNA levels for three 20S proteasome subunits were observed in the muscles from diabetic animals (Price et al., 1996).

Two different groups have also provided evidence that burn injury stimulated multiple proteolytic pathways in skeletal muscle, including the ubiquitin system (Fang et al., 1995; Naman, Ferland, Zerouala, Auclair, Goldberg, and Garrel unpublished observations). Finally, an increased expression of multiple components of the ubiquitin-proteasome pathway (ubiquitin, 14 kDa E_2, and 20S proteasome subunits) was also observed following trauma of the gastrocnemius muscle (Baracos and Attaix, unpublished observations).

Taken together these data strongly suggest that the activation of the ubiquitin-proteasome pathway was mainly responsible for the muscle wasting observed in several pathological conditions. Enhanced ubiquitin-dependent proteolysis is not due to anorexia that often prevails in such conditions, since this adaptation occurred in tumor-bearing (Baracos et al., 1995), septic (Voisin et al., 1996), and diabetic (Price et al., 1996) rats compared to pair-fed controls.

E. Aging

A very progressive loss of muscle proteins is well described in aged humans (Forbes, 1976) and rodents. This slow muscle wasting results from an imbalance between protein synthesis and degradation rates. However, mechanisms underlying changes in the regulation of skeletal muscle protein breakdown during aging are still very obscure. Dardevet et al. (1995) studied dexamethasone-induced muscle wasting in adult and old rats, since the increased incidence of disease states may result in periods of glucocorticoids hypersecretion in aging. Glucocorticoids stimulated muscle protein breakdown in the adult, but not in aged animals. The disappearance of the stimulatory effect of glucocorticoids on protein breakdown in aging occurred along with a loss of ability of dexamethasone to enhance the expression of the 14 kDa E_2 and of subunits of the 20 S proteasome. However, these experiments were performed with catabolic doses of glucocorticoids. To determine whether or not these data have some physiological significance, and because glucocorticoids are involved in the enhanced protein breakdown seen during starvation (see below), protein breakdown was recently studied in the muscles from aged fasted rats. Again no increased proteolysis was observed in such muscles (Meynial-Denis and Attaix, unpublished observations). Thus, if glucocorticoids play any role in the age-related muscle atrophy, this is unlikely to result from an activation of the ubiquitin-proteasome pathway.

III. UBIQUITIN-PROTEASOME PROTEOLYSIS IN HUMAN DISEASES

An obvious goal will be to examine whether mechanisms observed in animal models are active in human diseases. Mansoor et al. (1996) recently reported enhanced whole-body protein breakdown and 3-methylhistidine urinary excretion in head trauma patients compared with healthy volunteers. These variations were associated with increased mRNA levels for ubiquitin, 14 kDa E_2, and HC2 and HC8 20S proteasome subunits in muscle biopsies from head trauma patients. Similar observations also prevailed in muscle biopsies from cachectic cancer patients (Rallière, Tauveron, and Attaix, unpublished observations). It seems important to note that muscle wasting arises rapidly in such patients, as in acute animal models of muscle wasting. Thus, the ubiquitin-proteasome pathway could play an important role in muscle wasting in some human acute diseases. By contrast, in patients with Cushing's syndrome who exhibited muscle wasting due to high levels of glucocorticoids, no significant variations in mRNA levels for cathepsin D, m-calpain, ubiquitin, 14 kDa E_2, and proteasome subunits were observed in muscle (Rallière et al., 1997). Similarly, the mRNA levels for these genes were not altered in biopsies from Duchenne muscular dystrophy patients (Combaret et al., 1996). Finally, the expression of ubiquitin and the C2 20S proteasome subunit was unchanged in biopsies from patients with chronic renal failure after the correction of acidosis, although the mRNA levels for E_1 and the 14 kDa E_2 were downregulated (Roberts et al., 1996). The lack of detectable activation of some proteolytic genes in human muscle biopsies could be due to the very progressive loss of myofibrillar proteins that prevails in chronic diseases. Alternatively, some regulatory mechanisms may prevent increased expression of proteolytic genes in such situations, for avoiding excessive and rapid muscle wasting. These hypotheses must be confirmed by further experiments in chronic situations of muscle loss. Interestingly, however, an increased expression of only ubiquitin, but not of proteasome subunits, was also seen in the atrophying muscles from IL-6 transgenic mice (Tsujinaka et al., 1996).

IV. REGULATION BY HORMONES

A. Insulin

Insulin has a well-known antiproteolytic effect in muscle preparations (reviewed by Kettelhut et al., 1988). The hormone downregulated mRNA levels for the

14 kDa E_2 in cultured myoblasts, suggesting an inhibition of the ubiquitin-proteasome pathway (Wing and Banville, 1994). By contrast, although it is well documented that insulin inhibits *in vivo* protein breakdown at the whole body level, neither the tissues nor the proteolytic pathways on which insulin exerts its antiproteolytic effect are well characterized. The effects of insulin on mRNA levels for cathepsin D, m-calpain, and ubiquitin were determined *in vivo*, in skeletal muscle, skin, liver, and intestine (Larbaud et al., 1996). A six hour hyperinsulinemic, euglycemic, and hyperaminoacidemic clamp was performed in goats, a species in which insulin markedly inhibited whole-body protein breakdown under similar conditions (Tesseraud et al., 1993). Hyperinsulinemia and hyperaminoacidemia only depressed ubiquitin mRNA levels in fast-twitch and mixed skeletal muscles without any concomitant reduction in mRNA levels for the 14 kDa E_2 and 20S proteasome subunits. The reduced ubiquitin mRNA levels in skeletal muscle may represent a possible mechanism explaining the antiproteolytic effect of insulin *in vivo*. These data are in accordance with the recent demonstration that the inhibition of postabsorptive muscle protein breakdown by insulin is likely mediated by a nonlysosomal proteolytic pathway in humans (Barrett et al., 1995). They do not preclude that insulin may have an inhibitory effect on muscle proteolysis by alternative mechanisms. Among such possibilities, Duckworth et al. (1994) reported a direct inhibitory effect of insulin on the *in vitro* lysis of *N*-succinyl-Leu-Leu-Val-Tyr-7--amido-4-methylcoumarin by a purified complex including the 20S proteasome and the insulin degrading enzyme.

B. Insulinlike Growth Factor-I (IGF-I)

IGF-I also reduced 14 kDa E_2 mRNA levels in L6 myoblasts, but was 100-fold more potent than insulin (Wing and Bedard, 1996). IGF-I did not alter rates of transcription of the 14 kDa E_2 gene, but enhanced the rate of degradation of the 1.2 kb transcript. This effect was specific of that particular gene, since IGF-I did not affect mRNA levels for ubiquitin and proteasome subunits. Levels of IGF-1 fall with fasting and increase upon refeeding. These variations could be related to the increased, and decreased 14 kDa E_2 expression seen in muscle upon fasting, and refeeding, respectively (Wing and Banville, 1994). These studies provide a possible mechanism explaining the antiproteolytic properties of the hormone, which is clearly recognized as a stimulator of muscle growth. For example, IGF-I can restore growth to diabetic animals in the absence of insulin therapy (Scheiwiller et al., 1986).

C. Glucocorticoids

Glucocorticoids are required for the increase in muscle proteolysis seen in fasting (reviewed by Kettelhut et al., 1988), as well as in some pathological states. Wing and Goldberg (1993) reported that adrenalectomy partially prevented the increase in skeletal muscle proteolysis and totally suppressed the increase in ubiquitin mRNA levels normally seen in fasted rats. By contrast, injection of dexamethasone restored both responses. Glucocorticoids appeared essential for the activation of the ATP-dependent process, but not for the rise in lysosomal proteolysis in fasting. Glucocorticoids may also be required for increased expression of ubiquitin and 20S proteasome subunits in skeletal muscle following metabolic acidosis (Price et al., 1994). Acidification alone resulted in increased mRNAs of the ubiquitin-proteasome pathway in BC_3H1 myocytes, but both acidosis and glucocorticoids were required to stimulate protein degradation (Isozaki et al., 1996). Auclair et al. (1997) have shown that the activation of the ubiquitin-proteasome pathway by catabolic doses of corticosterone in rats was responsible for the loss of myofibrillar proteins. However, Dardevet et al. (1995) provided evidence that elevated protein breakdown rates and enhanced mRNA levels for the 14 kDa E_2 and 20S proteasome subunits were observed in dexamethasone-treated adult animals, but not in aged rats. Glucocorticoids were also recently reported to regulate ATP-ubiquitin-dependent proteolysis in sepsis (Tiao et al., 1996), and burn injury (Naman, Ferland, Zerouala, Auclair, Goldberg, and Garrel, unpublished observations). By contrast, the 20S proteasome C2 mRNA was unaffected by dexamethasone in cultured L8 myotubes (Hong and Forsberg, 1995).

D. Thyroid Hormones

Finally, there is some evidence that hypothyroidism may result in reduced ATP-dependent proteolysis by altering the content of 20S and 26S proteasomes in skeletal muscle (Tawa and Goldberg, 1994). Conversely, the administration of thyroid hormone stimulated ubiquitin-proteasome-dependent proteolysis in rat skeletal muscle (Tawa et al., 1997).

V. REGULATION BY CYTOKINES

Tumor necrosis factor-α (TNF-α) overproduced by activated macrophages promotes muscle catabolism in sepsis (Flores et al., 1989). A single intravenous injection of TNF-α resulted in an accumulation of ubiquitin conjugates in rat skeletal muscle (Garcia-Martinez et al., 1993). The important role of TNF-α in the activation of the ATP-ubiquitin-dependent proteolytic pathway was also demonstrated in tumor-bearing rats (Llovera et al., 1996). Anti-TNF treatment reverted increased expression of ubiquitin and 20S proteasome subunits in such animals. Xanthine derivatives such as pentoxifylline and torbafylline, which suppressed the muscle wasting in septic rats (Breuillé et al., 1993) presumably by inhibiting TNF production (Semmler et al., 1993), also blocked the activation of ubiquitin-proteasome proteolysis in cancer rats (Attaix et al., 1997). By contrast, IL-1, another cytokine that may signal increased proteolysis (Baracos et al., 1983) in synergy with TNF-α (Flores et al., 1989), had no effect on ubiquitin gene expression in septic muscle (Garcia-Martinez et al., 1995). A third cytokine that may contribute to signal muscle wasting is IL-6 (Goodman, 1994). IL-6 was shown to increase the expression of subunits of 20S (C2, C8) and 26S (S4) proteasome, as well as of cathepsins B and L in cultured myotubes (Ebisui et al., 1995). Furthermore, treatment with the IL-6 receptor antibody inhibited muscle atrophy, increased cathepsin B and L activities and increased expression of cathepsin and ubiquitin in IL-6 transgenic mice (Tsujinaka et al., 1996). However, it should be noted that no increased expression of 20S and 26S proteasome subunits was detected in the atrophying muscles from these transgenic animals. Cytokines are not the only catabolic factor(s) that may signal increased muscle protein breakdown in sepsis, cancer, trauma, and presumably AIDS. For example, Todorov et al. (1996) have recently identified a circulating proteoglycan that caused cancer cachexia both in mice and humans. Further experiments should test whether such factors act directly or indirectly on ubiquitin-proteasome proteolysis.

VI. SUBSTRATES OF UBIQUITIN-PROTEASOME PROTEOLYSIS IN MUSCLE

The best known substrates of the ubiquitin-proteasome pathway are short-lived or denatured, misfolded, and mislocalized proteins (Rechsteiner, 1991), but evidence was also provided for a role of this pathway in the turn-over of long-lived proteins (Rock et al. 1994). The ubiquitin system is activated in every catabolic condition so far studied in rodent (fasting, denervation atrophy, simulated weightlessness, sepsis, cancer, metabolic acidosis, burn injury, diabetes, glucocorticoid and thyroid hormone administration, and trauma), except muscular dytrophy (Combaret et al., 1996). This activation is suppressed when muscle wasting is stopped (e.g., upon refeeding following starvation, Medina et al., 1995, and in a late septic phase, Voisin et al., 1996). Furthermore, there is evidence for a downregulation of the ubiquitin-proteasome pathway when protein mass is preserved (e.g., in protein deficiency, Tawa et al., 1992) or improved (e.g., following chemotherapy, Tilignac, Temparis, and Attaix, unpublished observations). These observations indicate that ubiquitin-dependent proteolysis plays a major role in the breakdown of myofibrillar proteins, which are long-lived proteins. Other lines of evidence support this hypothesis. Firstly, ATP depletion (Kettelhut et al., 1988; Medina et al., 1991; Tiao et al., 1994) or the proteasome inhibitor lactacystin (Hobler et al., 1998) inhibits 3-methylhistidine release *in vitro* in striking contrast to inhibitors of lysosomal and Ca^{2+}-dependent proteolysis (Lowell et al., 1986; Furuno et al., 1990; Tiao et al., 1994). Secondly, a rise in ubiquitin conjugates was recently found in the myofibrillar fraction, suggesting that contractile proteins could be degraded by this pathway (Wing et al., 1995). Both actin (Ball et al., 1987) and myosin light chains (Laub and Jennissen, 1991) can be ubiquitylated. Thirdly, purified proteasome preparations can degrade myofibrillar proteins (Mykles and Haire, 1991, 1995; Solomon and Goldberg, 1996) or intact myofibrils (Taylor et al., 1995) in certain conditions.

VII. RATE LIMITING STEP

There are at least two possible rate limiting steps in the ubiquitin-proteasome pathway: substrate ubiquitylation and/or the degradation of proteins conjugated to ubiquitin.

A. Ubiquitylation of Protein Substrates

The most likely site of regulation of ubiquitin conjugation would be at the reaction mediated by E_2s. Wing and Banville (1994) suggested that substrate ubiquitylation by the 14 kDa E_2 was the rate limiting step in skeletal muscle, because the expression of this E_2 is up- and downregulated in muscles from fasted and refed rats, respectively. Furthermore, the expression of the 14-kDa E_2 increased in a variety of muscle wasting conditions, both in animal models (Temparis et al., 1994; Dardevet et al., 1995; Taillandier et al., 1996; Voisin et al., 1996), and humans (Mansoor et al., 1996). However, several observations strongly suggest that the 14 kDa E_2 is not the rate limiting point in the ubiquitin-dependent proteolysis. First, the accumulation of ubiquitylated proteins observed in muscles from fasted, denervated (Wing et al., 1995), and tumor-bearing (Llovera et al., 1994; Baracos et al., 1995) rats strongly argues that the proteolytic activities of the proteasome could be the rate limiting step. Tawa and colleagues (1997) recently reported an accumulation of ubiquitin-conjugates in diaphragm muscles incubated with the proteasome inhibitor MG132. It is noteworthy, however, that the accumulation of ubiquitin-conjugates can be delayed with respect to increased proteolysis, as in denervation atrophy (Wing et al., 1995) or was not seen in some models of muscle wasting (e.g., Yoshida sarcoma-induced cachexia, Combaret, Wing, and Attaix, unpublished observations). Furthermore, levels of ubiquitin-conjugates were even reduced in corticosterone-treated animals (Auclair et al., 1997). Second, no increase in mRNA levels for the 14 kDa E_2 was seen in the atrophying muscles from acidotic (Price et al., 1994) or acutely septic (Voisin et al., 1996) rats, although in both conditions the activation of a nonlysosomal, Ca^{2+}-independent, but ubiquitin-dependent proteolytic process was responsible for increased proteolysis. Conversely, increased expression of the 14 kDa E_2 did not necessarily correlate with enhanced protein breakdown (Temparis et al., 1994).

B. Proteolytic Activities of the Proteasome and Regulatory Components

It seems probable that the rate of degradation of ubiquitylated muscle proteins by the proteasome is the rate limiting step. The increased expression of 20S proteasome subunits that systematically correlates with enhanced protein breakdown in various muscle wasting conditions (Ilian and Forsberg, 1992; Mitch et al., 1994; Temparis et al., 1994; Baracos et al., 1995; Dardevet et al., 1995; Medina et al., 1995; Price et al., 1996; Taillandier et al., 1996; Voisin et al., 1996) support this hypothesis. Although it is not proved that increased mRNA levels for proteasome subunits result in altered proteolytic activities (Dahlmann et al., 1995), independent lines of evidence strongly suggested that such increased mRNA levels have some significance in three different catabolic states. First, the enhanced expression of 20S proteasome subunits correlated with increased abundance of the 27 kDa subunit in cancer (Baracos et al., 1995). Second, analysis of subunit C9 mRNA distribution profile in polyribosomes showed that this subunit entered active translation in simulated weightlessness (Taillandier et al., 1996). Finally, Bailey and colleagues (1996) reported an increase in transcribed C3 proteasome subunit mRNA in acidosis.

There is also growing evidence that increased mRNA levels for subunits of the 19S complex occurs in muscle wasting (Attaix et al., 1997). The 19S complex associates with the 20S proteasome to form the 26S proteasome that degrades ubiquitin conjugates. Dawson et al. (1995) first reported a dramatic increase in the expression of the ATPase subunit MS73 in abdominal intersegmental muscles of *Manduca sexta* during programmed cell death. Western analyses showed increases in several ATPases including MS73, MSS1, and mts2, but not TBP1. They were associated with large changes in the peptidase activities of the 26S proteasome. An enhanced content of the non-ATPase subunit S5a that binds ubiquitin conjugates (Deveraux et al., 1994) was also found to be specific of degenerating intersegmental muscles (Takayanagi et al., 1996). Interestingly, Jones et al. (1995) reported that four new subunits were synthesized that changed both the enzymatic properties and the conformation of the proteasome in such conditions. Table 3 shows that increased expression of some subunits of the 19S complex occurred in the atrophying muscles from hindlimb suspended and tumor-bearing rats. Increased mRNA levels for two ATPases (MSS1 and P45) and two non-ATPases (P112 and P31) subunits were observed in unweighted muscles. By contrast, only the mRNA level for MSS1, which belongs to the ATPase family, prevailed in the muscles from tumor bearing animals. Since the mRNA levels for MSS1 and S4, another ATPase subunit, did not change in the atrophying muscles from IL-6 transgenic mice (Tsujinaka et al., 1996), as was seen with P45 in cancer

Table 3. Abundance of mRNA Levels for Subunits of the 19S Complex in Two Different Muscle Wasting Conditions in Rats*

	Hindlimb Suspension	*Cancer*
ATPase subunit		
MSS1 (S7)	Increased	Increased
P45 (S8)	Increased	Unchanged
Non-ATPase subunit		
P112 (S1)	Increased	Unchanged
P31 (S13/14)	Increased	ND

Notes: * From Attaix et al., 1997. Increased ubiquitin-proteasome-dependent proteolysis was previously reported to be responsible for muscle atrophy in both hindlimb suspension (Taillandier et al., 1996) and cancer (Temparis et al., 1994).

ND, not determined.

cachexia (Table 3), the expression of these subunits of the 19S complex clearly depended on a given catabolic state. The increased mRNA levels for MSS1 in muscles from both cancer and unweighted rats strongly suggested that MSS1 may play a crucial role in protein breakdown. Analysis of MSS1 mRNA distribution profile in polyribosomes showed that this subunit entered active translation in unweighted muscles (Attaix et al., 1997). ATPase subunits provide energy to help inject the substrate into the proteolytic chamber of the 20S proteasome. They also provide energy for the assembly of the 26S proteasome, which occurs by an ATP-dependent association of the 20S proteasome with the 19S complex, and for the breakdown of ubiquitylated and some non-ubiquitylated protein substrates (Coux et al., 1996). However, additional experiments showed that the increased expression of MSS1 is regulated independently of 20S proteasome subunits, and did not result in accelerated proteolysis (Attaix et al., 1997).

PA28 (or 11S regulator) also associates with the 20S proteasome and stimulates its peptidase activities, without affecting the breakdown of proteins (Dubiel et al., 1992). This hexamer is composed of two different subunits: α and β (Ahn et al., 1995). Bovine PA28 modulated the peptidase activities of the lobster muscle proteasome (Mykles, 1996). A small but reproducible increase in mRNA levels for subunits α and β was observed in unweighted muscles (Attaix et al., 1997). PA28 may play a role either in the final cleavages of peptides arizing from protein breakdown, or/and in the breakdown of some abundant endogenous peptides in muscle (e.g., anserine and carnosine). By contrast, the mRNA level for subunit α was unchanged in the muscles from cachectic cancer animals (Attaix et al., 1997).

VIII. FUNCTIONAL ASSOCIATION BETWEEN PROTEOLYTIC SYSTEMS

A coordinate stimulation of the ATP-ubiquitin-dependent proteolytic pathway with either the Ca^{2+}-dependent (Temparis et al., 1994), or the lysosomal process (Wing and Goldberg 1993; Baracos et al., 1995), or both (Furuno et al., 1990; Mansoor et al., 1996; Taillandier et al., 1996; Voisin et al., 1996), prevails in different catabolic states. This is not unexpected, assuming that these pathways may serve to eliminate different classes of proteins. The lack of detectable activation of a given proteolytic system (e.g., the Ca^{2+}-activated process in fasted rats; Wing and Goldberg, 1993) or proteinase (e.g., cathepsin B in cancer rats; Temparis et al., 1994) in a particular wasting condition is more intriguing. Several explanations may account for these observations. First, the breakdown of myofibrillar and nonmyofibrillar proteins is regulated independently (Lowell et al., 1986; Furuno et al., 1990; Wing and Goldberg, 1993). Second, the methods used to detect an activation of the different proteinases may not be sensitive enough to detect small variations in proteolysis. Third, the breakdown rate of some classes of proteins may not be systematically increased in any catabolic state. Very different stimuli known to increase muscle proteolysis, for example cancer and disuse, may differentially affect the distribution pattern of individual muscle proteins.

Whether the ubiquitin-proteasome pathway may be functionally associated to either the Ca^{2+}-dependent or the lysosomal system is still an open question. The former hypothesis is not supported by any experimental data. In addition, experiments in mdx mice did not support an alteration of the ubiquitin pathway, but clearly indicated a selective activation of m-calpain (Combaret et al., 1996) consistent with Ca^{2+}-stimulated proteolysis (Turner et al., 1988; MacLennan et al., 1991). By contrast, the lysosomal and the ubiquitin pathways could constitute a single entity. For example, Gropper et al. (1991) have shown that ubiquitylated proteins can be detected in the lysosome and that the ubiquitin-activating enzyme, E_1, was required for stress-induced lysosomal degradation of proteins.

IX. POSSIBLE APPROACHES TO PREVENT MUSCLE WASTING IN CATABOLIC CONDITIONS

Muscle wasting may be advantageous during the initial phase of pathological states, since protein mobilization provides amino acids for acute phase protein synthesis and for energy metabolism. However, because of the vital role of muscle in locomotion and respiratory effort, sustained loss of skeletal muscle may have a very negative impact upon rehabilitation and/or survival of the organism. Thus, a final goal will be to prevent muscle wasting. A first approach is to use anabolic agents which inhibit protein breakdown. The β-agonist clenbuterol blocked presumably the activation of the ubiquitin pathway in cancer rats (Costelli et al., 1995). Another strategy is to suppress the production (or direct and indirect effects) of catabolic factors. An anti-TNF antibody (Llovera et al., 1996), and the glucocorticoid receptor antagonist RU 38486 (Tiao et al., 1996) were found to be effective in tumor-bearing, and septic rats, respectively. These compounds, however, cannot easily be used in humans. For example, β-agonists have adverse secondary effects, and anti-TNF antibodies are prohibitely expensive for routine use. Thus, there is a need for low-cost 4 drugs that have no adverse properties in humans. We recently identified pentoxifylline (Combaret et al., 1997a) and torbafylline (Attaix et al., 1997), two xandthine derivatives that block TNF production (Semmler et al., 1993), as such molecules. Both drugs significantly reduced the generalized muscle wasting of Yoshida sarcoma-bearing rats by inhibiting the activation of the ubiquitin-proteasome pathway. However, torbafylline was more potent than pentoxifylline to suppress enhanced protein breakdown in these rats (Attaix et al., 1997). A third possibility is to use pharmacological inhibitors of the proteasome, or of ubiquitin conjugation (Mitch and Goldberg, 1996). This approach is not straightforward since the ubiquitin-proteasome pathway is involved in many other essential cellular processes. However, all these strategies may clearly contribute to the development of new appropriate therapies in cachectic patients.

X. CONCLUDING REMARKS

Overwhelming evidence that the ubiquitin-proteasome pathway substantially participates in the breakdown of muscle proteins has been recently provided. However, these studies raise many questions. The precise identification of proteins that are ubiquitylated in skeletal muscle is of obvious importance. Equally important is the identification of signals in proteins that are involved in the selection of proteins for ubiquitin ligation and degradation. How the proteasome has access to myofibrillar proteins is totally unknown, and one clear rate-limiting step in their degradation is their dissociation from the myofibril (Solomon and Goldberg, 1996). The significance of variations in mRNA levels for multiple components of the ubiquitin-proteasome pathway, including the regulatory subunits of the 19S and 11S complexes, in muscle atrophy must be elucidated. In almost all instances of muscle atrophy it is still unclear whether these variations reflect increased transcription, alterations in mRNA processing and transport, or in degradation rates, and had any effect on proteolytic and/or peptidase activities. Although upstream sequences of several proteasomal genes have been described, potential regulatory elements have not yet been reported. Furthermore, there seems to be a transcriptional program that favors increased expression of all 20S proteasome subunits in muscle atrophy (Mitch and Goldberg, 1996). This program, however, did not coordinately regulate the expression of subunits of the 19S and 11S complexes (Attaix et al., 1997). Finally, the role of hormones and specific nutrients (i.e., amino acids) in the activation and/or inactivation of the ubiquitin system is also extremely poorly defined. Future experiments must be designed to elucidate the signaling pathway of hormones, cytokines, and newly identified catabolic factors that may directly or indirectly regulate this pathway.

ACKNOWLEDGMENTS

We are very grateful to our colleagues who provided us with preprints, and/or unpublished data. Studies in the laboratory of the authors were supported by the Association Française contre les Myopathies, the Association pour la Recherche sur le Cancer, the Conseil Régional d'Auvergne, Hoechst Aktiengesellschaft (Wiesbaden, Germany), the Institut National de la Recherche Agronomique, and the French Ministére de l'Education Nationale de la Recherche et de la Technologie.

REFERENCES

Ahn, J. Y., Tanahashi, N., Akiyama, K, Hisamatsu, H., Noda, C., Tanaka, K., Chung, C. H., Shibmara, N., Willy, P. J., Mott, J. D., Slaughter, C. A., & DeMartino, G. N. (1995). Primary structures of two homologous subunits of PA28, a γ-interferon-inducible protein activator of the 20S proteasome. FEBS Lett. 366, 37-42.

Attaix, D., Taillandier, D., Combaret, L., Ralliére, C., Larbaud, D., Aurousseau, E., & Tanaka, K. (1997). Expression of subunits of the 19S complex and of the PA28 activator in rat skeletal muscle. Mol. Biol. Rep. 24, 95-98.

Attaix, D., Taillandier, D., Temparis, S., Larbaud, D., Aurousseau, E., Combaret, L., & Voisin, L. (1994). Regulation of ATP-ubiquitin-dependent proteolysis in muscle wasting. Reprod. Nutr. Dev. 34, 583-597.

Auclair, D., Garrel, D. R., Zerouala, A. C., & Ferland, L. H. (1997). Activation of the ubiquitin pathway in rat skeletal muscle by catabolic doses of glucocorticoids. Am. J. Physiol. 272, C1007-C1016.

Bailey, J. L., Wang, W., England, B. K., Price, S. R., Ding, X., & Mitch, W. E. (1996). The acidosis of chronic renal failure activates muscle proteolysis in rats by augmenting transcription of genes encoding proteins of the ATP-dependent ubiquitin-proteasome pathway. J. Clin. Invest. 97, 1447-1453.

Ball, E., Karlik, C. C., Beall, C. J., Saville, D. L., Sparrow, J. C., Bullard, B., & Fyrberg, E. A. (1987). Arthrin, a myofibrillar protein of insect flight muscle, is an actin-ubiquitin conjugate. Cell 51, 221-228.

Baracos, V. E., DeVivo, C., Hoyle, D. H. R., & Goldberg, A. L. (1995). Activation of the ATP-ubiquitin-proteasome pathway in skeletal muscle of cachectic rats bearing a hepatoma. Am. J. Physiol. 268, E996-E1006.

Baracos, V. E, Rodemann, H. P, Dinarello, C. A., & Goldberg A. L. (1983). Stimulation of muscle protein degradation and prostaglandin E2 release by leukocytic pyrogen (interleukin-1). N. Engl. J. Med. 308, 553-558.

Barrett, E. J., & Gelfand, R. A. (1989). The in vivo study of cardiac and skeletal muscle protein turnover. Diabetes Metab. Rev. 5, 133-148.

Barrett, E. J., Jahn, L. A., Oiveras, D. M., & Friburg, D. A. (1995). Chloroquine does not exert insulin-like actions on human forearm muscle metabolism. Am. J. Physiol. 268, E820-E824.

Baumeister, W., Walz, J., Zühl, F., & Seemüller, E. (1998). The proteasome: Paradigm of a self-conpartmentalizing protease. Cell 92, 367-380.

Belkhou, R., Béchet, D., Cherel, Y., Galluser, M., Ferrara, M., & Le Maho, Y. (1994). Effects of fasting and thyroidectomy on cysteine proteinase activities in liver and muscle. Biochim. Biophys. Acta 1199, 195-201.

Botbol, V., & Scornick, O.A. (1997). Measurement of muscle protein degradation in live mice by accumulation of bestatin-induced peptides. Am. J. Physiol. 273, E1149-E1157.

Breuillé, D., Farges, M. C., Rosé, F., Arnal, M., Attaix, D., & Obled, C. (1993). Pentoxifylline decreases body weight loss and muscle protein wasting characteristics of sepsis. Am. J. Physiol. 265, E660-E666.

Carillo, S., Pariat, M., Jariel-Encontre, I., Poulat, F., Berta, P., & Piechaczyk, M. (1996). PEST motifs are not required for rapid calpain-mediated proteolysis of c-fos protein. Biochem. J. 313, 245-251.

Combaret, L., Taillandier, D., Voisin, L., Samuels, S. E., Boespflug-Tanguy, O., & Attaix, D., (1996). No alteration in gene expression of components of the ubiquitin-proteasome proteolytic pathway in dystrophin-deficient muscles. FEBS Lett. 393, 292-296.

Costelli., P., Garcia-Martinez, C., Llovera, M., Carbo, N., Lopez-Soriano, F. J., Agell, N., Tessitore, L., Baccino, F. M., & Argiles, J. M. (1995). Muscle protein waste in tumor-bearing rats is effectively antagonized by a β_2-adrenergic agonist (clenbuterol): role of the ubiquitin-energy-dependent pathway. J. Clin. Invest. 95, 2367-2372.

Coux, O., Tanaka, K., & Goldberg, A. L. (1996). Structure and functions of the 20S and 26S proteasomes. Annu. Rev. Biochem. 65, 801-847.

Croall, D. E., & DeMartino, G. N. (1991). Calcium-activated neutral protease (calpain) system: structure, function, and regulation. Physiol. Rev. 71, 813-847.

Dahlmann, B., & Kuehn, L. (1995). The 20S/26S proteasomal pathway of protein degradation in muscle tissue. Mol. Biol. Rep. 21, 57-62.

Dahlmann, B., Kuehn, L., & Reinauer, H. (1995). Studies on the activation by ATP of the 26S proteasome complex from rat skeletal muscle. Biochem. J. 309, 195-202.

Dardevet, D., Sornet, C., Taillandier, D., Savary, I., Attaix, D., & Grizard, J. (1995). Sensitivity and protein turnover response to glucorticoids are different in skeletal muscle from adult and old rats. Lack of regulation of the ubiquitin-proteasome proteolytic pathway in aging. J. Clin. Invest. 96, 2113-2119.

Dawson, S. P., Arnold, J. E., Mayer, N. J., Reynolds, S. E., Billett, M. A., Gordon, C., Colleaux, L., Kloetzel, P. M., Tanaka, K., & Mayer, R. J. (1995). Developmental changes of the 26S proteasome in abdominal intersegmental muscles of *Manduca sexta* during programmed cell death. J. Biol. Chem. 270, 1850-1858.

Deveraux, Q., Ustrell, V., Pickart, C., & Rechsteiner, M. (1994). A 26S protease subunit that binds ubiquitin conjugates. J. Biol. Chem. 269, 7059-7061.

Driscoll, J., & Goldberg, A. L. (1989). Skeletal muscle proteasome can degrade proteins in an ATP-dependent process that does not require ubiquitin. Proc. Natl. Acad. Sci. USA 86, 787-791.

Dubiel, W., Pratt, G., Ferrell, K., & Rechsteiner, M. (1992). Purification of an 11S regulator of the multicatalytic protease. J. Biol. Chem. 267, 22369-22377.

Dubiel, W., & Rechsteiner, M. (1998). The 19S regulatory complex of the 26S proteasome. In: Advances in Molecular and Cell Biology (Rivett, A. J., Ed.), Vol. 27 pp. 129-163. JAI Press, Greenwich, CT.

Duckworth, W. C., Bennett, R. G., & Hamel, F. G. (1994). A direct inhibitory effect of insulin on a cytosolic proteolytic complex containing insulin-degrading enzyme and multicatalytic proteinase. J. Biol. Chem. 269, 24575-24580.

Ebisui, C., Tsujinaka, T., Morimoto, T., Kan, K., Iijima, S., Yano, M., Kominami, E., Tanaka, K., & Monden, M. (1995). Interleukin-6 induces proteolysis by activating intracellular proteases (cathepsins B and L, proteasome) in C2C12 myotubes. Clin. Sci. 89, 431-439.

Fagan, J. M., Wajnberg, E. F., Culbert, L., & Waxman, L. (1992). ATP depletion stimulates calcium-dependent protein breakdown in chick skeletal muscle. Am. J. Physiol. 262, E637-E643.

Fagan, J. M., & Waxman, L. (1989). A novel ATP-requiring protease from skeletal muscle that hydrolyzes non-ubiquitinated proteins. J. Biol. Chem. 264, 17868-17872.

Fagan, J. M., Waxman, L., & Goldberg, A. L. (1987). Skeletal muscle and liver contain a soluble ATP+ubiquitin-dependent proteolytic system. Biochem. J. 243, 335-343.

Fang, C., Tiao, G., James, H., Ogle, C., Fischer, J. E., & Hasselgren, P. O. (1995). Burn injury stimulates multiple proteolytic pathways in skeletal muscle, including the ubiquitin-energy-dependent pathway. J. Am. Coll. Surg. 180, 161-170.

Flores, E. A., Bistrian, B. R., Pomposelli, J. J., Dinarello, C. A., Blackburn, G. L., & Istfan, N. W. (1989). Infusion of tumor necrosis factor/cachectin promotes muscle catabolism in the rat. A synergistic effect with interleukin 1. J. Clin. Invest. 83, 1614-1622.

Forbes, G. B. (1976). The adult decline in lean body mass. Hum. Biol. 48, 161-173.

Furuno, K., Goodman, M. N., & Goldberg, A. L. (1990). Role of different proteolytic systems in the degradation of muscle proteins during denervation atrophy. J. Biol. Chem. 265, 8550-8557.

Garcia-Martinez, C., Agell, N., Llovera, M., Lopez-Soriano, F. J., & Argiles, J. M. (1993). Tumour necrosis factor-α increases the ubiquitinization of rat skeletal muscle proteins. FEBS Lett. 323, 211-214.

Garcia-Martinez, C., Llovera, M., Agell, N., Lopez-Soriano, F.J., & Argiles, J.M. (1995). Ubiquitin gene expression in skeletal muscle is increased during sepsis: Involvement of TNF-α, but not IL-1. Biochem. Biophys. Res. Commun. 217, 839-844.

Goldspink, D. F., Morton, P. A. J., Loughna, P., & Goldspink, G. (1986). The effect of hypokinesia and hypodynamia on protein turnover and the growth of four skeletal muscles of the rat. Pflügers Arch. 407, 333-340.

Goll, D. E., Thompson, V. F., Taylor, R. G., & Zalewska, T. (1992). Is calpain activity regulated by membranes and autolysis or by calcium and calpastatin? Bioessays 14, 549-556.

Gonen, H., Stancovski, I., Shkedy, D., Hadari, T., Bercovich, B., Bengal, E., Mesilati S., Abu-Atoum, O., Schwartz, A. L., & Ciechanover, A. (1996). Isolation, characterization, and partial purification of a novel ubiquitin-protein ligase, E3. J. Biol. Chem. 271, 302-310.

Goodman, M. N. (1994). Interleukin-6 induces skeletal muscle protein breakdown in rats. Proc. Soc. Exp. Biol. Med. 205, 182-185.

Gropper, R., Brandt, R. A., Elias, S., Bearer, C. F., Mayer, A., Schwartz, A. L., & Ciechanover, A. (1991). The ubiquitin-activating enzyme, E_1, is required for stress-induced lysosomal degradation of cellular proteins. J. Biol. Chem. 266, 3602-3610.

Handley, P. M., Mueckler, M., Siegel, N. R., Ciechanover, A., & Schwartz, A. L. (1991). Molecular cloning, sequence, and tissue distribution of the human ubiquitin-activating enzyme E1. Proc. Natl. Acad. Sci. USA 88, 258-262.

Haas, A. L., Baboshina, O., Williams, B., & Schwartz, L. M. (1995). Coordinate induction of the ubiquitin conjugation pathway accompanies the developmentally programmed death of insect skeletal muscle. J. Biol. Chem. 270, 9407-9412.

Haas, A. L., & Riley, D. A. (1988). The dynamics of ubiquitin pools within skeletal muscle. In: The Ubiquitin System (Schlesinger, M., & Herskho, A. eds.), pp. 178-185. Cold Spring Harbor Laboratory Press, New York.

Hasselgren, P. O. (1993). Protein Metabolism in Sepsis. R. G. Landes Co., Austin.

Hershko, A., & Ciechanover, A. (1992). The ubiquitin system for protein degradation. Annu. Rev. Biochem. 61, 761-807.

Hong, D. H., Forsberg, N. E. (1995). Effects of dexamethasone on protein degradation and protease gene expression in rat L8 myotube cultures. Mol. Cell. Endocrinol. 108, 199-209.

Ilian, M. A., & Forsberg, N. E. (1992). Gene expression of calpains and their specific endogenous inhibitor, calpastatin, in skeletal muscle of fed and fasted rabbits. Biochem. J. 287, 163-171.

Isozaki, Y., Mitch, W. E., England, B. K., & Price, S. R. (1996). Protein degradation and increased mRNAs encoding proteins of the ubiquitin-proteasome proteolytic pathway in BC_3H1 myocytes require an interaction between glucocorticoids and acidification. Proc. Natl. Acad. Sci. USA 93, 1967-1971.

Johnson, P. (1990). Calpains (intracellular calcium-activated cysteine proteinases): structure-activity relationships and involvement in normal and abnormal cellular metabolism. Int. J. Biochem. 22, 811-822.

Jones, M. E. E., Haire, M. F., Kloetzel, P.-M., Mykles, D. L., & Schwartz, L. M. (1995). Changes in the stucture and function of the multicatalytic proteinase (proteasome) during programmed cell death in the intersegmental muscles of the hawkmoth, *Manduca sexta*. Dev. Biol. 169, 436-447.

Kettelhut, I. C., Wing, S. S., & Goldberg, A. L. (1988). Endocrine regulation of protein breakdown in skeletal muscle. Diabetes Metab. Rev. 4, 751-772.

Knecht, E., de Llano, U. M., Andreu, E. J., & Miralles, I. M. (1998). Pathways for the degradation of intracellular proteins within lysosomes in higher eukaryotes. In: Advances in Molecular and Cell Biology (Rivett, A. J., ed.), Vol. 27 pp. 201-234. JAI Press, Greenwich, CT.

Larbaud, D., Debras, E., Taillandier, D., Samuels, S. E., Temparis, S., Champredon, C., Grizard, J., & Attaix, D. (1996). Euglycemic hyperinsulinemia and hyperaminoacidemia decrease skeletal muscle ubiquitin mRNA in goats. Am. J. Physiol. 271, E505-E512.

Laub, M., & Jennissen, H. P. (1991). Ubiquitination of myosin light chains. Biol. Chem. Hoppe-Seyler 372, 704.

Li, J. B., & Goldberg, A. L. (1976). Effects of food deprivation on protein synthesis and degradation in rat skeletal muscles. Am. J. Physiol. 231, 441-448.

Li, J. B., Higgins, J. E., & Jefferson, L. S. (1979). Changes in protein turnover in skeletal muscle in response to fasting. Am. J. Physiol. 236, E222-E228.

Llovera, M., Carbo, N., Garcia-Martinez, C., Costelli, P., Tessitore, L., Baccino, F. M., Agell, N., Bagby, G. J., Lopez-Soriano, F. J., & Argiles, J. M. (1996). Anti-TNF treatment reverts increased muscle ubiquitin gene expression in tumour-bearing rats. Biochem. Biophys. Res. Commun. 221, 653-655.

Llovera, M., Garcia-Martinez, C., Agell, N., Lopez-Soriano, F. J., & Argiles, J. M. (1995). Muscle wasting associated with cancer cachexia is linked to an important activation of the ATP-dependent ubiquitin-mediated proteolysis. Int. J. Cancer 61, 138-141.

Llovera, M., Garcia-Martinez, C., Agell, N., Marzabal, M., Lopez-Soriano, F. J., & Argiles, J. M. (1994). Ubiquitin gene expression is increased in skeletal muscle of tumour-bearing rats. FEBS Lett. 338, 311-318.

Lowell, B. B., Ruderman, N. B., & Goodman, M. N. (1986). Evidence that lysosomes are not involved in the degradation of myofibrillar proteins in rat skeletal muscle. Biochem. J. 234, 237-240.

MacLennan, P. A., McArdle, A., & Edwards, R. H. T. (1991). Effects of calcium on protein turnover of incubated muscles from mdx mice. Am. J. Physiol. 260, E594-E598.

Mansoor, O., Beaufrère, B., Boirie, Y., Rallière, C., Taillandier, D., Aurousseau, E., Schoeffler, P., Arnal, M., & Attaix, D. (1996). Increased mRNA levels for components of the lysosomal, Ca^{2+}-activated and ATP-ubiquitin-dependent proteolytic pathways in skeletal muscle from head trauma patients. Proc. Natl. Acad. Sci., USA 93, 2714-2718.

Medina, R., Wing, S. S., & Goldberg, A. L. (1995). Increase in levels of polyubiquitin and proteasome mRNA in skeletal muscle during starvation and denervation atrophy. Biochem. J. 307, 631-637.

Medina, R., Wing, S. S., Haas, A., & Goldberg, A. L. (1991). Activation of the ubiquitin-ATP-dependent proteolytic system in skeletal muscle during fasting and denervation atrophy. Biomed. Biochim. Acta 50, 347-356.

Medina, R., Wing, S. S., Kettelhut, I., & Goldberg, A. L. (1992). Regulation of different proteolytic systems in muscle by insulin and food intake. In: Protein Metabolism in Diabetes Mellitus (Nair, K. S., ed.), pp. 111-123. Smith Gordon, London.

Melloni, E., & Pontremoli, S. (1991). The calpain-calpastatin system: structural and functional properties. J. Nutr. Biochem. 2, 467-476.

Millward, D. J., Bates, P. C., Grimble, G. K., Brown, J. G., Nathan, M., & Rennie, M. J. (1980). Quantitative importance of non-skeletal-muscle sources of N-methylhistidine in urine. Biochem. J. 190, 225-228.

Mitch, W. E., & Goldberg, A. L. (1996). Mechanisms of muscle wasting. The role of the ubiquitin-proteasome pathway. N. Engl. J. Med. 335, 1897-1905.

Mitch, W. E., Medina, R., Grieber, S., May, R. C., England, B. K., Price, S. R., Bailey J. L., & Goldberg, A. L. (1994). Metabolic acidosis stimulates muscle protein degradation by activating the adenosine triphosphate-dependent pathway involving ubiquitin and proteasomes. J. Clin. Invest. 93, 2127-2133.

Molinari, M., Anagli, J., & Carafoli, E. (1995). PEST sequences do not influence substrate susceptibility to calpain proteolysis. J. Biol. Chem. 270, 2032-2035.

Mortimore, G. E., Kadowaki, M., & Heydrick, S. J. (1992). The autophagic pathway in liver: its regulation and role in macromolecular turnover. In: Protein Metabolism in Diabetes Mellitus (Nair, K. S., ed.), pp.125-138. Smith Gordon, London.

Mykles, D. L. (1993). Lobster muscle proteasome and the degradation of myofibrillar proteins. Enzyme Protein 47, 220-231.

Mykles, D. L. (1996). Differential effects of bovine PA 28 on six peptidase activities of the lobster muscle proteasome (multicatalytic proteinase). Arch. Biochem. Biophys. 325, 77-81.

Mykles, D. L., & Haire, M. F. (1991). Sodium dodecyl sulfate and heat induce two distinct forms of lobster muscle multicatalytic proteinase: the heat-activated form degrades myofibrillar proteins. Arch. Biochem. Biophys. 288, 543-551.

Mykles, D. L., & Haire, M. F. (1995). Branched-chain-amino acid-preferring peptidase activity of the lobster multicatalytic proteinase (proteasome) and the degradation of myofibrillar proteins. Biochem. J. 306, 285-291.

Pepato, M. T., Migliorini, R. H., Goldberg, A. L., & Kettelhut, I. C. (1996). Role of different proteolytic pathways in degradation of muscle protein from streptozotocin-diabetic rats. Am. J. Physiol. (1996). 271, E340-E347.

Pisters, P. W. T., & Brennan, M. F. (1990). Amino acid metabolism in human cancer cachexia. Annu. Rev. Nutr. 10, 107-132.

Price, S. R., Bailey, J. L., Wang, X., Jurkovitz, C., England, B. K., Ding, X., Phillips, L. S., & Mitch, W. E. (1996). Muscle wasting in insulinopenic rats results from activation of the ATP-dependent, ubiquitin-proteasome proteolytic pathway by a mechanism including gene transcription. J. Clin. Invest. 98, 1703-1708.

Price, S. R., England, B. K., Bailey, J. L., Van Vreede, K., & Mitch, W. E. (1994). Acidosis and glucocorticoids concomitantly increase ubiquitin and proteasome subunit mRNA levels in rat muscles. Am. J. Physiol. 267, C955-C960.

Rallière, C., Tauveron, I., Taillandier, D., Guy, L., Boiteux, J.-P., Giraud, B., Attaix, D., & Thiéblot, P. (1997). Glucocorticoids do not regulate the expression of proteolytic genes in skeletal muscle from Cushing syndrome patients. J. Clin. Endocrinol. Metab. 82, 3161-3164.

Rechsteiner, M. (1991). Natural substrates of the ubiquitin proteolytic pathway. Cell 66, 615-618.

Richard, I., Broux, O., Allamand, V., Fougerousse, F., Chiannilkulchai, N., Bourg, N., Brenguier, L., Devaud, C., Pasturaud, P., Roudaut, C., Hillaire, D., Passos-Bueno, M.-R., Zatz, M., Tischfield, J. A., Fardeau, M., Jackson, C. E., Cohen, D., & Beckmann, J. S. (1995). Mutations in the proteolytic enzyme calpain 3 cause limb-girdle muscular dystrophy type 2A. Cell 81, 27-40.

Rivett, A. J. (1993). Proteasomes: multicatalytic proteinase complexes. Biochem. J. 291, 1-10.

Roberts, R. G., Redfern, C. P. F., & Goodship, T. H. J. (1996). Changes in the expression of proteolytic enzymes following correction of acidosis in humans with renal failure. In: Abstract Book of the 4th International Symposium on Amino Acid/Protein Metabolism in Health and Disease. Padova, Italy, April 12-13, Abstract 1.

Rock, K. L., Gramm, C., Rothstein, L., Clark, K., Stein, R., Dick, L., Hwang, D., & Goldberg, A. L. (1994). Inhibitors of the proteasome block the degradation of most cell proteins and the generation of peptides presented on MHC class I molecules. Cell 78, 761-771.

Rogers, S., Wells, R., & Rechsteiner, M. (1986). Amino acids sequences common to rapidly degraded proteins: the PEST hypothesis. Science 234, 364-368.

Sakamoto, K., Sorimachi, H., Kinbara, K., Tezuka, M., Amano, S., Yoshizawa, T., Sugita, H., Ishiura, S., & Suzuki, K. (1994). Quantification of calpain-related molecules by specific PCR amplification and its application to human muscular dystrophy. Biomed. Res. 15, 337-346.

Samuels, S. E., Taillandier, D., Aurousseau, E., Cherel, Y., Le Maho, Y., Arnal, M., & Attaix, D. (1996). Gastrointestinal tract protein synthesis and mRNA levels for proteolytic systems in adult fasted rats. Am. J. Physiol. 271, E232-E238.

Scheiwiller, E., Guller, H.-P., Merryweather, J., Scandella, C., Maerki, W., Zapf, J., & Froesch, E. R. (1986). Growth restoration of insulin-deficient diabetic rats by recombinant human insulin-like growth factor I. Nature (London) 323, 169-171.

Semmler, J., Gebert, U., Eisenhut, T., Moeller, J., Schönharting, M. M., Alléra, A., & Endres, S. (1993). Xanthine derivatives: comparison between suppression of tumour necrosis factor-α production and inhibition of cAMP phosphodiesterase activity. Immunology 78, S20-525.

Smith, L. K., Bradshaw, M., Croall, D. E., & Garner, C. W. (1993). The insulin receptor substrate (IRS-1) is a PEST protein that is susceptible to calpain degradation in vitro. Biochem. Biophys. Res. Commun. 196, 767-772.

Solomon, V., & Goldberg, A.L. (1996). Importance of the ATP-ubiquitin-proteasome pathway in the degradation of soluble and myofibrillar proteins in rabbit muscle extracts. J. Biol. Chem. 271, 26690-26697.

Sorimachi, H., Imajoh-Ohmi, S., Emori, Y., Kawasaki, H., Ohno, S., Minami, Y., & Suzuki, K. (1989). Molecular cloning of a novel mammalian calcium-dependent protease distinct from both m- and mu type: specific expression of the mRNA in skeletal muscle. J. Biol. Chem. 264, 20106-20111.

Sorimachi, H., Ishiura, S., & Suzuki, K. (1997). Structure and physiological function of calpains. Biochem. J. 328, 721-732.

Sorimachi, H., Toyama-Sorimachi, N., Ishiura, S., & Suzuki, K. (1993). Identification and localization of a novel muscle-specific calpain, p94. In: Proteolysis and Protein Turnover (Bond, J. S., & Barrett, A. J., eds.), pp. 45-49. Portland Press, London.

Spencer, M. J., Croall, D. E., & Tidball, J. G. (1995). Calpains are activated in necrotic fibers from mdx dystrophic mice. J. Biol. Chem. 270, 10909-10914.

Sugden, P. H., Fuller, S. J. (1991). Regulation of protein turnover in skeletal and cardiac muscle. Biochem. J. 273, 21-37.

Taillandier, D., Aurousseau, E., Meynial-Denis, D., Béchet, D., Ferrara, M., Cottin, P., Ducastaing, A., Bigard, X., Guezennec, C.-Y., Schmid, H.-P., & Attaix, D. (1996). Coordinate activation of lysosomal, Ca^{2+}-activated and ATP-ubiquitin-dependent proteinases in the unweighted rat soleus muscle. Biochem. J. 316, 65-72.

Takayanagi, K., Dawson, S., Reynolds, S. E., & Mayer, R. J. (1996). Specific developmental changes in the regulatory subunits of the 26S proteasome in intersegmental muscles preceding eclosion in *Manduca sexta*. Biochem. Biophys. Res. Commun. 228, 517-523.

Tanaka, K., & Tanahashi, N. (1998). The 20S proteasome: Subunits and functions. In: Advances in Molecular and Cell Biology (Rivett, A. J., Ed.), Vol. 27 pp. 105-128. JAI Press, Greenwich, CT. In press.

Tawa, N. E., Jr., & Goldberg, A. L. (1994). Protein and amino acid metabolism in muscle. In: Myology (Engel, A. G., & Franzini-Armstrong, C., eds.), pp. 683-707, Mc Graw-Hill New York.

Tawa, N. E., Jr., Kettelhut, I. C., & Goldberg, A. L. (1992). Dietary protein deficiency reduces lysosomal and nonlysosomal ATP-dependent proteolysis in muscle. Am. J. Physiol. 263, E326-E334.

Tawa, Jr., N.E., Odessey, R., & Goldberg, A.L. (1997). Inhibitors of the proteasome reduce the accelerated proteolysis in atrophying rat skeletal muscles. J. Clin. Invest. 100, 197-203.

Taylor, R. G., Tassy, C., Briand, M., Robert, N., Briand, Y., & Ouali, A. (1995). Proteolytic activity of proteasome on myofibrillar structures. Mol. Biol. Rep. 21, 71-73.

Temparis, S., Asensi, M., Taillandier, D., Aurousseau, E., Larbaud, D., Obled, A., Béchet, D., Ferrara, M., Estrela, J. M., & Attaix, D. (1994). Increased ATP-ubiquitin-dependent proteolysis in skeletal muscles of tumor-bearing rats. Cancer Res. 54, 5568-5573.

Tesseraud, S., Grizard, J., Debras, E., Papet, I., Bonnet, Y., Bayle, G., & Champredon, C. (1993). Leucine metabolism in lactating and dry goats: effect of insulin and substrate availability. Am. J. Physiol. 265, E402-E413.

Thomason, D. B., & Booth, F. W. (1990). Atrophy of the soleus muscle by hindlimb unweighting J. Appl. Physiol. 68, 1-12.

Tiao, G., Fagan, J. M., Roegner, V., Lieberman, M., Wang, J. J., Fischer, J. E., & Hasselgren, P. O. (1996). Energy-ubiquitin-dependent muscle proteolysis during sepsis is regulated by glucocorticoids. J. Clin. Invest. 97, 339-348.

Tiao, G., Fagan, J. M., Samuels, N., James, J. H., Hudson, K., Lieberman, M., Fischer, J. E., & Hasselgren, P. O. (1994). Sepsis stimulates nonlysosomal, energy-dependent proteolysis and increases ubiquitin mRNA levels in rat skeletal muscle. J. Clin. Invest. 94, 2255-2264.

Tischler, M., Desautels, M., & Goldberg, A. L. (1982). Does leucine, leucyltRNA or some metabolite of leucine regulate protein synthesis and degradation in skeletal and cardiac muscle? J. Biol. Chem. 257, 1613-1621.

Tischler, M. E., Rosenberg, S., Satarug, S., Henriksen, E. J., Kirby, C. R., Tome, M., & Chase, P. (1990). Different mechanisms of increased proteolysis in atrophy induced by denervation or unweighting of rat soleus muscle. Metabolism 39, 756-763.

Todorov, P., Cariuk, P., McDevitt, T., Coles, B., Fearon, K., & Tisdale, M. (1996). Characterization of a cancer cachectic factor. Nature (London) 379, 739-742.

Tsujinaka, T., Fujita, J., Ebisui, C., Yano, M., Kominami, E., Suzuki, K., Tanaka, K., Katsume, A., Ohsugi, Y., Shiosaki, H., & Monden, M. (1996). Interleukin 6 receptor antibody inhibits muscle atrophy and modulates proteolytic systems in interleukin 6 transgenic mice. J. Clin. Invest. 97, 244-249.

Turner, P. R., Westwood, T., Regen, C. M., & Steinhardt, R. A. (1988). Increased protein degradation results from elevated free calcium levels found from mdx mice. Nature (London) 335, 735-738.

Voisin, L., Breuillé, D., Combaret, L., Pouyet, C., Taillandier, D., Aurousseau, E., Obled, C., & Attaix, D. (1996). Muscle wasting in a rat model of long lasting sepsis results from the activation of lysosomal, Ca^{2+}-activated and ubiquitin-proteasome proteolytic pathways. J. Clin. Invest. 97, 1610-1617.

Wassner, S. J., & Li, J. B. (1982). N-methylhistidine release: Contributions of rat skeletal muscle, gastrointestinal tract and skin. Am. J. Physiol. 243, E293-E297.

Waterlow, J. C., Garlick, P. J., & Millward, D. J. (1978). Protein Turnover in Mammalian Tissues and in the Whole Body. Elsevier North Holland, Amsterdam.

Wefes, I., Mastrandrea, L. D., Haldeman, M., Koury, S. T., Tamburlin, J., Pickart, C. M., & Finley, D. L. (1995). Induction of ubiquitin-conjugating enzymes during terminal erythroid differentiation. Proc. Natl. Acad. Sci. USA 92, 4982-4986.

Wilkinson, K. D. (1998). Cellular regulation by ubiquitin-dependent processes. In: Advances in Molecular and Cell Biology (Rivett, A. J., ed.) Vol. 27 pp. 71-104. JAI Press, Greenwich, CT. In press.

Wing, S. S., & Banville, D. (1994). 14-kDa ubiquitin-conjugating enzyme: structure of the rat gene and regulation upon fasting and by insulin. Am. J. Physiol. 267, E39-E48.

Wing, S. S., & Bedard, N. (1996). Insulin-like growth factor I stimulates degradation of a mRNA transcript encoding the 14 kDa ubiquitin-conjugating enzyme. Biochem. J. 319, 455-461.

Wing, S. S., Chiang, A. L., Goldberg, A. L., & Dice, J. F. (1991). Proteins containing peptide sequences related to lys-phe-glu-arg-gln are selectively depleted in liver and heart but not skeletal muscle of fasted rats. Biochem. J. 275, 165-169.

Wing, S. S., & Goldberg, A. L. (1993). Glucocorticoids activate the ATP-ubiquitin-dependent proteolytic system in skeletal muscle during fasting. Am. J. Physiol. 264, E668-E676.

Wing, S. S., Haas, A. L., & Goldberg, A. L. (1995). Increase in ubiquitin-protein conjugates concomitant with the increase in proteolysis in rat skeletal muscle during starvation and atrophy denervation. Biochem. J. 307, 639-645.

Wing, S. S., & Jain, P. (1995). Molecular cloning, expression and characterization of a ubiquitin conjugation enzyme ($E2_{17kB}$) highly expressed in rat testis. Biochem. J. 305, 125-132.

Young, V. R., & Munro, H. N. (1978). Nτ-methylhistidine (3-methylhistidine) and muscle protein turnover: an overview. Fed. Proc. 37, 2291-2300.

PROTEOLYSIS IN APOPTOSIS:
ENZYMES AND SUBSTRATES

Grant G.F. Mason and Jill Beyette

Advances in Molecular and Cell Biology
Volume 27, pages 267-290.

ISBN: 0-7623-0387-5

I. INTRODUCTION

Apoptosis or programmed cell death is a genetically controlled and highly regulated process through which multicellular organisms remove unwanted cells, and might be considered a specific form of terminal differentiation (Schwartz, 1991). It also occurs in response to a wide range of toxic stimuli. Apoptosis is important during embryogenesis, for homeostasis of tissue mass, and for prevention of disease (Raff, 1992; Vaux and Strasser, 1996). If apoptosis occurs too much the results are impaired developement or degenerative disease. On the other hand too little may also result in impaired development as well as tumor formation and sustained viral or bacterial infection. Unlike necrosis which results from plasma membrane injury and causes cells to swell, apoptosis is characterized by cell shrinkage, the chromatin becomes pyknotic and may form crescent shaped bodies along the nuclear envelope, the nucleus may break up, the cell produces processes or buds (blebbing) which often contain nuclear fragments, and the DNA is cleaved into intranucleosomal pieces giving rise to characteristic 185 bp fragments (Majno and Joris, 1995).

Early evidence that proteases may play a role in apoptosis came from studies of granzyme B, a component of cytotoxic T cell and natural killer cell granules. It was shown that another constituent of these granules perforin (a pore-forming protein) in combination with granzyme B alone was sufficient to induce apoptosis (Shi et al., 1992). It has been shown more recently that many protease inhibitors of different specificities can inhibit programmed cell death (Kumar and Harvey, 1995). It has now become obvious that a number of proteases play regulatory roles in apoptosis. Whether they function as executors of cell death has yet to be elucidated. This review focuses on the major classes of proteases that have been ascribed roles in apoptosis and presents the evidence for their involvment in programmed cell death.

II. THE ICE FAMILY OF PROTEASES AND THEIR SUBSTRATES

A. *Caenorhabditis elegans* Death Genes

During development of the nematode *Caenorhabditis elegans* 1,090 cells are generated of which 131 subsequently undergo apoptosis (Sulston and Horvitz, 1977; Sulston et al., 1983). A growing number of genes have been identified that play a role in cell death in *C. elegans*. Three genes, *ced-3, ced-4,* and *ced-9*, have been shown to play vital roles in the initiation or execution of cell death while others are required for the phagocytosis and removal of dead cells (Hengartner and Horvitz, 1994). Both *ced-3* and *ced-4* are required for cell death to occur. In recessive mutants for these genes, with few exceptions, all cells that normally die during development survive (Ellis and Horvitz, 1986). These extra cells do not undergo subsequent division but differentiate and appear to have normal function; function that would normally be carried out by other cells. *Ced-9*, on the other hand, antagonizes the function of *ced-3* and *ced-4* by protecting cells from apoptosis. Mutants deficient in *ced-9* function die early in development due to the death of cells that would normally survive (Hengartner et al., 1992).

The *ced-4* gene is known to encode a novel 63 kDa protein and, like *ced-3*, to be expressed largely during the time of development when apoptosis occurs (Yuan and Horvitz, 1992). Only very recently has a mammalian homologue been found (Zou et al., 1997). Ced-4 and its mammalian counterpart Apaf-1 appear to promote the activation of ced-3 and the caspases, respectively (reviewed in Hengarten, 1997). However, earlier cloning of the *ced-3* gene revealed that the deduced ced-3 protein is 503 amino acids in length and is similar (29% identical in amino acid sequence) to the human interleukin-1beta (IL-1β)-converting enzyme (ICE) (Yuan et al., 1993), a cysteine protease responsible for processing the inactive precursor of IL-1β to the active cytokine (Thornberry et al., 1992). The overexpression of *ICE*, like that of *ced-3*, in mammalian cells causes apoptosis (Miura et al., 1993), while overexpression of *ced-9* or its homologues such as *Bcl-2* prevent apoptosis (reviewed in Nuñez and Clarke, 1994), suggesting a similarity of function and an evolutionary conservation of the apoptotic mechanism.

B. Interleukin-1β-Converting Enzyme

IL-1β, is a major effector of inflammatory and autoimmune reactions, responsible for the increased synthesis of other cytokines and prostaglandins (reviewed in Dinarello, 1991), and is produced as an inactive 31 kDa precursor (proIL-1β) requiring processing to the 17 kDa active cytokine. ICE, first purified (Black et al., 1989; Kostura et al., 1989) and subsequently cloned (Cerretti et al., 1992; Thornberry et al., 1992) from monocytic cells, was shown to cleave proIL-1β between Asp 116 and Ala 117 to the active form.

ICE is a heteromeric enzyme composed of 10 kDa (p10) and 20 kDa (p20) subunits, both of which are required for catalytic activity (Thornberry et al., 1992; Miller et al., 1993a). ICE has an unusual substrate specificity in that it displays a very strong preference for Asp in the P_1 and small hydrophobic residues in the P_1' positions (Thornberry et al., 1992). Sequence analysis showed that both subunits arise from a common 45 kDa precursor (p20, residues 120–297; p10 residues 317–404) by cleavage of Asp-X bonds where X equals Asn 120, Ser 298, and Ala 317. Active ICE has been shown to be able to cleave its own precursor but the precursor appears to be unable to undergo autoprocessing (Thornberry et al., 1992). ICE, however, does not cleave all proteins with Asp-X bonds. To date, the only known physiological substrates of ICE are human and murine proIL-1β.

Based on inhibition studies and structural information (see below), ICE has been classified as a novel cysteine protease. A number of agents such as *p*-chloromercurobenzoate, iodoacetamide (Miller et al., 1993b), and iodoacetate (Thornberry et al., 1992) which interact with cysteine, and diethylpyrocarbanate, which interacts with histidine, have been shown to inactivate ICE in a competitive manner (Wilson et al., 1994), suggesting a role for cysteine and histidine in the catalytic mechanism. Indeed, inhibition by [^{14}C]iodoacetate was used to identify the catalytic cysteine as the final residue of the heptapeptide Val-Ile-Ile-Ile-Gln-Ala-Cys derived from p20. However, ICE is insensitive to other classical inhibitors of cysteine proteases such as E64 and antipain but is inhibited by the serine protease inhibitors N-tosyl-L-phenylalanine chloromethyl ketone and 3,4-dichloroisocoumarin (Wilson et al., 1994).

The best peptide substrate for ICE is Ac-Tyr-Val-Ala-Asp-NH-CH_3, having a relative V_{max}/K_m comparable to proIL-1β (Thornberry et al., 1992). It is the basis for specific competitive inhibitors such as the aldehyde inhibitor Ac-Tyr-Val-Ala-Asp-CHO which prevents the release of IL-1β from activated human blood monocytes. Thus, ICE provides a potential target site for anti-inflammatory agents.

The crystal structure of ICE bound to the tetrapeptide aldehyde inhibitor Ac-Tyr-Val-Ala-Asp-CHO has been solved (Walker et al., 1994; Wilson et al., 1994) and reveals that ICE is a tetramer composed of two p10/p20 dimers. The p10/p20 subunits of each heterodimer are very closely associated, with the p20 almost surrounding the p10 to form a single domain. The oligomerization sites appear to be on the p10 between residues 318–322 and 386–396, 13 of which are conserved between the human and mouse enzymes (Nett et al., 1992; Thornberry et al., 1992; Wilson et al., 1994). Mutation of His 322, which lies 35 Å from the catalytic site abrogates ICE activity.

The active site comprises residues from both p10 and p20. The aldehyde inhibitor used was found to be covalently bound at Cys 285, in agreement with the labeling study above. Mutagenesis studies confirmed that this is the catalytic cysteine. The imidazole side-chain of His 237 is in close proximity and is believed, along with Gly 238, to also play a role in catalysis. Both subunits contribute to specific substrate recognition with Arg 179, Gln 283, and Arg 341 forming a binding pocket for the carboxylate side-chain of the P_1 aspartic acid. Mutation of His 237 or Arg 179 also abolishes or greatly decreases enzymatic activity (Wilson et al., 1994).

The distance from the p20 COOH-terminus to the p10 NH_2-terminus is 67 Å, a distance greater than could be bridged by the 19 amino acids lost during the processing of p45. This suggests that, unless gross conformational changes take place following processing, a p10/p20 dimer is derived from two different p45 precursors.

While the studies above provide evidence that ICE plays an important role in the apoptotic mechanism, evidence exists to suggest that this may not be the whole story. For instance, the only known substrate of ICE is IL-1β, yet a number of cell lines stably secrete IL-1β without undergoing apoptosis. Mice deficient in ICE develop normally and respond normally to most apoptosis-promoting stimuli, yet do not produce active IL-1β (Kuida et al., 1995; Li et al., 1995). In an *in vitro* model of apoptosis using extracts from chicken DU249 cells committed to apoptosis there was no cleavage of proIL-1β, the ICE substrate (Lazebnik et al., 1994). However, the extracts did contain an enzyme activity which cleaved the nuclear enzyme poly (ADP-ribose) polymerase (PARP) to an 85 kDa product that is characteristic of apoptosis in intact cells. Cleavage was seen to occur at a tetrapeptide sequence, Glu-Val-Asp ↓ Gly, identical to one of the ICE cleavage sites in proIL-1β with Asp at the P_1 position. Inhibitor studies ruled out the possibility of this cleavage being due to granzyme B and the enzyme activity was named prICE. prICE, however does not cleave proIL-1β nor does ICE cleave PARP (Lazebnik et al., 1994). Thus, proteases other than ICE probably play roles in the apoptotic mechanism.

C. ICE-Like Proteases

Relationship to ICE and Ced-3

A growing number of proteins related to ced-3 and ICE have now been identified and comprise a novel family of cysteine proteases, the ICE-like proteases. Ten family members have been characterized so far and have been divided by phylogenetic analysis into three subfamilies according to their degree of identity with ICE and ced-3 (see Table 1) (Alnemri et al., 1996; Chinnaiyan and Dixit, 1996; Duan et al., 1996a).

As many members of this family of proteases have multiple names, a new nomenclature for the human enzymes has been put forward (Alnemri et al., 1996; see Table 1). It is proposed to call the members of this protease family "caspases" (C = cysteine proteases, aspase = cleavage after asp). The individual members are referred to by arabic numerals based on the order of original publication (Table 1). This review will use the original names with which most people are familiar at the present time.

Table 1. The Family of ICE-Like Proteases

Nedd2 Subfamily	
Nedd2/Ich-1	caspase-2
ICE-LAP6	caspase-9
ICE Subfamily	
ICE	caspase-1
ICE_{rel}II/TX/Ich-2	caspase-4
ICE_{rel}III/TY	caspase-5
Ich-3	
Ced-3 Subfamily	
Ced-3	
CPP32/Yama/Apopain	caspase-3
ICE-LAP3/Mch-3/CMH-1	caspase-7
FLICE/MACH	caspase-8
Mch-2	caspase-6

The Nedd2 subfamily comprises Nedd2, the product of a developmentally down-regulated mouse gene identified by subtraction cloning (Kumar et al., 1994), its human homologue Ich-1 (Wang et al., 1994), and ICE-LAP6 (Duan et al., 1996b) (see Table 1). The *Nedd2/Ich-1* genes encode proteins which share 29% and 27% identity with ICE and 31% and 28% identity with ced-3, respectively. *ICE-LAP6* encodes a protein which shares 31% sequence identity to ced-3 and approximately 26% to ICE.

The ICE subfamily comprises ICE (Cerretti et al., 1992; Thornberry et al., 1992), ICE_{rel}II/TX/Ich-2 (Faucheu et al., 1995; Kamens et al., 1995; Munday et al., 1995), ICE_{rel}III/TY (Munday et al., 1995; Faucheu et al., 1996), and Ich-3 (Wang et al., 1996a) (see Table 1). These subfamily members all share a greater degree of identity to each other and ICE than ced-3 or Nedd2/Ich-1. The *ICE_{rel}II/TX/Ich-2* gene encodes a protein that shares 53% identity with ICE, 27% with Ich-1, and 26% with ced-3. ICE_{rel}III/TY shares 51% identity with ICE, 27% with Ich-1, 24% with ced-3, and 74% with ICE_{rel}II/TX/Ich-2 (Faucheu et al., 1995, 1996; Kamens et al., 1995; Munday et al., 1995). The Ich-3 protein shares approximately 45% identity with ICE, 60% with TX, 26% with ced-3, and 30% with Ich-1.

The ced-3 subfamily comprises ced-3 (Yuan et al., 1993), CPP32/Yama/Apopain (Fernandes-Alnemri et al., 1994; Nicholson et al., 1995; Tewari et al., 1995a), Mch-2 (Fernandes-Alnemri et al., 1995a), ICE-LAP3/Mch-3/CMH-1 (Fernandes-Alnemri et al., 1995b; Duan et al., 1996a; Lippke et al., 1996), and FLICE/MACH (Boldin et al., 1995; Muzio et al., 1996). The members of this subfamily share a greater degree of identity to each other or ced-3 than ICE or Nedd2/Ich-1. The *CPP32/Yama/Apopain* gene encodes a protein which shares 35% identity with ced-3, 30% identity with ICE, and 30% identity with Nedd2. Mch-2α shares 38% identity with CPP32/Yama/Apopain, 35% with ced-3, 29% with Ich-1, and 29% with ICE. *ICE-LAP3/Mch3/CMH-1* encodes a protein which shares > 50% identity with CPP32, approximately 35% identity with ced-3, and < 30% identity with members of the other subfamilies.

Structural Features of the ICE-Like Proteases

All these ICE-like proteases appear to share common structural features. With the exception of ICE-LAP6 and FLICE, all have the conserved active site QACRG pentapeptide. This amino acid sequence is replaced by QAC<u>G</u>G in ICE-LAP6 and with QAC<u>Q</u>G in FLICE. The central cysteine of this pentapeptide has been shown in all cases to be required for catalytic activity. Also conserved among these proteases are the amino acid residues that form the P_1 binding pocket conferring the specificity for aspartic acid in the P_1 site characteristic of these proteases. The residues forming the P_2–P_4 sites are less well conserved, presumably allowing for different substrate preferences among these proteases. All the members of this family of proteases contain several Asp residues in putative cleavage sites that would generate subunits similar to the p10/p20 subunits of ICE. Indeed, it is in these putative subunit regions that the greatest degree of sequence conservation is observed. All of the ICE-like proteases are synthesized as inactive zymogens and have an N-terminal pro-domain that may be long (up to over 100 amino acids) as in the members of the ICE and Nedd2 subfamilies. The Ced-3 subfamily members, with the exception of FLICE, have short pro-domains. The long pro-domain of FLICE has specific functional significance as detailed below. While these proteases appear structurally and enzymatically to be very similar to ICE none has been reported to efficiently cleave the ICE substrate pro-IL-1β at physiological concentrations.

Evidence for the Involvment of ICE-Like Proteases in Apoptosis

ICE-like proteases appear to be expressed in most tissues and cell lines studied and it would appear that most, if not all, might be expressed in a particular cell type (Fernandes-Alnemri et al., 1995b). Overexpression of these proteases in a variety of mammalian and insect cells produces apoptosis. The degree of apoptosis varies somewhat depending on the protease and cell type and may depend on the efficiency of cleavage of the pro-domain (Munday et al., 1995). Mutation of the presumed catalytic cysteine residue prevents apoptosis. While there is evidence that some of these proteases may undergo autoprocessing when ectopically expressed (Faucheu et al., 1995, 1996; Lippke et al., 1996) it cannot be ruled out that other proteases are responsible for the initial pro-domain cleavage. Indeed, when members of this family have been purified as full length zymogens they do not undergo autoprocessing, whereas expressed ectopically they do cause apotosis (Kumar et al., 1994; Fernandes-Alnemri et al., 1995b; Tewari et al., 1995a). Recently, ICE-LAP3 has been shown to be activated by cathepsin G at a Gln-Ala bond (Zhou and Salvesen, 1997).

Some of the ICE-like proteases have been reported to be expressed as multiple forms arising from splicing variants including ICE itself, Ich-1, Mch2, Mch3 and CPP32 (Fernandes-Alnemri et al., 1994, 1995a,b; Wang et al., 1994; Alnemri et al., 1995). The *ICE* gene encodes at least five different transcripts. Two of these, ICEδ, which lacks the cleavage sites between the p20 and p10 subunits, and ICEε, in which the pro-region and the p20 subunit are deleted, are inactive. ICEε can assemble with ICEα p20 to form an inactive complex. Indeed, coexpression of ICEε and ICEδ with ICEα in sf9 cells resulted in an extended survival of the cells in a manner analogous to that seen with Bcl-2 (Alnemri et al., 1995). Thus, ICEε at least appears to be able to act as an inhibitor of ICE activity. The *Ich-1* gene generates two transcripts. In addition to the form described above, designated Ich-1_L, there is a splice variant encoding a protein that terminates 21 amino aacids after the active site pentapeptide and thus lacks the equivalent of ICEs p10 subunit, Ich-1_S. In contrast to the case for Ich-1_L, the heterologous expression of Ich-1_S does not cause apoptosis. Indeed, expression of Ich-1_S in rat-1 fibroblasts protects them from apoptosis due to serum deprivation (Wang et al., 1994). Mch2 also gives rise to two isoforms, Mch2α the full length protein described above and Mch2β which lacks amino acids 14–102 of the α form and thus lacks approximately half the p20 subunit equivalent of ICE

(Fernandes-Alnemri et al., 1995a). Mch2β appears to be inactive as a protease and does not cause apoptosis when expressed heterologously. Whether it acts as an inhibitor of apoptosis is not known (Fernandes-Alnemri et al., 1995a). *Mch3* also gives rise to a splice variant Mch3β which because of insertion/deletion changes causes a change in the reading frame after amino acid 148 of Mch3α. This protein does not contain the active site QACRG and is inactive (Fernandes-Alnemri et al., 1995b). Thus, splice variants cause these genes to promote or prevent apoptosis and therefore control of mRNA splicing may act as a regulatory check-point in apoptosis (Wang et al., 1994).

D. Substrates of the Ice-Like Proteases

Poly(ADP-Ribose) Polymerase

The 116 kDa nuclear protein poly(ADP-ribose)polymerase, PARP, was observed to be cleaved to produce a characteristic 85 kDa fragment early in apoptosis induced by a variety of agents and γ-radiation (Kaufmann, 1989; Kaufmann et al., 1993). As mentioned above, an ICE-like activity known as prICE was found to be responsible for the cleavage of PARP in apoptotic cell extracts (Lazebnik et al., 1994). The ICE inhibitor Ac-Tyr-Val-Ala-Asp-chloromethyl ketone blocks prICE-induced apoptosis in a cell-free system (Lazebnik et al., 1993) suggesting that prICE activity lies towards the begining of an apoptotic pathway. CPP32 was the first ICE-like protease to be shown to be able to cleave PARP and was, indeed, purified as the PARP-cleaving protease (Nicholson et al., 1995; Tewari et al., 1995a). Attenuation of apoptotic changes could be achieved by the use of an inhibitor of CPP32-mediated PARP cleavage, Ac-Asp-Glu-Val-Asp-CHO, but not by the ICE inhibitor Ac-Tyr-Val-Ala-Asp-CHO. When CPP32 was depleted from cytosolic fractions of apoptotic osteosarcoma cells, these PARP cleavage-deficient extracts were unable to cause significant apoptotic changes to healthy nuclei. Addition of CPP32 but not of ICE to these extracts caused apoptotic changes to the nuclei (Nicholson et al., 1995). However, in the absence of a cytosolic fraction CPP32 alone could not cause apoptotic changes in the nuclei. Thus, it was concluded that CPP32 was necessary but not sufficient for apoptosis. It is of interest to note here that although major nuclear changes are characteristic of apoptosis these experiments and the immunolocalization of ICE-LAP3 to the cytoplasmic compartment (Duan et al., 1996a) suggest that the apoptotic effector machinery resides within the cytoplasm.

With regard to PARP cleavage it is interesting to note that CrmA, which has been shown to inhibit apoptosis in many systems (Gagliardini et al., 1994; Wang et al., 1994; Boudreau et al., 1995; Tewari and Dixit, 1995), was seen to only weakly inhibit PARP cleavage by CPP32 (Nicholson et al., 1995) in contrast to the findings of Tewari et al. (1995a). Indeed, CrmA was observed to have a 10,000-fold preference for ICE (Nicholson et al., 1995). It is more likely that CPP32 is part of a cascade mechanism whereby apoptotic signals cause activation of other proteases which in turn activate CPP32 which may in turn activate others leading to apoptosis and PARP cleavage. Other ICE-like proteases have been shown to cleave PARP including CMH-1, Mch2, Mch3, ICE-LAP6, and FLICE. TX, Nedd2 and ICE have been reported also to cleave PARP albeit at nonphysiological concentrations (Gu et al., 1995).

Interestingly, Mch3 can be cleaved by CPP32 but not vice versa. The cleaved, activated Mch3 can then cleave PARP. Thus, while CPP32 activity may be responsible for early PARP cleavage, Mch3 may be involved at later stages. PARP-null mice, however, develop normally and thus PARP cleavage cannot be essential for apoptosis (Wang et al., 1995a). PARP cleavage may thus simply remove one mechanism of DNA repair rather than directly lead to cell death. Indeed another enzyme involved in DNA repair, DNA-dependent protein kinase, has been shown recently to be cleaved during apoptosis (Song et al., 1996). Alternatively, Earnshaw (1995) has suggested that PARP is cleaved to conserve cellular nicotinamide-adenine dinucleotide and ATP stores required for the execution of apoptosis.

Lamins

The cleavage of nuclear lamins has been shown to be an early event in apoptosis (Neamati et al., 1994) resulting in nuclear disassembly. Lamins A, B1/B2, and C are cleaved by an ICE-like protease activity of which Mch2 is a constituent (Lazebnik et al., 1995; Orth et al., 1996a). However, two pieces of evidence point to the lamin and PARP cleaving enzymes being distinct. Firstly, although the ICE inhibitor Ac-Tyr-Val-Ala-Asp-chloromethyl ketone inhibited both PARP cleavage and lamin cleavage in cell-free extract the latter cleavage was approximately 100-fold more sensitive to inhibition. Secondly, toslylysine-chloromethyl ketone was shown to inhibit lamin cleavage but not PARP cleavage in the same system (Lazebnik et al., 1995).

Sterol Regulatory Element Binding Proteins

Sterol regulatory element binding proteins SREBP-1 and -2 have been shown to be cleaved *in vitro* by CPP32 at an Asp between the basic helix-loop-helix leucine zipper domain and the first transmembrane domain (Wang et al., 1995b). In Chinese hamster ovary (CHO) cells induced to undergo apoptosis by staurosporine it was observed that CPP32 was activated and that under these conditions SREBP-1/2 were cleaved to transcriptionally active fragments in a manner distinct from sterol-regulated activation (Wang et al., 1996b).

70 kDa U-1 Protein

The 70 kDa protein subunit of the U-1 small nuclear riboprotein (U-1snRNP) is cleaved during apoptosis induced by a number of agents (Casciola-Rosen et al.,

1994; Tewari et al., 1995b). The cleavage is inhibitable by CrmA (Tewari et al., 1995b) but not by Ac-Tyr-Val-Ala-Asp-CHO (Casciola-Rosen et al., 1994) , suggesting that an ICE-like protease other than ICE itself is responsible.

Others

Protein kinase Cδ has been shown to be cleaved at an Asp residue in apoptotic cells by a mechanism blocked by Bcl2 and Bcl-x_L. Cleavage is also blocked by Ac-Tyr-Val-Ala-Asp-CHO (Emoto et al., 1995). A component of the microfilament system, Gas2, has been shown to be cleaved during apoptosis and this cleavage is dependent on an Asp residue at position 279 (Brancolini et al., 1995). D4 guanine nucleotide dissociaton inhibitor for the Ras-related Rho family of GTPases is reported to be a substate of CPP32/Yama/Apopain (Na et al., 1996).

E. Ordering of the ICE-Like Proteases in the Apoptotic Pathway

At present it is unknown if the ICE-ike proteases act in a single sequential pathway in concert with other effectors of cell death or, most likely, whether they act in a number of parallel pathways which can interact at a number of levels. The ordering of ICE-like proteases within these pathways will be important for understanding the execution mechanism of apoptosis (Discussed in Cohen, 1997; Savesen and Dixit, 1997). Recently using a cell-free system to determine the effect of various ICE-like proteases of the Ced-3 subfamily on a number of apoptotic indicators some evidence for ordering these proteases in the apoptotic pathway has been observed (Orth et al., 1996b). Using this system it was shown that members of this family have distinct activities in the process of apoptosis. For instance, Mch2 and native extracts were shown to induce chromatin condensation and the formation of apoptotic bodies whereas Yama or ICE-LAP3 caused the condensation of chromatin at the nuclear rim but little apoptotic body formation. Further observations of the proteases activities towards other substrates, including other members of the Ced-3 family, was consistent with a pathway in which a death stimulus resulted in the activation of Mch2. Once activated, Mch2 could cleave lamin A and could activate pro-Yama and pro-ICE-LAP3 amplifying the range of death substates by allowing these activated proteases to more efficiently cleave PARP and the U1-70 kDa protein.

Good evidence has come from the cloning of FLICE/MACH to place these proteases at the head of the proteolytic cascade in the case of Fas- and tumor necrosis factor (TNF)-receptor mediated apoptosis. The TNF receptor 1, TNFR1, and Fas/Apo-1/CD95 receptors are the only members of the TNFR superfamily of receptors that initiate apoptosis on binding of their respective ligands. Both these receptor molecules contain an intracellular domain known as the death domain. This domain consists of some 70 amino acids and is required for the ability to induce apoptosis (Itoh and Nagata, 1993; Tartaglia et al., 1993). Three other death domain containing molecules have been isolated and shown to associate with the activated receptors via death domain interactions: Fas-associated protein with death domain (FADD)/MORT1 and receptor interacting protein (RIP) in the case of Fas (Boldin et al., 1995; Chinnaiyan et al., 1995; Stanger et al., 1995), and TNFR1-associated death domain protein (TRADD) and RIP in the case of TNFR1 (Hsu et al., 1995; Stanger et al., 1995). FADD, TRADD, and RIP all cause apoptosis when expressed ectopically. Deletion analysis of FADD/MORT1 suggests that although its death domain is required for association with Fas it is not required for apoptosis. Apoptosis requires an N-terminal region called the death effector domain (DED) (Chinnaiyan et al., 1995). The death domains on these receptors thus act as recruitment centers for other death domain containing proteins. FADD/MORT1 binds directly to Fas. It can however also bind to TNFR1 through binding to TRADD (Hsu et al., 1996). Thus, FADD/MORT1 appears to be the point of convergence between the two pathways.

FLICE/MACH, while sharing identity with the ICE-like proteases, also contains within its N-terminal sequence two 60 amino acid stretches with homology to the DED of FADD which mediates its association with FADD/MORT1. These data would suggest that upon activation of the receptors FADD/MORT1 is recruited to the death domain regions and in turn recruits FLICE/MACH through their DEDs. Cleavage and activation of FLICE/MACH as an ICE-like protease is then presumed to follow (Boldin et al., 1996; Muzio et al., 1996). Several dominant negative isoforms of MACH exist and block both TNFR1- and Fas-induced apoptosis implying a role for MACH/FLICE in both receptor pathways (Boldin et al., 1996).

III. GRANZYME B

The requirement of ICE for Asp at the P_1 position is shared by granzyme B, a structurally unrelated serine protease which plays an important role in cell killing by cytotoxic T lymphocytes (CTLs) and natural killer (NK) cells (Heusel et al., 1994; reviewed in Shresta et al., 1995). CTLs and NK cells kill their target cells by either a calcium-independent (involving the Fas/Apo1/CD95 receptor) or a calcium-dependent mechanism. The latter mechanism involves the release of the contents of the cytotoxic effector cells cytolytic granules close to the target cells. As mentioned earlier these contain the pore-forming protein perforin and granzyme proteases. Perforin in the presence of calcium oligomerizes to form pores in the target cell membrane allowing entry of granzyme B. This results in apoptosis of the target cell with all the characteristic hallmarks such as DNA fragmentation. Perforin alone does not cause this. A number of ICE-like proteases have now been shown to be cleaved and activated by granzyme B including, CPP32/Yama/Apopain (Darmon et al., 1995; Martin et al., 1996; Quan et al., 1996), CMH-1 (Gu et al., 1996), Ich-3 (Wang et al., 1996a), and ICE-LAP6 (Duan et al., 1996b). FLICE/MACH has also been shown to be cleaved by granzyme B and could possibly mediate apoptosis signaled via this molecule as well as via the Fas and TNF receptors (Muzio et al., 1996).

IV. CALPAINS

A. Structure and Activation

Calpains are one of the two major classes of intracellular, non-lysosomal proteinases (for reviews see Croall and DeMartino, 1991; Suzuki et al., 1995). Calpains are ubiquitous in animal cells, particularly in vertebrates, where they have been characterized extensively. Calpains do not require ATP for activity, and they do not degrade proteins to completion, but instead cleave substrates by limited proteolysis. Two major isoforms of calpain, μ-calpain and m-calpain, are each composed of two subunits. A 30 kDa subunit is shared between the two isoforms which also comprise similar but distinct 80 kDa subunits. The larger subunits contain the catalytic domain. μ-calpain requires micromolar, whereas m-calpain requires millimolar, levels of Ca^{2+} for activity. A third isozyme which requires Ca^{2+} levels intermediate between μ- and m-calpains has been discovered recently in chicken (Sorimachi et al., 1995) and is known as μ/m calpain.

Autolysis is an important feature of the calpains in which the amino terminal portion of each subunit is cleaved, producing enzymes with higher Ca^{2+} sensitivities. Autolysis lowers the *in vitro* Ca^{2+} requirement to near physiological levels in the case of μ-calpain (down to 0.1 μM Ca^{2+}), but autolyzed m-calpain still requires in the range of 10 μM Ca^{2+}. Recent evidence, however, demonstrates that calpains become activated upon dissociation of the 80 and 30 kDa subunits, and that both the 80 kDa subunit procalpain and its 76 kDa autolyzed counterpart contain similar biochemical properties and Ca^{2+} sensitivities (reviewed in Suzuki et al., 1995), the 30 kDa subunit acting as an inhibitor of the 80 kDa subunit (Yoshizawa, 1995a). Levels of Ca^{2+} necessary for enzyme activity cause subunit dissociation (Yoshizawa et al., 1995b). The calpains may then become inactivated by either subunit reassociation in the absence of Ca^{2+}, or degraded. Autolysis appears to cause permanent subunit separation, even in the absence of Ca^{2+}; autolysis may accelerate subunit dissociation, promoting activation and possibly predisposing the enzyme to degradation (Suzuki et al., 1995).

In mammals, the calpain family of proteases has expanded recently with the discovery of tissue-specific calpains; skeletal muscle contains nCL-1 (p94 calpain), and three other isoforms (nCL-2, nCL-2', and nCL-3) have been recently identified in smooth muscle (reviewed in Suzuki et al., 1995). These share high protein sequence homology with the well-characterized μ- and m-calpains, and an absolute requirement for Ca^{2+} for activity (with the exception of nCL-2', an alternate splice product of nCL-2 that is missing the Ca^{2+}-binding domain). In addition to the mammalian enzymes, several calpain-like proteinases have been characterized from invertebrates that share sequence homology and biochemical properties (reviewed in Beyette and Mykles, in press).

Calpains have an extensive list of substrates *in vitro* (see Croall and DeMartino, 1991), including proto-oncoproteins, growth factor and hormone receptors, protein kinases, and cytoskeletal proteins. It is because these latter proteins have been implicated as proteolytic substrates during cell death that interest in the calpains was kindled with respect to apoptosis. Furthermore, an increase in intracellular Ca^{2+} during apoptosis, required for the process of DNA fragmentation (Wyllie, 1980; McConkey et al., 1989, 1991), could involve calpain activation as a required or additional step.

B. Calpains in Apoptosis

Contradictory evidence, based on the effect of inhibitors, exists as to whether inhibition of calpains upregulates or downregulates apoptosis. Several groups have found evidence that calpain activation is a necessary event for triggering apoptosis in different cell types and in response to certain inducing agents (Bruno et al., 1992; Sarin et al., 1994; Squier et al., 1994). The results of Squier et al. (1994) showed that calpains were necessary for activation in three out of four models of apoptosis. Dexamethasone-treated mouse thymocytes showed an increase in calpain activity one hour after treatment, well before the appearance of DNA fragmentation or morphological characteristics of apoptosis. Concurrently, μ-calpain was autolyzed to its 76 kDa form. When thymocytes were preincubated with two different inhibitors of calpains (calpain inhibitor I and MDL-28170), amounts of DNA fragmentation and apoptosis were decreased for both dexamethasone-treated and γ-irradiated cells.

The above are examples of the "induction" model of apoptosis in which macromolecular synthesis is required for apoptosis to proceed (Cohen, 1991). In the alternate "release" model macromolecular synthesis is not required, indeed inhibition of macromolecular synthesis induces apoptosis. In this model it is envisioned that cells are already programmed to die but that apoptosis is held in check by a short-lived inhibitor. Such a model system is the cyclohexamide induced apoptosis in bone marrow metamyelocytes. Calpain inhibitor 1 and MDL–28170 were both observed to inhibit apoptosis in this model (Squier et al., 1994). Thus, calpains act downstream of the postulated inhibitor of apoptosis. Calpains, however, were unable to prevent the extremely rapid cell death induced by cytotoxic T lymphocytes. However, in this case, cell killing very likely is induced downstream of calpain activity in the apoptotic pathway.

The above results indicate a role for calpains in the upregulation of apoptosis. However, other groups have demonstrated just the opposite effect in different cell systems. In human monoblast U937 cells that were preincubated with γ-interferon prior to elicitation of apoptosis with TNF-α, μ-calpain also was autolyzed to its 76 kDa active form in only apoptotic cell fractions (Kikuchi and Imajoh-Ohmi, 1995). This autocatalytic processing was inhibited in the presence of two calpain inhibitors (acetyl-leucyl-leucyl-norleucinal, or ALLnL, a calpain-specific inhibitor; and E64d, an irreversible inhibitor of cysteine proteinases). However, these two inhibitors increased DNA fragmentation and the proportion of apoptotic monoblasts from 36% to 64%, while having no apparent cytotoxic effects, and causing no apoptosis in the absense of TNF-α. A similar result was shown in human prostate adenocarcinoma cells (Zhu et al., 1995). In these cells, calpain inhibition (by calpain inhibitor I and carboxybenzoyl-L-leucyl-L-leucyl-L-tyrosine-diazomethyl ketone) resulted in apoptosis. The use of calpain inhibitors also resulted in a prolonged increase in AP-1 activity in the prostate adenocarcinoma cells. The expression of AP-1 was shown to increase in the prostate during apoptosis induced by castration in the rodent (Isaacs et al., 1992), and may be involved in apoptosis (Colotta et al., 1992; Smeyne et al., 1993).

The studies that have been carried out with calpain inhibitors must be evaluated conservatively because many of these inhibitors such as E64d and calpain inhibitor 1 are not specific for calpains. Recently, an inhibitor directed towards the calcium-binding sites of calpain was shown to inhibit apoptosis in dexamethasone-treated thymocytes (Squier and Cohen, 1997). In the cases where calpain inhibitors induced apoptosis (Kikuchi and Imajoh-Ohmi, 1995; Zhu et al., 1995), inhibitors were added to proliferating cells. The human monoblast U937 cells are induced to proliferate when cultured in the presence of IFN-γ (Imajoh-Ohmi et al., 1995). There are many observations of proliferating and carcinomic or tumorigenic cells being more sensitive to apoptosis-inducing factors; these cell systems may already be triggered for apoptosis, and in these cases calpains may play a different, and opposing, role.

Recently, it has been shown that calpain and serine protease inhibitors can cause apoptosis in HL-60 cells (Lu and Mellgren, 1996). Two calpain inhibitors, benzyloxycarbonyl-Leu-Leu-Tyr diazomethyl ketone and Ac-Leu-Leu-Nle-CHO were found to be unable to inhibit apoptosis due to cyclohexamide but were found to induce apoptosis in the absence of cyclohexamide. The serine protease inhibitors 3,4-dichloroisocoumarin and tosyl-Phe-chloromethyl ketone (TPCK) were also found to induce apoptosis in the absence of cyclohexamide but were found to inhibit apoptosis induced by cyclohexamide in these cells. The thiol protease inhibitor benzyloxycarbonyl-Leu-Val-Gly diazomethyl ketone was found to neither promote or inhibit apoptosis. Inhibitors of the ICE-like proteases did not induce apoptosis in HL-60 cells but did inhibit apoptosis caused by cyclohexamide and the serine and calpain inhibitors. These results are consistent with the pivotal role of the ICE-like proteases in apoptosis while suggesting that calpains and serine proteases sensitive to 3,4 dichloroisocoumarin and TPCK are not required for the induction of apoptosis in HL-60 cells (Lu and Mellgren, 1996).

V. PROTEASOMES

A. Introduction

The 26S proteasome complex, of which the 20S proteasome provides five or more distinct proteolytic activites, is present in high levels in the cytoplasm and/or nuclei of all eukaryotic and some prokaryotic cells, comprising up to 1% of intracellular proteins (reviewed by Tanaka and Tanahashi, 1997, and Dubiel and Rechsteiner, 1997, in this volume). In an ATP-dependent reaction, the 26S proteasome degrades abnormal, regulatory, and short-lived proteins that have been tagged by ubiquitin for proteolysis (for review, refer to Wilkinson, 1997, in this volume). In addition to the proteinase, ATP-ubiquitin-dependent proteolysis by the 26S proteinase requires a ubiquitin activating enzyme (E1), one of several ubiquitin conjugating enzymes (E2), and sometimes a ubiquitin-protein ligase (E3) which probably facilitates substrate recognition by the 26S proteinase.

Regulation of protein degradation by the 26S proteasome may occur by many different mechanisms. Activators and inhibitors of the 20S and 26S complexes exist (reviewed in this volume by Dubiel and Rechsteiner, 1997), and are currently under characterization in several laboratories. In addition, other possible means of modulating 26S proteasome degradation of a particular substrate(s) include changes in subunit content (Mason and Rivett, 1994) or phosphorylation (Mason et al., 1996), changes in intracellular proteasome location, ubiquitinylation of substrates, and regulation of E1, E2, and E3 enzymes.

B. Proteasomes and Apoptosis

A number of recent publications have begun to examine relationships between ubiquitin and/or proteasome function and apoptosis. The recent development of a proteasome-specific inhibitor lactacystin (Fenteany et al., 1995), has greatly facilitated elucidation of proteasome involvement during different cell processes. Earlier studies suffered from problems with the selectivity of the proteasome inhibitors used. Lactacystin treatment of human monocyte U937 cells induced to apoptosis by TNF-α has demonstrated a possible role of proteasomes in the degradation of the ICE-like protease CPP32 (Fujita et al., 1996). Lactacystin and other proteasome inhibitors enhanced cleavage of the CPP32 substrate PARP *in vivo*, and CPP32-like activity increased in cells treated with proteasome inhibitors compared to those without, whereas the cysteine proteinase inhibitor E64d had no effect. The proteasome apparently protected cells from TNF-induced apoptosis by degrading CPP32, its activating enzyme and/or other ICE-like proteases activated by CPP32. These results are in agreement with those of Imajoh-Ohmi and colleagues (1995), who observed that lactacystin induced apoptosis when proliferating human monoblast U937 cells were cultured in its presence. Five μM lactacystin, a concentration necessary to induce neurite outgrowth (Fenteany et al., 1994), was observed to cause cell nuclei condensation and chromatin fragmentation.

Proteasome function is essential during the cell cycle (Kominami et al., 1995; Pagano et al., 1995) and lactacystin has been shown to inhibit cell cycle progression through inhibition of the proteasome (Katagiri et al., 1995). Proteasome inhibition by lactacystin has been shown to induce neuroblastoma differentiation (Omura et al., 1991). Thus, proteasome inhibition during cell proliferation may direct cells to differentiation or apoptosis (Imajoh-Ohmi, 1995). However, as with the calpains, contradictory evidence exists as to the function of proteasomes in apoptosis. Proteasome inhibitors were observed to inhibit apoptosis and PARP cleavage in thymocytes induced to die by a number of agents such as γ-radiation, glucocorticoids, or the phorbol ester, phorbol 12-myristate, 13-acetate (PMA) (Grimm et al., 1996). However, while lactacystin could inhibit apoptosis induced by PMA it did not inhibit apoptosis caused by PMA and the calcium ionophore A23187. Calpain inhibitors were also shown to be ineffective in inhibiting apoptosis due to PMA and A23187. Thus, the increase in intracellular Ca^{2+} must initiate apoptosis by some other mechanism or act at a point in the pathway after that requiring proteasomes. The kinase inhibitor staurosporine has also been shown to induce apoptosis in a number of cell types and can induce apoptosis in the presence of protease inhibitors. Thus, inhibition of proteasomes causes inhibition of apoptosis by only a specific group of initiators.

Similar findings have been observed in neurons stimulated to undergo apoptosis by withdrawl of nerve growth factor (NGF). In these cells proteasome inhibitors, including lactacystin, were shown to be able to inhibit apoptosis due to NGF deprivation and to prevent PARP cleavage, suggesting that the role of the proteasome in apoptosis lies upstream to that of the ICE-like proteases (Sadoul et al., 1996). This ordeing was confirmed by showing that proteasome inhibitors inhibit the processing of pro-IL-1β and ICE in lipopolysaccharide-activated macrophages without inhibiting the activity of ICE directly (Sadoul et al., 1996). However, a recent report suggests that the proteasome is responsible for an ICE-like protease activity in P19 embryonal carcinoma cells induced to differentiate by retinoic acid (Kobayashi et al., 1996). Indeed, the authors report that purified proteasome from mouse liver can cleave the ICE-like proteinase substrate peptide Ac-Tyr-Val-Ala-Asp-7-amino-4-methylcoumarin, although not very efficiently.

The contradictory actions of proteasome inhibition above may be explained by the proliferative state of the cell and by the mechanism of induction of apoptosis. It is apparent that inhibition of proteasome function in proliferating cells blocks the cell cycle and as a result may direct cells into apoptosis. However, in nondividing cells where short exposure to proteasome inhibitors is not particularly cytotoxic proteasomes can play a role in apoptosis induced by other agents (Grimm et al., 1996).

The proteasome plays a major role in the degradation of many short-lived proteins in the cell, some of which are known to influence apoptosis themselves, including p53 and NF-κB (Palombella et al., 1994; Shkedy et al., 1994; Beg and Baltimore, 1996; Lowe, 1996; Van Antwerp et al., 1996; Wang et al., 1996c). Thus, it is likely that depending on which apoptosis-inducing agent is used there will be differing effects with the use of proteasome, and indeed other protease, inhibitors.

Our recent studies (Beyette et al., 1998) have shown that the activities of proteasomes against a number of substrates decreased in thymocytes induced to apoptosis by dexamethasone and that this decrease was reversed by the caspase inhibitor benzyloxycarbonyl-Val-Ala-Asp (OMe)-fluoromethylketone. This caspase inhibitor was also observed to inhibit one of the activities of the proteasome.

VI. SUMMARY

In conclusion, while it is apparent that the ICE-like proteases play a pivotal role in apoptosis induced by a variety of agents, the exact ordering of these proteases in the apoptotic pathway is not fully understood. Whether all the ICE-like proteases play a role in any particular pathway or whether specific proteases participate in specific pathways has yet to be elucidated. The role of other proteases covered above is even less well understood. It is very likely that these proteases all work upstream of the ICE-like proteases in the induction of apoptosis but may also play roles downstream in the execution of apoptosis. The upstream effect of these other proteases will reflect the roles played by their substrates in the many signaling pathways leading to apoptosis.

REFERENCES

Alnemri, E. S., Fernandes-Alnemri, T., & Litwack, G. (1995). Cloning and expression of four novel isoforms of human interleukin-1β converting enzyme with different apoptotic activities. J. Biol. Chem. 270, 4312-4317.

Alnemri, E. S., Livingston, D. J., Nicholson, D. W., Salvesen, G., Thornberry, N. A., Wong, W. W., & Yuan, J. (1996). Human ICE/CED-3 protease nomenclature. Cell 87, 171.

Beg, A. A., & Baltimore, D. (1996). An essential role for NF-κB in preventing TNF-α-induced cell death. Science 274, 782-784.

Beyette, J., & Myckles, D. (1997). Crustacean Ca^{2+} dependent proteinases. In: The Pharmacology of Calpains (Wang, K. K. K. G., & Yuen, P.-W., eds.). In press.

Beyette, J., Mason, G.G.F., Murray, R.Z., Cohen, G.M., & Rivett, A.J. (1998). Proteasome activities decrease during dexamethasone-induced apoptosis of thymocytes. Biochem. J. (In press).

Black, R. A., Kronheim, S. R., & Sleath, P. R. (1989). Activation of interleukin-1 by a co-induced protease. FEBS. Lett. 247, 386-390.

Boldin, M. P., Varfolomeev, E. E., Pancer, Z., Mett, I. L., Camonis, J. H., & Wallach, D. (1995). A novel protein that interacts with the death domain of Fas/APO1 contains a sequence motif related to the death domain. J. Biol. Chem. 270, 7795-7798.

Boldin, M. P., Goncharov, T. M., Goltsev, Y. V., & Wallach, D. (1996). Involvement of MACH, a novelMORT1/FADD-interacting protease, in Fas/APO-1- and TNF receptor-induced cell death. Cell 85, 803-815.

Boudreau, N., Sympson, C. J., Werb, Z., & Bissell, M. J. (1995). Suppression of ICE and apoptosis in mammary epithelial cells by extracellular matrix. Science 267, 891-893.

Brancolini, C., Benedetti, M., & Schneider, C. (1995). Microfilament reorganization during apoptosis: the role of Gas2, possible substrate for ICE-like proteases. EMBO J. 14, 5179-5190.

Bruno, S., Lassota, P., Giaretti, W., & Darzynkiewicz, Z. (1992). Apoptosis of rat thymocytes triggered by prednisolone, camptothecin, or teniposide is selective to G_0 cells and is prevented by inhibitors of proteases. Onco. Res. 4, 29-35.

Casciola-Rosen, L. A., Miller, D. K., Anhalt, G. J., & Rosen A. (1994). Specific cleavage of the 70-kDa protein component of small nuclear ribonucleoprotein is a characteristic biochemical feature of apoptotic cell death. J. Biol. Chem. 269, 30757-30760.

Cerretti, D. P., Kozlosky, C. J., Mosley, B., Nelson, N., Van Ness, K., Greenstreet, T. A., March, C. J., Kronheim, S. R., Druck, T., Cannizzaro, L. A., Huebner, K., & Black, R. A. (1992). Molecular cloning of the interleukin-1β converting enzyme. Science 256, 97-100.

Chinnaiyan, A. M., O Rourke, K., Tewari, M., & Dixit, V. M. (1995). FADD, a novel death domain-containing protein, interacts with the death domain of Fas and initiates apoptosis. Cell 81, 505-512.

Chinnaiyan, A. M., & Dixit, V. M. (1996). The cell-death machine. Curr. Biol. 6, 555-562.

Cohen, G.M. (1997). Caspases: The executioners of apoptosis. Biochem. J. 326, 1-16.

Cohen, J.J. (1991). Progammed cell-death in the immune system. Adv. Immunol. 50, 55-85.

Colotta, F., Polentrarutti, N., Sironi, M., & Mantovani, A. (1992). Expression and involvement of c-*fos* and c-*jun* protooncogenes in programmed cell death induced by growth factor deprivation in lymphoid cell lines. J. Biol. Chem. 267, 18278-18283.

Croall, D. E., & DeMartino, G. N. (1991). Calcium-activated neutral protease (calpain) system: Structure, function, and regulation. Physiol. Rev. 71, 813-847.

Darmon, A. J., Nicholson, D. W., & Bleackley, R. C. (1995). Activation of the apoptotic proteases CPP32 by cytotoxic T-cell-derived granzyme B. Nature (London) 377, 446-448.

Dinarello, C. A. (1991). Interleukin-1 and interleukin-1 antagonism. Blood 77, 1627-1652.

Duan, H., Chinnaiyan, A. M., Hudson, P. L., Wing, J. P. He, W.-W., & Dixit, V. M. (1996a). ICE-LAP3, a novel mammalian homologue of the *Caenorhabditis elegans* cell death protein Ced-3 is activated during Fas- and tumor necrosis factor-induced apoptosis. J. Biol. Chem. 271, 1621-1625.

Duan, H., Orth, K., Chinnaiyan, A. M., Poirier, G. G., Froelich, C. J., He, W.-W., & Dixit, V. M. (1996b). ICE-LAP6, a novel member of the ICE/Ced-3 gene family, is activated by the cytotoxic T cell protease Granzyme B. J. Biol. Chem. 271, 16720-16724.

Dubiel, W., & Rechsteiner, M. (1997). The 19S regulatory complex of the 26S proteeasome. In:

Earnshaw, W. C. (1995). Apoptosis: Lessons from *in vitro* systems. Trends Cell Biol. 5, 217-220.

Ellis, H.M., & Horvitz, H. R. (1986). Genetic control of programmed cell death in the nematode *C. elegans*. Cell 44, 817-829.

Emoto, Y., Manome, Y., Meinhardt, G., Kisaki, H., Kharbanda, S., Robertson, M., Ghayur, T., Wong, W. W., Kamen, R., Weichselbaum, R., & Kufe, D. (1995). Proteolytic activation of protein kinase C δ by an ICE-like protease in apoptotic cells. EMBO J. 14, 6148-6156.

Faucheu, C., Diu, A., Chan, A. W. E., Blanchet, A.-M., Miossec, C., Hervé, F., Collard-Dutilleul, V., Gu, Y., Aldape, R. A., Lippke, J. A., Rocher, C., Su, M. S.-S., Livingston, D. J., Hercend, T., & Lalanne, J.-L. (1995). A novel human protease similar to the interleukin-1β converting enzyme induces apoptosis in transfected cells. EMBO J. 14, 1914-1922.

Faucheu, C., Blanchet, A.-M., Collard-Dutilleu, V., Lalanne, J.-L., & Diu-Hercend, A. (1996). Identification of a cysteine protease closely related to interleukin-1β-converting enzyme. Eur. J. Biochem. 236, 207-213.

Fenteany, G., Standaert, R. F., Reichard, G. A., Corey, R. E. J., & Schreiber, S. L. (1994). A β-lactone related to lactacystin induces neurite outgrowth in a neuroblastoma cell line and inhibits cell cycle progression in an osteosarcoma cell line. Proc. Natl. Acad. Sci. USA 91, 3358-3362.

Fenteany, G., Standaert, R. F., Lane, W. S., Choi, S., Corey, E. J., & Schreiber, S. L. (1995). Inhibition of proteasome activities and subunit-specific amino-terminal threonine modification by lactacystin. Science 268, 726-731.

Fernandes-Alnemri, T., Litwack, G., & Alnemri, E.S. (1994). CPP32, a novel human apoptotic protein with homology to *Caenorhabditis elegans* cell death protein Ced-3 and mammalian interleukin-1β-converting enzyme. J. Biol. Chem. 269, 30761-30764.

Fernandes-Alnemri, T., Litwack, G., & Alnemri, E. S. (1995a). *Mch2*, a new member of the apoptotic *Ced-3/Ice* cysteine protease gene family. Cancer Res. 55, 2737-2742.

Fernandes-Alnemri, T., Takahashi, A., Armstrong, R., Krebs, J., Fritz, L., Tomaselli, K. J., Wang, L., Yu, Z., Croce, C. M., Salvesen, G., Earnshaw, W. C., Litwack, G., & Alnemri, E. S. (1995b). Mch3, a novel human apoptotic cysteine protease highly related to CPP32. Cancer Res. 55, 6045-6052.

Fujita, E., Mukasa, T., Tsukahara, T., Arahata, K., Omura, S., & Momoi, T. (1996). Enhancement of CPP-32-like activity in theTNF-treated U937 cells by the proteasome inhibitors. Biochem. Biophys. Res. Comm. 224, 74-79.

Gagliardini, V., Fernandez, P.-A., Lee, R. K. K., Drexler, H. C. A., Rotello, R. J., Fishman, M. C., & Yuan, J.-Y. (1994). Prevention of vertebrate neuronal death by the *CrmA* gene. Science 263, 826-828.

Grimm, L. M., Goldberg, A. L., Poirier, G. G., Schwartz, L. M., & Osborne, B. A. (1996). Proteasomes play an essential role in thymocyte apoptosis. EMBO J. 15, 3835-3844.

Gu, Y., Sarnecki, C., Aldape, R. A., Livingston, D. J., & Su, M. S. S. (1995). Cleavage of poly(ADP-ribose)polymerase by interleukin-1-β converting-enzyme and its homologs TX and Nedd-2. J. Biol. Chem. 270, 18715-18718.

Gu, Y., Sarnecki, C., Fleming, M. A., Lippke, J. A., Bleackley, R. C., & Su, M. S.-S. (1996). Processing and activation of CMH-1 by granzyme B. J. Biol. Chem. 271, 10816-10820.

Hengartner, M.O. (1997). CED-4 is a stranger no more. Nature (London) 388, 714-715.

Hengartner, M. O., & Horvitz, H. R. (1994). Programmed cell death in *C. elegans*. Curr. Opin. Genet. Dev. 4, 581-586.

Hengartner M. O. Ellis, R. E., & Horvitz, H.R. (1992). *Caenorhabditis elegans* gene ced-9 protects cells from programmed cell death. Nature (London) 356, 494-499.

Heusel, J. W., Wesseischmidt, R. L., Shresta, S., Russel, J. H., & Ley, T. J. (1994). Cytotoxic lymphocytes require granzyme B for the rapid induction of DNA fragmentation and apoptosis in allogeneic target cells. Cell 76, 977-987.

Hsu, H., Xiong, J., & Goeddel, D. V. (1995). The TNF receptor 1-associated protein TRADD signals cell death and NF-κB activation. Cell, 81, 495-504.

Hsu, H., Shu, H.-B., Pan, M.-P., & Goeddel, D. V. (1996). TRADD-TRAF2 and TRADD-FADD interactions define two distinct TNF receptor-1 signal transduction pathways. Cell 84, 299-308.

Imajoh-Ohmi, S., Kawaguchi, T., Sugiyama, S., Tanaka, K., Omura, S., & Kikuchi, H. (1995). Lactacystin, a specific inhibitor of the proteasome, induces apoptosis in human monoblast U937 cells. Biochem. Biophys. Res. Comm. 217, 1070-1077.

Isaacs, J. T., Lundmo, P. I., Berges, R., Martikainen, P., Kyprianou, N., & English, H. F. (1992). Androgen regulation of programmed death of normal and malignant prostatic cells. J. Andro. 13, 457-464.

Itoh, N., & Nagata, S. (1993). A novel protein domain required for apoptosis. Mutational analysis of human Fas antigen. J. Biol. Chem. 268, 10932-10937.

Kamens, J., Paskind, M., Hugunin, M., Talanian, R. V. Allen, H., Banach, D., Bump, N., Hackett, M., Johnston, C. G., Li, P., Mankovich, J. A., Terranova, M., & Ghayur, T. (1995). Identification and characterization of ICH-2, a novel member of the interleukin-1β-converting enzyme family of cysteine proteases. J. Biol. Chem. 270, 15250-15256.

Katagiri, M., Hayashi, M., Matsuzaki, K., Tanaka, H., & Omura, S. (1995). The neuritogenesis inducer lactacystin arrests cell cycle at both G0/G1 and G2 phases in Neuro 2a cells. J. Antibiotics 48, 344-346.

Kaufmann, S.H. (1989). Induction of endonucleocytic DNA cleavage in human acute myelogenous leukemia cells by etoposide, camptohecin, and other anticancer drugs: a cautionary note. Cancer Res. 49, 5870-5878.

Kaufmann, S. H., Desnoyers, S., Ottaviano, Y., Davidson, N. E., & Poirier, G. G. (1993). Specific proteolytic cleavage of poly(ADP-ribose)polymerase: an early marker of chemotherapy-induced apoptosis. Cancer Res. 53, 3976-3985.

Kikuchi, H., & Imajoh-Ohmi, S. (1995) Activation and possible involvement of calpain, a calcium-activated cysteine protease, in down-regulation of apoptosis of human monoblast U937 cells. Cell Death Diff. 2, 195-199.

Kobayashi, T., Shinozaki, A., Momoi, T., Arahata, K., & Tsukahara, T. (1996). Identification of an interleukin-1β converting enzyme-like activity that increases upon treatment of P19 cells with retinoic acid as the proteasome. J. Biochem. 120, 699-704.

Kominami, K., DeMartino, G. N., Moomaw, C. R., Slaughter, C. A., Shimbara, N., Fujimuro, M., Yokosawa, H., Hisamatsu, H., Tanahashi, N., Shimizu, Y., Tanaka, K., & Toh-E, A. (1995). Nin1p, a regulatory subunit of the 26S proteasome, is necessary for activation of Cdc28p kinase of *Saccharomyces cerevisiae*. EMBO J. 14, 3105-3115.

Kostura, M. J., Tocci, M. J., Limjuco, G., Chin, J., Cameron, P., Hillman, A. G., Chartrain, N. A., & Schmidt, J. A. (1989). Identification of a monocyte specific pre-interleukin-1β convertase activity. Proc. Natl. Acad. Sci. USA, 86, 5227-5231.

Kuida, K., Lippke, J. A., Ku, G., Harding, M. W., Livingston, D. J., & Su, M. S. S. (1995). Altered cytokine export and apoptosis in mice deficient in interleukin-1β converting enzyme. Science 267, 2000-2003.

Kumar, S., Kinoshita, M., Noda, M., Copeland, N. G., & Jenkins, N. A. (1994). Induction of apoptosis by the mouse *Nedd2* gene, which encodes a protein similar to the product of the *Caenorhabditis elegans* cell death gene *ced-3* and the mammalian IL-1β-converting enzyme. Genes Dev. 8, 1613-1626.

Kumar, S., & Harvey, N. L. (1995). Role of multiple cellular proteases in the execution of programmed cell death. FEBS Lett. 375, 169-173.

Lazebnik, Y. A., Cole, S., Cooke, C. A., Nelson, W. G., & Earnshaw, W. C. (1993). Nuclear events of apoptosis in-vitro in cell-free mitotic extracts - a model system for analysis of the active phase of apoptosis. J. Cell Biol. 123, 7-22.

Lazebnik, Y. A., Kaufmann, S. H., Desnoyers, S., Poirier, G. G., & Earnshaw, W. C. (1994). Cleavage of poly(ADP-ribose)polymerase by a proteinase with properties like ICE. Nature (London) 371, 346-347.

Lazebnik, Y. A., Takahashi, A., Moir, R. D., Goldman, R. D., Poirier, G. G., Kaufmann, S. H., & Earnshaw, W. C. (1995). Studies of the lamin proteinase reveal multiple parallel biochemical pathways during apoptotic execution. Proc. Natl. Acad Sci. USA 92, 9042-9046.

Li, P., Allen, H., Banerjee, S., Franklin, S., Herzog, L., Johnson, C .(1995). Mice deficient in IL-1β converting enzyme are defective in production of mature IL-1β and resistant to endotoxic shock. Cell 80, 401-411.

Lippke, J. A., Gu, Y., Sarnecki, C., Caron, P. R., & Su, M. S.-S. (1996). Idntification and characterization of CPP32/*Mch2* homolog 1, a novel cysteine protease similar to CPP32. J. Biol. Chem. 271, 1825-1828.

Lowe, S. W. (1996). The role of p53 in apoptosis. In: Apoptosis in Normal Development and Cancer (Sluyser, M., ed.), pp. 97-125. Taylor & Francis, London.

Lu, Q., & Mellgren, L.L. (1996). Calpain inhibitors and serine-protease inhibitors can produce apoptosis in HL-60 cells. Arch. Biochem. Biophys. 334, 175-181.

Majno, G., & Joris, I. (1995). Apoptosis, oncosis, and necrosis: an overview of cell death. Am. J. Pathol. 146, 3-15.

Martin, S. J., Amarante-Mendes, G. P., Shi, L., Chuang, T.-H., Casiano, C. A., O Brien, G. A., Fitzgerald, P., Tan, E.M., Bokoch, G. M. Greenberg, A. H., & Green, D. R. (1996). The cytotoxic cell ptotease granzyme B initiates apoptosis in a cell-free system by proteolytic processing and activation of the ICE/CED-3 family protease, CPP32, via a novel two-step mechanism. EMBO J. 15, 2407-2416.

Mason, G. G. F., & Rivett, A. J. (1994). Proteasomes: The changing face of proteolysis. Chem. Biol. 1, 197-199.

Mason, G. G. F., Henbil, K. B., & Rivett, A. J. (1996). Phosphorylation of proteasomes in mammalian cells: Identification of two phosphorylated subunits and the effect of phosphorylation on activity. Eur. J. Biochem. 238, 453-462.

McConkey, D. J., Hartzell, P., Jondal, M., & Orrenius, S. (1989). Inhibition of DNA fragmentation in thymocytes and isolated thymocyte nuclei by agents that stimulate protein kinase C. J. Biol. Chem. 264, 13399-13402.

McConkey, D. J., Aguilar-Santelises, M., Hartzell, P., Eriksson, I., Mellstedt, H., Orrenius, S., & Jondal, M. (1991). Induction of DNA fragmentation in chronic B-lymphocytic leukemia cells. J. Immunol. 146, 1072-1076.

Miller, D. K., Ayala, J. M., Egger, L. A., Raju, S. M., Yamin, T.-T., Ding, G. J.-F., Gaffney, E. P., Howard, A. D., Palyha, O. C., Rolando, A. M., Salley, J. P., Thornberry, N. A., Weidner, J. R., Williams, K. T., Chapman, K. T., Jackson, J., Kostura, M. J., Limjuco, G., Molineaux, S. M., Mumford, R. A., & Calaycay, J. R. (1993a). Purification and characterization of active human interleukin 1β-converting enzyme from THP.1 monocytic cells. J. Biol. Chem. 268, 18062-18069.

Miller, D. K., Calaycay, J. R., Chapman, K. T., Howard, A. D. Kostura, M. J., Molineaux, S. M., & Thornberry, N. A. (1993b).The IL-1-β converting enzyme as a therapeutic target. Ann. N.Y. Acad. Sci. 696, 133-148.

Miura, M., Zhu, H., Rotello, R., Hartwieg, E. A., & Yuan, J. (1993). Induction of apoptosis in fibroblasts by IL-1β-converting enzyme, a mammalian homolog of the *C. elegans* cell death gene *ced-3*. Cell 75, 653-660.

Munday, N. A., Vaillancourt, J. P., Ali, A., Casano, F. J., Miller, D. K., Molineaux, S. M., Yamin, T.-T., Yu, V. L., & Nicholson, D. W. (1995). Molecular cloning and pro-apoptotic activity of ICE_{rel}II and ICE_{rel}III, members of the ICE/CED-3 family of cysteine proteases. J. Biol. Chem. 270, 15870-15876.

Muzio, M., Chinnaiyan, A. M., Kischkel, F. C., O'Rourke, K., Shevchenko, A., Ni, J., Scaffidi, C., Bretz, J. D., Zhang, M., Gentz, R., Mann, M., Krammer, P. H., Peter, M. E., & Dixit, V. M. (1996). FLICE, a novel FADD-homologous ICE/CED-3-like protease, is recruited to the CD95 (Fas/APO-1) death-inducing signalling complex. Cell 85, 817-827.

Na, S., Chuang, T.-H., Cunningham, A., Turi, T. G., Hanke, J. H., Bokoch, G. M., & Danley, D. E. (1996). D4-GDI, a substrate of CPP32, is proteolyzed during Fas-induced apoptosis. J. Biol. Chem. 271, 11209-11213.

Neamati, N., Fernandez, A., Wright, S., Kiefer, J., & McConkey, D. J. (1994). Degradation of lamin B1 precedes oligonucleosomal DNA fragmentation in apoptotic thymocytes and isolated thymocyte nuclei. J. Immunol. 154, 1693-1700.

Nett, M. A., Cerretti, D. P., Berson, D. R., Seavitt, J., Gilbert, D. J., Jenkins, N. A., Copeland, N. G., Black, R. A., & Chaplin, D. D. (1992). Molecular cloning of the murine IL-1β converting enzyme cDNA. J. Immunol. 149, 3254-3259.

Nicholson, D. W., Ali, A., Thornberry, N. A., Vaillancourt, J. P., Ding, C. K., Gallant, M., Gareau, Y., Griffin, P. R., Labelle, M., Lazebnik, Y. A., Munday, N. A., Sayaparaju, M. R., Smulson, M. E., Yamin, T. T., Yu, V. L., & Miller, D. K. (1995). Identification and inhibition of the ICE/CED-3 protease necessary for mammalian apoptosis. Nature (London) 376, 37-43.

Nuñez, G., & Clarke, M. F. (1994). The Bcl-2 family of proteins: regulators of cell death and survival. Trends in Cell Biol. 4, 399-403.

Omura, S., Fujimoto, T., Otoguro, K., Matsuzaki, K., Moriguchi, R., Tanaka, Y., & Sasaki, Y. (1991). Lactacystin, a novel microbial metabolite, induces neuritogenesis of neuroblastoma cells. J. Antibiotics 44, 113-116.

Orth, K., Chinnaiyan, A. M., Garg, M., Froelich, J., & Dixit, V. M. (1996a). The ced-3/ICE-like protease Mch2 is activated during apoptosis and cleaves the death substrate lamin A. J. Biol. Chem. 271, 16443-16446.

Orth, K., O Rourke, K, Salvesen, G. S., & Dixit, V. M. (1996b). Molecular ordering of apoptotic mammalian CED-3/ICE-like proteases. J. Biol. Chem. 271, 20977-20980.

Pagano, M., Tam, S. W., Theodoras, A. M., Beer-Romero, P., Del Sal, G., Chau, V., Yew, P. R., Draetta, G. F., & Rolfe, M. (1995). Role of the ubiquitin-proteasome pathway in regulating abundance of the cyclin-dependent kinase inhibitor p27. Science 269, 682-685.

Palombella, V. J., Rando, O. J., Goldberg, A. L., & Maniatis, T. (1994). The ubiquitin-proteasome pathway is required for processing the NF- B1 precursor protein and the activation of NF-κB. Cell 78, 773-785.

Quan, L. T., Tewari, M., O Rourke, K., Dixit, V. M., Snipas, S. J., Poirier, G. G., Ray, C., Pickup, D. J., & Salvesen, G. S. (1996). Proteolytic activation of the cell death protease Yama/CPP32 by granzyme B. Proc. Natl. Acad. Sci. USA 93, 1972-1976.

Raff, M. C. (1992). Social controls on cell-survival and cell-death. Nature (London) 356, 397-400.

Sadoul, R., Fernandez, P.-A., Quiquerez, A.-L., Martinou, I., Maki, M., Schröter, M., Becherer, J. D., Irmler, M., Tschopp, J., & Martinou, J.-C. (1996). Involvement of the proteasome in the programmed cell death of NGF-deprived sympathetic neurons. EMBO J. 15, 3845-3852.

Salvesen, G.S., & Dixit, V.M. (1997). Caspases: Intracellular signaling by proteolysis. Cell 91, 443-446.

Sarin, A., Clerici, M., Blatt, S. P., Hendrix, C. W., Shearer, G. M., & Henkart, P. A. (1994). Inhibition of activation-induced programmed cell death and restoration of defective immune responses of HIV+ donors by cysteine protease inhibitors. J. Immuno. 153, 862-872.

Schwartz, L. M. (1991). The role of cell death genes during development. BioEssays 13, 389-395.

Shi, L., Kam, C. M., Powers, J. C., Aebersold, R., & Greenberg, A. H. (1992). Purification of three cytotoxic lymphocyte granule serine proteases. J. Exp. Med. 176, 1521-1529.

Shkedy, D., Gonen, H., Bercovich, B., & Ciechanover, A. (1994). Complete reconstitution of conjugation and subsequent degradation of the tumor supressor protein p53 by purified components of the ubiquitin proteolytic system. FEBS Lett. 348, 126-130.

Shresta, S., Heusel, J.W., Macivor, D.M., Wesselschmidt, R.L., Russell, J.H., & Ley, T.J. (1995). Granzyme B plays a critical role in cytotoxic lymphocyte-induced apoptosis. Immunol. Rev. 146, 211-221.

Smeyne, R. J., Vendrell, M., Hayward, M., Baker, S. J., Miao, G. G., Schilling, K., Robertson, L. M., Curran, T. D., & Morgan, J. I. (1993). Continuous c-Fos expression precedes programmed cell-death *in-vivo*. Nature (London) 363, 166-169.

Song, Q., Lees-Miller, S. P., Kumar, S., Zhang, N., Chan, D. W., Smith, G. C. M., Jackson, S. P., Alnemri, E. S., Litwack, G., Khanna, K. K., & Lavin, M. F. (1996). DNA-dependent protein kinase catalytic subunit: a target for an ICE-like protease in apoptosis. EMBO J. 15, 3238-3246.

Sorimachi , H., Tsukahara, T., Okada-Ban, M., Sugita, H., Ishiura, H., & Suzuki, K. (1995). Identification of a third ubiquitous calpain species. Chicken muscle expresses four distinct calpains. Biochim. Biophys. Acta 1261, 381-393.

Squier, M.K.T., & Cohen, J.J. (1997). Calpain, an upstream regulator of thymocyte apoptosis. J. Immunol. 158, 3690-3697.

Squier, M. K. T., Miller, A. C. K., Malkinson, A. M., & Cohen, J. J. (1994). Calpain activation in apoptosis. J. Cell Physiol. 159, 229-237.

Stanger, B. Z., Leder, P., Lee, T.-H., Kim, E., & Seed, B. (1995). RIP: A novel protein containing a death domain that interacts with Fas/APO-1 (CD95) in yeast and causes cell death. Cell 81, 513-523.

Sulston, J.E., & Horvitz, H.R. (1977). Post-embryonic cell lineages of the nematode *Caenorhabditis elegans*. Dev. Biol. 82, 110-156.

Sulston, J. E., Schierenberg, E., White, J. G., & Thomson, N (1983). The embryonic cell lineage of the nematode *Caenorhabditis elegans*. Dev. Biol. 100, 64-119.

Suzuki, K., Sorimachi, H., Yoshizawa, T., Kinbara, K., & Ishiura, S. (1995). Calpain: Novel family members, activation, and physiological function. Biol. Chem. Hoppe-Seyler 376, 523-529.

Tanaka, K., & Tanahashi, N. (1997). The 20S proteasome. Subunits and functions. In:

Tartaglia, L. A., Ayres, T. M., Wong, G. H. W., & Goeddel, D. V. (1993). A novel domain within the 55 kD TNF receptor signals death. Cell 74, 845-853.

Tewari, M., & Dixit, V. M. (1995). Fas- and TNF-induced apoptosis is inhibited by the poxvirus *crmA* gene product. J. Biol. Chem. 270, 3255-3260.

Tewari, M., Quan, L. T., O'Rourke, K., Desnoyers, S., Zeng, Z., Beidler, D.R., Poirier, G. G., Salvesen, G. S., & Dixit, V. M. (1995a). Yama/CPP32β, a mammalian homolog of CED-3, is a CrmA-inhibitable protease that cleaves the death substrate poly(ADP-ribose) polymerase. Cell 81, 801-809.

Tewari, M., Beidler, D. R., & Dixit, V. M. (1995b). CrmA-inhibitable cleavage of the 70-kDa protein component of the U1 small nuclear ribonucleoprotein during Fas- and tumor necrosis factor-induced apoptosis. J. Biol. Chem. 270, 18738-18741.

Thornberry, N. A., Bull, H. G., Calaycay, J. R., Chapman, K. T., Howard, A. D., Kostura, M. J., Miller, D. K., Molineaux, S. M., Weidner, J. R., Aunins, J., Elliston, K. O., Ayala, J. M., Casano, F. J., Chin, J., Ding, G. J.-F., Egger, L. A., Gaffney, E. P., Limjuco, G., Palyha, O. C., Raju, S M., Rolando, A. M., Salley, J. P., Yamin, T.-T., Lee, T. D., Shively, J. E., MacCross, M., Mumford, R. A., Schmidt, J. A., & Tocci, M. J. (1992). A novel heterodimeric cysteine protease is required for interleukin-1b processing in monocytes. Nature (London) 356, 768-774.

Van Antwerp, D. J., Martin, S. J., Kafri, T., Green, D. R., & Verma, I. M. (1996). Suppression of TNF-α-induced apoptosis by NF-κB. Science 274, 787-789.

Vaux, D. L., & Strasser, A. (1996). The molecular biology of apoptosis. Proc. Natl. Acad. Sci. USA 93, 2239-2244.

Walker, N. P. C., Talanian, R. V., Brady, K. D., Dang, L. C., Bump, N. J., Ferenz, C. R., Franklin, S., Ghayur, T., Hackett, M. C., Hammill, L. D., Herzog, L., Hugunin, M., Houy, W., Mankovich, J. A., McGuiness, L., Orlewicz, E., Paskind, M., Pratt, C. A., Reis, P., Summani, A., Terranova, M., Welch, J. P., Xiong, L., Möller, A., Tracey, D.E., Kamen, R., & Wong, W. W. (1994). Crystal structure of the cysteine protease interleukin-1β-converting enzyme: A $(p20/p10)_2$ homodimer. Cell 78, 343-352.

Wang, L., Miura, M., Bergeron, L., Zhu, H., & Yuan, J. (1994). *Ich-1*, an *Ice/ced-3*-related gene, encodes both positive and negative regulators of programmed cell death. Cell 78, 739-750.

Wang, Z.-Q., Auer, B., Stingl, L., Berghammer, H., Haidacher, D., Schweiger, M., & Wagner, E. F. (1995a). Mice lacking ADPRT and poly(ADP-ribosyl)ation develop normally but are susceptible to skin disease. Genes Dev. 9, 509-520.

Wang, X., Pai, J.-T., Wiedenfeld, E. A. Medina, J. C., Slaughter, C. A., Goldstein, J. A., & Brown, M. S. (1995b). Purification of an interleukin-1β converting enzyme-related cysteine protease that cleaves sterol regulatory element-binding proteins between the leucine zipper and transmembrane domains. J. Biol. Chem. 270, 18044-18050.

Wang, S., Miura, M. Jung, Y.-k., Zhu, H., Gagliardini, V., Shi, L., Greenberg, A. H., & Yuan, J. (1996a). Identification and characterization of Ich-3, a member of the interleukin-1β converting enzyme (ICE)/Ced-3 family and an upstream regulator of ICE. J. Biol. Chem. 271, 20580-20587.

Wang, X., Zelenski, N. G., Yang, J., Sakai, J., Brown, M. S., & Goldstein, J. L. (1996b). Cleavage of sterol regulatory element binding proteins (SREBPs) by CPP32 during apoptosis. EMBO J. 15, 1012-1020.

Wang, C.-Y., Mayo, M. W., & Baldwin Jr., A. S. (1996c).TNF- and cancer therapy-induced apoptosis: potentiation by inhibition of NF-κB. Science 274, 784-787.

Wilkinson, K. D. (1997). Cellular regulation by ubiquitin-dependent processes. In:

Wilson, K. P., Black, J.-A. F., Thomson, J. A., Kim, E. E., Griffith, J. P., Navia, M. A., Murcko, M. A., Chambers, S. P., Aldape, R. A., Raybuck, S. A., & Livingston, D. J. (1994). Structure and mechanism of interleukin-1β converting enzyme. Nature (London) 370, 270-274.

Wyllie, A. H. (1980). Glucocorticoid-induced thymocyte apoptosis is associated with endogenous nuclease activation. Nature (London) 284, 555-556.

Yoshizawa, T., Sorimachi, H., Tomioka, S., Ishiura, S., & Suzuki, K. (1995a). A catalytic subunit of calpain possesses full proteolytic activity. FEBS Lett. 358, 101-103.

Yoshizawa, T., Sorimachi, H., Tomioka, S., Ishiura, S., & Suzuki, K. (1995b). Calpain dissociates into subunits in the presence of calcium ions. Biochem. Biophys. Res. Comm. 208, 376-383.

Yuan, J., & Horvitz, H. R. (1992). The *Caenorhabditis elegans* cell death gene *ced-4* encodes a novel protein and is expressed during the period of extensive programmed cell death. Development 116, 309-320.

Yuan, J., Shaham, S., Ledoux, S., Ellis, H. M., & Horvitz, H. R. (1993). The *C.elegans* cell death gene ced-3 encodes a protein similar to mammalian interleukin-1b-converting enzyme. Cell 75, 641-652.

Zhou, Q., & Salvesen, G.S. (1997). Activation of pro-caspase-7 by serine proteases includes a non-canonical specificity. Biochem. J. 324, 361-364.

Zou, H., Henzel, W.J., Liu, X.S., Lutschg, A., & Wang, X.D. (1997). Apaf-1, a human protein homologous to *C. elegans* CED-4 participates in cytochrome-c–dependent activation of caspase-3. Cell 90, 405-413.

Zhu, W., Murtha, P. E., & Young, C. Y. F. (1995). Calpain inhibitor-induced apoptosis in human prostate adenocarcinoma cells. Biochem. Biophys. Res. Comm. 214, 1130-1137.

INDEX

www.ingramcontent.com/pod-product-compliance
Ingram Content Group UK Ltd.
Pitfield, Milton Keynes, MK11 3LW, UK
UKHW020242290926
15470UKWH00002B/695

9 780762 30387